Powder Technology

Handling and Operations,
Process Instrumentation,
and Working Hazards

Powder Technology

Handling and Operations, Process Instrumentation, and Working Hazards

edited by

Hiroaki Masuda
Ko Higashitani
Hideto Yoshida

CRC Press
Taylor & Francis Group
Boca Raton London New York

CRC Press is an imprint of the
Taylor & Francis Group, an **informa** business

This material was previously published in *Powder Technology Handbook, Third Edition.* © CRC Press LLC, 2006.

CRC Press
Taylor & Francis Group
6000 Broken Sound Parkway NW, Suite 300
Boca Raton, FL 33487-2742

First issued in paperback 2019

© 2007 by Taylor & Francis Group, LLC
CRC Press is an imprint of Taylor & Francis Group, an Informa business

No claim to original U.S. Government works

ISBN-13: 978-1-4200-4412-6 (hbk)
ISBN-13: 978-0-367-38981-9 (pbk)

Library of Congress Cataloging-in-Publication Data

Powder technology: Handling and operations, process instrumentation, and working hazards / [edited by] Hiroaki Masuda and Ko Higashitani.
    p. cm.
Includes bibliographical references and index.
ISBN 1-4200-4412-5 (acid-free paper)
1. Masuda, Hiroaki, 1943- II. Higashitani, Ko, 1944-

TP156.P3P645 2006
620'.43--dc22
                                                       2006050485

Visit the Taylor & Francis Web site at
http://www.taylorandfrancis.com

and the CRC Press Web site at
http://www.crcpress.com

Preface

Particulate, or powder, technology is a fundamental engineering field that deals with a variety of particles, from submicroscale grains and aggregates to multi-phase colloids.

The applications of powders and particles are rapidly expanding into more diverse technologies, from the information market—including mobile phones, copy machines, and electronic displays—to pharmaceuticals, biology, cosmetics, food and agricultural science, chemicals, metallurgy, mining, mechanical engineering, and many other fundamental engineering fields. Fueling some of the latest developments, nanoparticles are the focus of promising research leading to more effective applications of various particles and powders.

Drawing from the recently published third edition of the acclaimed *Powder Technology Handbook*, this book concentrates its coverage on powder/particle handling methods and unit operations. This independent volume examines the purpose and considerations involved in different processes—including planning, equipment, measurements, and modeling techniques. It focuses on integrated strategies for finding the optimal solutions to problems in any context.

Substantially revised, updated, and expanded, this volume highlights new information on combustion and heating, electrostatic powder coating, and simulation. It also reflects recent data on the health effects caused by the inhalation of fine particles, along with ways to minimize harmful exposure.

While the book consolidates some sections from the last edition for ease of use, it also incorporates the innovative work and vision of new, young authors to present a broader and fully up-to-date representation of the technologies.

We hope this volume will serve as a strong guide for understanding the key aspects of industrial processing for particles and powders and encourage readers to apply the knowledge in this book to real applications, particularly those involving novel particles.

Special acknowledgment is given to all contributors and also to all the original authors whose valuable work is cited in this handbook. We would also like to acknowledge Dr. Matsusaka of Kyoto University, for his collaboration and editing, and we are grateful to our editorial staff at Taylor & Francis Books for their careful editing and production work.

Hiroaki Masuda

Ko Higashitani

Hideto Yoshida

Editors

Hiroaki Masuda is a professor in the Department of Chemical Engineering at Kyoto University, Japan. He is a member of the Society of Chemical Engineers, Japan and the Institute of Electrostatics (Japan), among other organizations, and is the president of the Society of Powder Technology, Japan. His research interests include electrostatic characterization, adhesion and reentrainment, dry dispersion of powder, and fine particle classification. Dr. Masuda received his Ph.D. degree (1973) in chemical engineering from Kyoto University, Japan.

Ko Higashitani graduated from the Department of Chemical Engineering, Kyoto University, Japan in 1968. He worked on hole pressure errors of viscoelastic fluids as a Ph.D. student under the supervision of Professor A. S. Lodge in the Department of Chemical Engineering, University of Wisconsin–Madison, USA. After he received his Ph.D. degree in 1973, he moved to the Department of Applied Chemistry, Kyushu Institute of Technology, Japan, as an assistant professor, and then became a full professor in 1983. He joined the Department of Chemical Engineering, Kyoto University, in 1992. His major research interests now are the kinetic stability of colloidal particles in solutions, such as coagulation, breakup, adhesion, detachment of particles in fluids, and slurry kinetics. In particular, he is interested in measurements of particle surfaces in solution by the atomic force microscope and how the surface microstructure is correlated with interaction forces between particles and macroscopic behavior of particles and suspensions.

Hideto Yoshida is a professor in the Department of Chemical Engineering at Hiroshima University, Japan. He is a member of the Society of Chemical Engineers, Japan and the Society of Powder Technology, Japan. His research interests include fine particle classification by use of high-performance dry and wet cyclones, standard reference particles, particle size measurement by the automatic-type sedimentation balance method, and the recycling process of fly-ash particles. Dr. Yoshida received his Ph.D. degree (1979) in chemical engineering from Kyoto University, Japan.

Contributors

Charles S. Campbell
School of Engineering, Aerospace
 and Mechanical Engineering
University of Southern California
Los Angeles, California, USA

Toyohisa Fujita
Department of Geosystem Engineering
 Graduate School of Engineering
University of Tokyo
Tokyo, Japan

Kuniaki Gotoh
Department of Applied Chemistry
Okayama University
Okayama, Japan

Jusuke Hidaka
Department of Chemical
Engineering and Materials Science
Faculty of Engineering
Doshisha University
Kyotanabe, Kyoto, Japan

Ko Higashitani
Department of Chemical Engineering
Kyoto University
Katsura, Kyoto, Japan

Hajime Hori
Department of Environmental Health
 Engineering
University of Occupational and
 Environmental Health
Kitakyushu, Fukuoka, Japan

Kengo Ichiki
Department of Mechanical Engineering
The Johns Hopkins University
Baltimore, Maryland, USA

Hironobu Imakoma
Department of Chemical Science and
Engineering
Kobe University, Nada-ku
Kobe, Japan

Eiji Iritani
Department of Chemical Engineering
Nagoya University
Nagoya, Chikusa-ku, Japan

Chikao Kanaoka
Ishikawa National College of Technology
Tsubata, Ishikawa, Japan

Yoichi Kanda
Department of Chemical Engineering
Kyoto University
Katsura, Kyoto, Japan

Yoshiteru Kanda
Department of Chemistry and Chemical
 Engineering
Yamagata University
Yonezawa, Yamagata, Japan

Hisao Makino
Central Research Institute
 of Electric Power Industry
Yokosuka, Kanagawa, Japan

Hiroaki Masuda
Department of Chemical Engineering
Kyoto University
Katsura, Kyoto, Japan

Shuji Matsusaka
Department of Chemical Engineering
Kyoto University
Katsura, Kyoto, Japan

Kei Miyanami
Department of Chemical Engineering
Osaka Prefecture University
Sakai, Osaka, Japan

Makio Naito
Joining and Welding Research Institute
Osaka University
Ibaraki, Osaka, Japan

Morio Okazaki
Department of Chemical Engineering
Kyoto University
Kyoto, Japan

Jun Oshitani
Department of Applied Chemistry
Okayama University
Okayama, Japan

Isao Sekiguchi
Department of Applied Chemistry
Chuo University
Bunkyo-ku, Tokyo, Japan

Yoshiyuki Shirakawa
Department of Chemical Engineering
  and Materials Science
Doshisha University
Kyotanabe, Kyoto, Japan

Minoru Sugita
Ohsaki Research Institute, Inc.
Tokyo, Japan

Minoru Takahashi
Ceramics Research Laboratory
Nagoya Institute of Technology
Tajimi, Aichi, Japan

Isamu Tanaka
Department of Environmental Health
  Engineering
University of Occupational and Environmental
  Health
Kitakyushu, Fukuoka, Japan

Tatsuo Tanaka
Hokkaido University West
Sapporo, Japan

Toshitsugu Tanaka
Department of Mechanical Engineering
Osaka University
Suita, Osaka, Japan

Ken-ichiro Tanoue
Yamaguchi University
Ube, Yamaguchi, Japan

Yuji Tomita
Department of Mechanical and Control
  Engineering
Kyushu Institute of Technology
Kitakyushu, Fukuoka, Japan

Shigeki Toyama
Nagoya University
Nagoya, Japan

JunIchiro Tsubaki
Department of Molecular Design and
  Engineering
Nagoya University
Nagoya, Japan

Hirofumi Tsuji
Yokosuka Research Laboratory
Central Research Institute of Electric
  Power Industry
Yokosuka, Kanagawa, Japan

Yutaka Tsuji
Department of Mechanical Engineering
Osaka University
Suita, Osaka, Japan

Hiromoto Usui
Department of Chemical Science and Engineering
Kobe University, Nada-ku
Kobe, Japan

Satoru Watano
Department of Chemical Engineering
Osaka Prefecture University
Sakai, Osaka, Japan

Richard A. Williams
Institute of Particle Science and Engineering
School of Process, Environmental, and
  Materials Engineering
University of Leeds
West Yorkshire, United Kingdom

Hiroshi Yamato
Department of Environmental
  Health Engineering
University of Occupational
  and Environmental Health
Kitakyushu, Fukuoka, Japan

Hideto Yoshida
Department of Chemical Engineering
Hiroshima University
Higashi-Hiroshima, Japan

Hiroki Yotsumoto
National Institute of Advanced Industrial
  Science and Technology
Tsukuba, Ibaraki, Japan

Shinichi Yuu
Ootake R. & D. Consultant Office
Fukuoka, Japan

Contents

PART II   Process Instrumentation

PART III   *Working Atmospheres and Hazards*

Part I

Powder Handling and Operations

1.1 Crushing and Grinding

Tatsuo Tanaka
Hokkaido University West, Sapporo, Japan

Yoshiteru Kanda
Yamagata University, Yonezawa, Yamagata, Japan

1.1.1 INTRODUCTION

Comminution is the oldest mechanical unit operation for size reduction of solid materials and an important step in many processes where raw materials are converted into intermediate or final products. The purposes of comminution are to reduce the size, to increase the surface area, and to free the useful materials from their matrices, and recently it has been involved in modification of the surface of solids, preparation of the composite materials, and recycling of useful components from industrial wastes. Comminution has a long history, but it is still difficult to control particle size and its distribution. Hence, fundamental analysis and optimum operation have been investigated.

A demand for fine or ultrafine particles is increasing in many kinds of industries. The energy efficiency of comminution is very low, and the energy required for comminution increases with a decrease in feed or produced particle size. Research and development to find energy-saving comminution processes have been performed.

1.1.2 COMMINUTION ENERGY

In design, operation, and control of comminution processes, it is necessary to correctly evaluate the comminution energy of solids. In general, the comminution energy (i.e., the size reduction energy) is expressed by a function of a particle size.[1]

Laws of Comminution Energy

Rittinger's Law

Rittinger assumes that the energy consumed is proportional to the produced fresh surface. Because the specific surface area is inversely proportional to the particle size, the specific comminution energy E/M is given by Equation 1.1:

$$\frac{E}{M} = C_R \left(S_p - S_f \right) = C_R' \left(x_p^{-1} - x_f^{-1} \right) \tag{1.1}$$

where S_p and S_f are the specific surface areas of product and feed, respectively, x_p and x_f are the corresponding particle sizes, and C_R and C_R' are constants which depend on the characteristics of materials.

3

Kick's Law

Kick's law assumes that the energy required for comminution is related only to the ratio of the size of the feed particle to the product particle:

$$\frac{E}{M} = C_k \ln\left(\frac{x_f}{x_p}\right) = C_k' \ln\left(\frac{S_p}{S_f}\right)$$

(1.2)

where C_K and C_K' are constants.

Equation 1.2 can be derived by assuming that the strength is independent of the particle size, the energy for size reduction is proportional to the volume of particle, and the ratio of size reduction is constant at each stage of size reduction.

Bond's Law[2]

Bond suggests that any comminution process can be considered to be an intermediate stage in the breakdown of a particle of infinite size to an infinite number of particles of zero size. Bond's theory states that the total work useful in breakage is inversely proportional to the square root of the size of the product particles, directly proportional to the length of the crack tips, and directly proportional to the square root of the formed surface:

$$W = W_i\left(\frac{10}{\sqrt{P}} - \frac{10}{\sqrt{F}}\right) = C_B'\left(S_p^{1/2} - S_f^{1/2}\right)$$

(1.3)

where $W(kWh/t)$ is the work input and F and P are the particle size in microns at which 80% of the corresponding feed and product passes through the sieve. $W_i(kWh/t)$ is generally called Bond's work index. The work index is an important factor in designing comminution processes and has been widely used.

Holmes's Law[3]

Holmes proposes a modification to Bond's law, substituting an exponent r, in place of 0.5 in Equation 1.3 as follows:

$$W = W_i\left(\frac{10}{P^r} - \frac{10}{F^r}\right)$$

(1.4)

Values of r which Holmes determined for materials are tabulated in Table 1.1.[4]

1.1.3   CRUSHING OF SINGLE PARTICLES

In principle, the mechanism of size reduction of solids is based on the fracture of a single particle and its accumulation during comminuting operations.

Fracture Properties of Solids

In a system composed of an elastic sphere gripped by a pair of rigid parallel platens, the load-deformation curve can be predicted by the theories of Hertz as summarized by Timoshenko and Goodier.[5] The

TABLE 1.1   Values of r Determined by Holmes

Materials	Holmes Exponent (r)
Amygdaloid	0.25
Malartic	0.40
Springs	0.53
Sandstone	0.66
Morenci	0.73
East Malartic	0.42
Chino Nevada Consolidated	0.65
Real del Monte	0.57
La Luz	0.34
Kelowna Exploratory	0.39
Utah Copper	0.50

elastic strain energy, E (J), input to a sphere up to the instant of fracture is given by the integral of the load acting through the deformation:

$$E = 0.832 \left(\frac{1-v^2}{Y} \right)^{2/3} x^{-1/3} p^{5/3} \tag{1.5}$$

where Y (Pa) is Yougth's modulus, v (-), Poisson's ratio, x (m) the diameter of the sphere (particle size), and P (N) is the fracture load. In this system, the compression strength of the sphere, S, is given by Hiramatsu et al.,[6] and the specific fracture energy E/M (J/kg) is given by

$$\frac{E}{M} = C_1 \rho^{-1} \pi^{2/3} \left(\frac{1-v^2}{Y} \right)^{2/3} S^{5/3} \tag{1.6}$$

where p (kg/m³) is density.

The relationship between the specific fracture energy and the strength for quartz and marble is shown in Figure 1.1.[7]

On the other hand, when two spherical particles, 1 and 2, collide with each other, the maximum stress, S_{max}, generated inside the particles is expressed by a function of particle size, x, relative velocity, v (m/s), and mechanical properties[8]:

$$S_{max} = C_2 \left(\frac{m_1 m_2}{m_1 + m_2} \right)^{1/5} v^{2/5} \left(\frac{2}{x_1} + \frac{2}{x_2} \right)^{3/5} \left(\frac{1-v_1^2}{Y_1} + \frac{1-v_2^2}{Y_2} \right)^{-4/5} \tag{1.7}$$

where m_1 (kg) and m_2 are the mass of the particles, and C_2 is a constant. The subscripts 1 and 2 denote two particles.

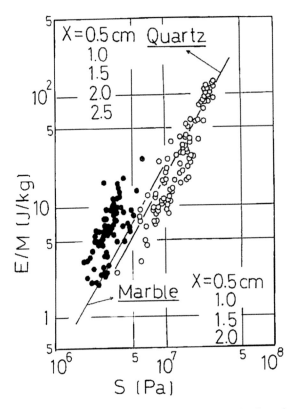

FIGURE 1.1 Relationship between strength S and specific fracture energy E/M.

Variation of Fracture Energy with Particle Size

Strength is a structure-sensitive property and changes with specimen volume. From a statistical consideration of the distribution of the presence of minute flaws,[9] Weibull[10] and Epstein[11] showed that the mean strength of the specimen, S, is proportional to the ($-1/m$) power of the specimen volume, V (m^3):

$$S = \left(S_0 V_0^{1/m}\right) V^{-1/m} \tag{1.8}$$

where S_0 (Pa) is the strength of unit volume V_0 (m^3), and m is Weibull's coefficient of uniformity. Experimental data lines determined by the least squares method for bolosilicate glass and quartz are shown in Figure 1.2[12] From Equation 1.6 and Equation 1.8, the relationship between specific fracture energy or fracture energy of a single particle, E, and particle size x is obtained as fallows:

$$\frac{E}{M} = C_3 (6)^{5/3m} \rho^{-1} \pi^{(2m-5)/3m} \left(\frac{1-v^2}{Y}\right)^{2/3} \left(S_0 V_0^{1/m}\right)^{5/3} x^{-5/m} \tag{1.9}$$

$$E = C_4 (6)^{5/3m} \pi^{(5m-5)/3m} \left(\frac{1-v^2}{Y}\right)^{2/3} \left(S_0 V_0^{1/m}\right)^{5/3} x^{(3m-5)/m} \tag{1.10}$$

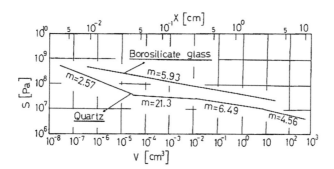

FIGURE 1.2 Variation of strength S with volume V of specimen.

The calculated result for quartz is shown in Figure 1.3.[12] It is important to note that the specific fracture energy increases rapidly for smaller particle size (less than approximately 500 μm), namely, the requirement of large amounts of energy in fine or ultrafine grinding can be presumed. The strength and the specific fracture energy increase also with an increase in loading rate.[13]

Crushing Resistance and Grindability

The importance of crushing resistance or grindability of solid materials and energy efficiency of comminuting equipment have been recognized in determining comminution processes in a variety of industries. Grindability is obtained from a strictly defined experiment. The two typical methods are the following.

Hardgrove Grindability Index (JIS M 8861,1993)

The machine to measure the grindability consists of a top-rotating ring with eight balls 1 in. in diameter. A load of 64 φ 0.5 lb is applied on the top-rotating ring. Fifty grams of material sieved between 1.19 and 0.59 mm is ground for the period of 60 revolutions. The Hardgrove grindability index, H.G.I., is defined as

$$H.G.I. = 13 + 6.93w \tag{1.11}$$

where w (g) is the mass of ground product finer than 75 μm.

Bond's Work Index (JIS M 4002,1976)

Bond's work index W_i, defined in Equation 1.3,[14] is given by.

$$W_i = \frac{1.1 \times 44.5}{P_1^{0.23} G_{bp}^{0.82} \left(10/\sqrt{P'} - 10/\sqrt{F}\right)} \tag{1.12}$$

where P_1 is the sieve opening in micron for test grindability, G_{bp} (g/rev) is the ball mill grindability, P' is the product size in microns (80% of product finer than size P_1 passes), and F is the feed size in microns (80% of feed passes). A standard ball mill is 12 in. (305 mm) in internal diameter and 12 in. in internal length, charged with 285 balls, as tabulated in Table 1.2. The lowest limit of the total mass of balls is 19.5 kg. The amount of feed material is 700 cm³ bulk volume, composed of

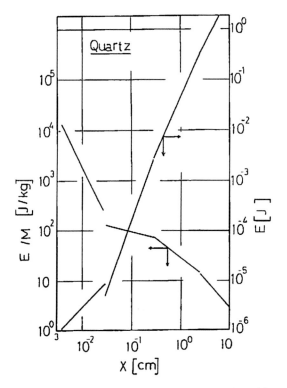

FIGURE 1.3   Relationship between size x and specific fracture energy E/M, or fracture energy E.

TABLE 1.2   Composition of Steel Balls for Measurement by Bond's Work Index

Diameter (mm)	No. of Balls
36.5	43
30.2	67
25.4	10
19.1	71
15.9	94
Sum	285

particles finer than 3360 μm. The mill is rotated a number of times so as to yield a circulating load of 250% at 70 rev/min, where the circulating load is defined as the component ratio of the oversize to the undersize. The process is continued until the net mass of undersize produced per revolution becomes constant $G_{b,p}$ in Equation 1.12. Table 1.3 shows the work index measured by wet process.[15] In fine grinding, when P in Equation 1.3 is smaller than 70 μm, the work index, W_i, is multiplied by a factor f to account for the increased work input. The factor f is found from the following empirical equation[16]:

$$f = \frac{P+10.3}{1.145P} \, (P \leq 70\mu m)$$
(1.13)

TABLE 1.3 Average Work Indexes

Material	Number tested	Average Specific gravity	Work index
All materials tested	1211		14.42
Andesite	6	2.84	18.25
Barite	7	4.50	4.73
Basalt	3	2.91	17.10
Bauxite	4	2.20	8.78
Cement clinker	14	3.15	13.56
Cement raw material	19	2.67	10.51
Coke	7	1.31	15.18
Copper ore	204	3.02	12.73
Diorite	4	2.82	20.90
Dolomite	5	2.74	11.27
Emery	4	3.48	56.70
Feldspar	8	2.59	10.80
Ferro-chrome	9	6.66	7.64
Ferro-manganese	5	6.32	8.30
Ferro-silicon	13	4.41	10.01
Flint	5	2.65	26.16
Fluorspar	5	3.01	8.91
Gabbro	4	2.83	18.45
Glass	4	2.58	12.31
Gneiss	3	2.71	20.13
Gold ore	197	2.81	14.93
Granite	36	2.66	15.05
Graphite	6	1.75	43.56
Gravel	15	2.66	16.06
Gypsumrock	4	2.69	6.73
Iron ore			
Hematite	56	3.55	12.93
Hematite-specular	3	3.28	13.84
Oolitic	6	3.52	11.33
Magnetite	58	3.88	9.97
Taconite	55	3.54	14.60
Lead ore	8	3.45	11.73
Lead-zinc ore	12	3.54	10.57
Limestone	72	2.65	12.54
Manganese ore	12	3.53	12.20
Magnesite	9	3.06	11.13
Molybdenum ore	6	2.70	12.80

(*Continued*)

TABLE 1.3 (Continued)   Average Work Indexes

Material	Number tested	Average Specific gravity	Work index
Nickel ore	8	3.28	13.65
Oilshale	9	1.84	15.84
Phosphate rock	17	2.74	9.92
Potash ore	8	2.40	8.05
Pyrite ore	6	4.06	8.93
Pyrrhotite ore	3	4.04	9.57
Quartzite	8	2.68	9.58
Quartz	13	2.65	13.57
Rutile ore	4	2.80	12.68
Shale	9	2.63	15.87
Silica sand	5	2.67	14.10
Silicon carbide	3	2.75	25.87
Slag	12	2.83	9.39
Slate	2	2.57	14.30
Sodium silicate	3	2.10	13.50
Spodumene ore	3	2.79	10.37
Syenite	3	2.73	13.13
Tin ore	8	3.95	10.90
Titanium ore	14	4.01	12.33
Trap rock	17	2.87	19.32
Zinc ore	12	3.64	11.56

Bond[16] proposed a relationship between the work index, W_i, and the Hardgrove grindability index (H.G.I.):

$$W_i = \frac{435}{\left(\text{H.G.I.}\right)^{0.91}} \tag{1.14}$$

Grindability in Fine Grinding

When the particle sizes of the products are submicron or micronized particles, it will be difficult to estimate the comminution energy by Equation 1.3, Equation 1.12, and Equation 1.13.

Bond[2] had proposed Equation 1.15 for measurement of W_i before Equation 1.12.

$$W_i = 1.1 \times 16 \left(\frac{P_1}{100}\right)^{0.5} \cdot G_{b.p}^{-0.82} \tag{1.15}$$

Equation 1.15 is simpler than Equation 1.12. There was not a great difference[17] between W_i calculated by Equation 1.12 and W_i calculated by Equation 1.15. Figure 1.4 shows the relationship

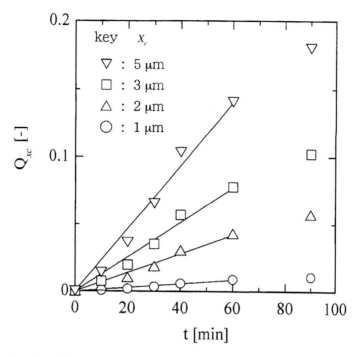

FIGURE 1.4 Relationship between grinding time t and mass fraction finer than size x_c, Q_{xc}.

between the mass fraction $Q_{x_c}(-)$ finer than the particle size (μm) and grinding time (min) in a ball mill grinding of silica glass.[18] In the early stage of grinding, a zero-order increasing rate is applicable, as shown in following equations.

$$Q_{x_c} = k_{x_c} t \tag{1.16}$$

$$W_{x_c} = Q_{x_c} \cdot W_s = k_{x_c} \cdot W_s \cdot t \tag{1.17}$$

where W_{x_c} is the mass of product finer than a size x_c, and W_s is the mass of the feed. From Equation 1.15 through Equation 1.17, the following equations can be obtained:

$$W_i \propto P_1^{0.5} \cdot G_{b_p}^{-0.82} \propto x_c^{0.5} \cdot \left(k_{x_c} \cdot W_s \right)^{-0.82} \tag{1.18}$$

$$W_{i_c} = x_c^{0.5} \cdot \left(k_{x_c} \cdot W_s \right)^{-0.82} \tag{1.19}$$

where W_{x_c} is proportional to W_i, which was proposed by Bond. W_i could be estimated by the examination of the zero-order increasing rate constant of the mass fraction less than a sieving size using an arbitrary ball mill.

Figure 1.5 shows the relationship between sieving size, x_c, and W_{i_c} for silica glass.[18] It was presumed that the work index could be approximately constant to a sieving size of 20 μm and increased in the range of a size less than 20 μm. It was also found that large amounts of energy are necessary to produce fine or ultrafine particles.

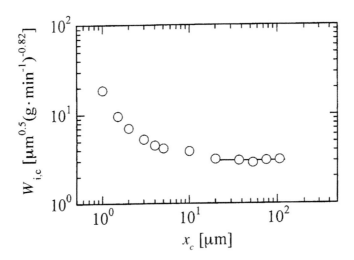

FIGURE 1.5   Relationship between sieving size x_c and corresponding work index W_{1_c}.

1.1.4   KINETICS OF COMMINUTION

A pulverizing machine is to be designed and operated by pursuing the comminution process of particle assemblage with the elapsed time. The size distribution after fracture of a single particle is derived by applying a stochastic theory by Gilvarry,[19] Gaudin and Meloy,[20] or Broadbent and Callcot.[21] These theories are based on the idea that some microcracks preexisting in the solid body are activated with stress and absorb elastic strain energy, so that the cracks develop rapidly and collide with each other to yield fragments of distributed sizes. The undersize cumulative fraction $B(\gamma, x)$ produced from a single particle of size γ is called a breakage function and is written as follows:

Gilvarry:

$$B(\gamma,x) = 1 - \exp\left[-\left(\frac{x}{c_1}\right) - \left(\frac{x}{c_2}\right)^2 - \left(\frac{x}{c_3}\right)^3 \right] \tag{1.20}$$

Gaudin and Meloy:

$$B(\gamma,x) = 1 - \left[1 - \left(\frac{x}{\gamma}\right) \right]^j \tag{1.21}$$

Broadbent and Callcot:

$$B(\gamma,x) = \frac{1 - \exp(-x/\gamma)}{1 - \exp(-1)} \tag{1.22}$$

where x is the particle size, c_1, c_2, and c_3 are constants, and j is the number of fragments or 10. They may be approximated as

$$B(\gamma,x) = \left(\frac{x}{\gamma}\right)^m \tag{1.23}$$

where m is a constant.[22] The derivative $\partial B/\partial x$ is called a distribution function appearing later.

With respect to the mass–size balance in batch grinding, the mass increment of a component of size x during a differential time interval dt is expressed by removal of a portion of the component due to selective grinding and by production of the same component due to selectively grinding a portion of all the coarser particles followed by the distribution to the noted size range, as illustrated in Figure 1.6:

$$\frac{\partial^2 D(x,t)}{\partial t\,\partial x} = -\frac{\partial D(x,t)}{\partial x} S(x,t)$$
$$+ \int_{x}^{x_m} \frac{\partial D(\gamma,t)}{\partial \gamma} S(\gamma,t) \frac{\partial B(\gamma,x)}{\partial x} dy$$

(1.24)

where $S(\gamma,t)$ is the probability density for particles of size γ to be selected for grinding and it is called a selection function or rate function, t is the grinding time, and X_m is the maximum size present. Thus, the rate of comminution of particle assemblage is determined by the size distribution after crushing a single particle and by the probability that each particle is selected for crushing within a certain time. Assuming Equation 1.23 for $B(\gamma,x)$ and the empirical relationship

$$S(x,t) = Kx^n$$

(1.25)

for $S(x,t)$ (Bowdish, 1960), the oversize cumulative fraction, $R(x,t) = 1 - D(x,t)$, is obtained by integration of Equation 1.24 using a rate constant, K:

$$m = n:$$

$$R(x,t) = R(x,0)\,\exp{-Kx^n t}$$

(1.26)

$$m \neq n:$$

$$R(x,t) \approx R(x,0)\,\exp[-(\mu Kx^n t)^v]$$

(1.27)

where $R(x,0)$ is the initial size distribution and can be regarded as unity if the grinding time t is long enough. μ and v are determined only by the value of m/n, as indicated in Figure 1.7. The size distribution is found to vary with grinding time in the Rosin–Rammler type, which was first confirmed experimentally by Chujo[23] using a ball mill. The selection function or the rate constant has been experimentally determined by Shoji and Austin[24] for ball milling, as shown in Figure 1.6, in which

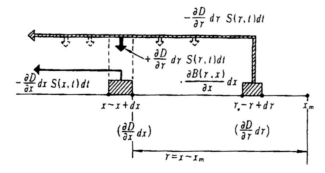

FIGURE 1.6 Explanation of mass balance in comminution process. [From Austin, L.G., and Klimpel, R.R., Ind. Eng. Chem., 56, no. 11, 18–29, 1964.]

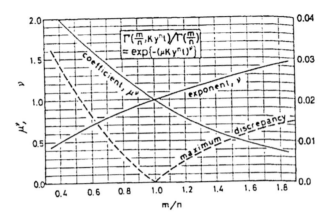

FIGURE 1.7   Approximation of Equation 1.25 by exponential function.

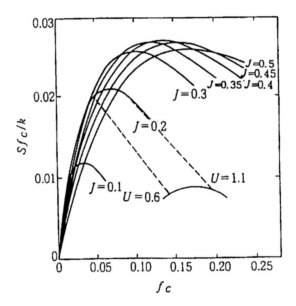

FIGURE 1.8   Variation of relative absolute rate of breakage with powder and ball filling for dry grinding.

$S(x)$ is depicted in relation to f_c, the fractional holdup of particles in a mill, to J, the ball filling degree, and U, the particle filling in the interstice of balls. The grinding rate is reported by Tanaka[25] for ultrafine grinding based on the current experimental works using very small beads in ball, vibration, planetary, and stirred milling, as

$$S(x) \propto \rho \left(\frac{x}{d} \right) \exp\left(-\frac{d_m}{d} \right) \qquad (1.28)$$

where d is the beads' diameter, d_m the optimum beads' diameter as a function of the beads density ρ, colliding speed, size, and strength of material crushed.

1.1.5   GRINDING OPERATIONS

To reduce the strength or the toughness of the material crushed, wet process grinding, the addition of grinding aids, and cryogenic grinding are proposed. The additives are likely to reduce the surface energy, leading to facilitating grinding of the particles. In addition to this, newly formed surfaces are active enough for stronger chemical bonding, so that surface modification can be expected by the grinding operation.

Furthermore, to prepare the size distribution required for product quality, a multipass of ground material through a mill and a classifier are necessary, as noted below.

Internal Classification System

The classification mechanism is assembled in a grinding mill. For example, an air-swept mill adopts internal classification by flowing fluids such as air through the mill, where particles finer than a critical size x_c are removed from the machine immediately after grinding. Then the following mass balance holds[26]:

$$F\left(\frac{dR_p(x)}{dx}\right) = H\int_{x_c}^{x_m}\left(\frac{dR(\gamma)}{d\gamma}\right)S(\gamma)\left(\frac{\partial B(\gamma,x)}{\partial x}\right)d\gamma \qquad (1.29)$$

where F is the feed to a continuous grinding machine, H is the holdup of particles in the machine, and the subscript p denotes product. Using Equation 1.23 and Equation 1.25 along with the conditions

$$\int_{x_c}^{x_m}\left(\frac{dR(\gamma)}{d\gamma}\right)d\gamma = 1; \quad \int_0^{x_c}\left(\frac{dR_p(x)}{dx}\right)dx = 1$$

the ideal size distribution of the ground material is given in the case where Equation 1.30 is applicable and sorting by a clean-cut classification is assumed:

$$\frac{F}{KH} = x_c^n \qquad (1.30)$$

$$R_p(x) = 1 - \left(\frac{x}{x_c}\right)^n \qquad (1.31)$$

Closed-Circuit Grinding System

In contrast to the preceding system, a closed-circuit system is characterized by an external classifier involved in the system. The ground material, D, is continuously sent to a classifier, where only the fine component is removed as the finished product, P. The coarse material is recirculated to the mill, as shown in Figure 1.9. Increasing the circulating load T, it is possible to avoid overcrushing so as to increase the grinding capacity. Controlling the size distribution of the finished product is also possible to some extent.

In a clean-cut classifier[27], the cumulative oversize fraction or the ground particles, R_D, is obtained from Equation 1.26 for the average residence time t_0 ($= H/F$) in the mill as

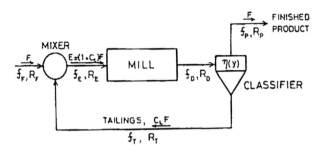

FIGURE 1.9 Typical connection of closed-circuit grinding system.

$$R_D(x) = \exp\left(-\frac{Kt_0 x^n}{1 + CL}\right)$$ (1.32)

where the average residence time is given by the overall residence time t_0 ($= H/F$) in this circuit and the circulating ratio CL as

$$i_{av} = \frac{H}{E} = \frac{H}{F(1 - CL)} = \frac{t_0}{1 - CL}$$ (1.33)

$$CL = \frac{T}{P} = \frac{T}{F} = \frac{R_D(x_c)}{1 - R_D(x_c)}$$ (1.34)

hence, the cutoff size x_c is given by combining Equation 1.32 and Equation 1.34 as

$$Kt_0 x_c^n = \ln\left[1 + \left(\frac{1}{CL}\right)\right]^{1 + CL}$$ (1.35)

The characteristic classification size x_c^* of x_c corresponding to infinite CL is also obtained from Equation 1.34 and Equation 1.35, as

$$x_c^{*n} = \left(\frac{F}{KH}\right)$$ (1.36)

As the clean-cut size distribution of product, $R_p(x)$, is written by

$$R_p(x) = \frac{R_D(x) - R_D(x_c)}{1 - R_D(x_c)}$$ (1.37)

the following are obtained using the equations above:

$$R_p\left(\frac{x}{x_c^*}\right) = (1+CL)\exp\left(-\frac{\left(x/x_c^*\right)^n}{1+CL}\right) - CL \tag{1.38}$$

$$\left(\frac{x_c}{x_c^*}\right)^n = (1+CL)\ln\left[1+\left(\frac{1}{CL}\right)\right] \tag{1.39}$$

The interrelationship between the above two equations is graphed in Figure 1.10. The chart is used such that a desired size ratio (x_{30}/x_{70}) of product at $R_p = 30\%$ and 70%, for example, is fitted to the horizontal distance between the two corresponding curves. Hence, x_c^* is calculated from the value on the abscissa using either x_{30} or x_{70}; then, x_c and CL are read from the ordinate on both sides. The flow system of the closed circuit can be designed by use of CL and F for the transportation equipments and x_c for the classifier. The mill design should be made on basis of H in Equation 1.36 for a known grinding rate constant K as well as F and x_c^*.

In the case of a non-clean-cut classifier,[28] classification performance can be expressed by two parameters, A and S, in a mathematical model of the partial classification efficiency $\eta(x)$ as follows:

$$\eta(x) = \left\{1+\exp\left[\left(\frac{4S}{A}\right)\left[1-\left(\frac{x}{x_c}\right)^A\right]\right]\right\}^{-1} \tag{1.40}$$

where $A = 1.5$ and $S = 1.0$ for air separators, and $A = 1.0$ and $S = 0.5$ for hydrocyclones. Choosing a reference curve in Figure 1.11 to be fitted to the desired cumulative undersize distribution of the finished product, the size corresponding to 1.0 on the abscissa is equal to x_c^* (for $v = 1$ in Equation 1.27), and the cumulative undersize $D_p(1)$ on the ordinate is connected with CL for specific values of S and A in Figure 1.12. Then, the value of CL indicates x_c/x_c^* in Figure 1.13; thus, the cutoff size, x_c, is obtained from x_c^*. Figure 1.14 illustrates some partial classification efficiency curves modeled by use of the two parameters S and A.[28]

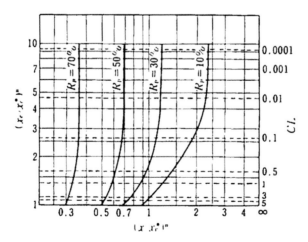

FIGURE 1.10 Collinear chart for ideal classification. [From Furuya, M., Nakajima, Y., and Tanaka, T., *Ind. Eng. Chem. Process. Des. Dev.*, 10, 449–456, 1971.]

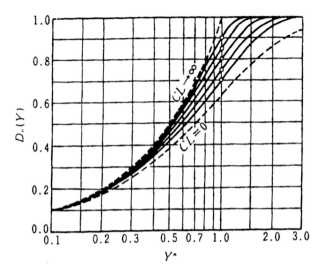

FIGURE 1.11 Reference curve for size distribution of product. (30) [From Furuya, M., Nakajima, Y., and Tanaka, T., *Ind. Eng. Chem. Process. Des. Dev.*, 12, 18–23, 1973.]

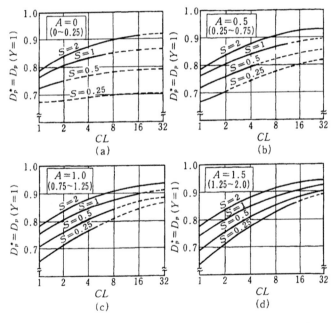

FIGURE 1.12 Relationship among classification parameters, circulating ratio, and undersize fraction of product at $x = x_c^*$. [From Furuya, M., Nakajima, Y., and Tanaka, T., *Ind. Eng. Chem. Process. Des. Dev.*, 12, 18–23, 1973.]

Alteration of the size distribution of the finished product is possible by some combinations of mills and classifiers, as well as feed positions. The analysis is given by Tanaka,[29] taking advantage of Figure 1.10. Grinding capacity increases in general with increasing CL due to the reduction of overcrushed fine particles. In this sense, it is worthwhile noting that when CL in Equation 1.38 tends to infinity, $R_p(x)$ becomes the same as Equation 1.31, the internal classification mechanism.

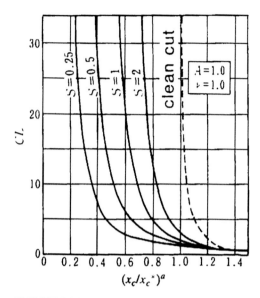

FIGURE 1.13 Relationship among classification parameters, circulating ratio, and $(x_c/x_c^*)^n$.

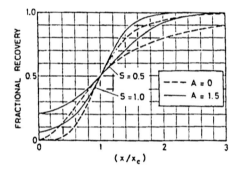

FIGURE 1.14 Fractional recovery curves calculated from the mathematical model (Equation 1.35 and Equation 1.36).

1.1.6 CRUSHING AND GRINDING EQUIPMENT

A major objective of comminution is to liberate minerals for concentration processes. Another objective is to produce particles of a required sizes. Comminution processes generally consist of several stages in series. Various types of crushing and grinding equipments have been used industrially as a mechanical way of producing particulate solids. The working phenomena in these types of equipment are complex, and different principles are adopted in the loading, such as compression, shear, cutting, impact, and friction; in the mechanism of force transmission or the mode of motion of grinding media, such as rotation, reciprocation, vibration, agitation, rolling, and acceleration due to fluids; and in the operational method, such as dry or wet system, batch or continuous operation, association of internal classification or drying and so on. But, in practice, it is most common to classify comminution processes into four stages by the particle size produced. Although the sizes are not clear cut, they are called the primary (first), intermediate (second), fine (third), and ultrafine (fourth) stages, according to the size of ground product. On the basis of the above classification, typical equipment types and their structure and characteristics are mentioned briefly below.

feed

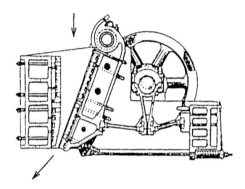

product

FIGURE 1.15 Jaw crusher.

Crushers

Crushers are widely used as the primary stage to produce particles finer than about 10 cm in size. They are classified as jaw, gyratory, and cone crushers based on compression, cutter mill and shredder based on shear, and hammer crusher based on impact.

Jaw Crusher

The jaw crusher shown in Figure 1.15 consists essentially of two crushing surfaces, inclined to each other. Material is crushed between a fixed and a movable plate by reciprocating pressure until the crushed products become small enough to fall through the narrowest gap between the crushing surfaces.

Gyratory Crusher

The essential features of a gyratory crusher are a solid cone on a revolving shaft, placed within a hollow shell which may have vertical or conical sloping sides, as shown in Figure 1.16. Material is crushed when the crushing surfaces approach each other, and the crushed products fall through the discharging chute.

Hammer Crusher, Swing-Hammer Crusher, and Impactor

These are used either as a one-step primary crusher or as a secondary crusher for products from a primary crusher. Pivoted hammers are mounted on a horizontal shaft, and crushing takes place by the impact between the hammers and breaker plates. A cylindrical grating or screen can be placed beneath the rotor. Materials are reduced to a size small enough pass through the bars of the grating or screen. Hammers are symmetrically designed. The size of product can be regulated by changing the spacing of the grate bars or the opening of the screen, and also by lengthening or shortening the hammer arms.

Intermediate Crushers

Intermediate crushers produce particles finer than about 1 cm. The roller mill, crushing roll, disintegrator, screw mill, edge runner, stamp mill, pin mill, and so on belong to this category. Roll

feed

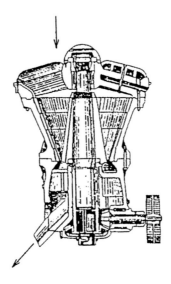

product

FIGURE 1.16 Gyratory crusher.

feed

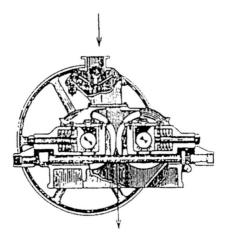

product

FIGURE 1.17 Crushing roll.

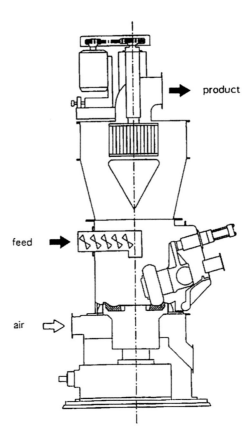

FIGURE 1.18 Roller mill.

crushers are of many types. They consist of at least one cylinder rotating on its principal axis, which nips material with two surfaces to compress and break the material into pieces. Figure 1.17 shows a typical crushing roll. It consists of two cylinders mounted on horizontal shafts, which are driven in opposite directions. The distance between the cylinders (rollers) is usually made adjustable. The size of feed materials is determined by the diameter of the cylinders, the required size of products, and the angle of nip. Recently, high-pressure roller mills (100–200 MPa) are available to comminute finely brittle materials.[30] Figure 1.18 shows a typical roller mill. Roller mills have been actively used for preparation of fine particles. Material is fed to the center of the horizontally rotating table, conveyed to its circumference by centrifugal force, ground by several units of rollers on the concave table, and moved toward the circumference.

Fine-Grinding Equipment

Fine-grinding equipment produces particles finer than about 10 μm. There are many kinds of machines in this category. They are roughly classified into three types: ball-medium type, medium agitating type, and fluid-energy type.

In a ball-medium type, the grinding energy is transferred to materials through media such as balls, rods, and pebbles by moving the mill body. Based on the mode of motion of the mill body, ball-medium mills are classified as tumbling ball mills, vibration mills, and planetary mills.

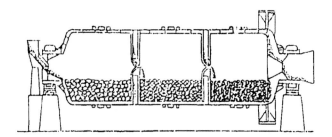

FIGURE 1.19 Compartment mill.

1. A tumbling mill or a ball mill is most widely used in both wet and dry systems, in batch and continuous operations, and on small and large scales. The optimum rotational speed is usually set at 65 to 80% of critical speed, N_c (rpm), when the balls are attached to the wall due to centrifugation:

$$N_c = \frac{42.3}{\sqrt{D_m}} \tag{1.41}$$

where D_m is the mill diameter in meters. It is desirable to reduce the ball size in correspondence with the smaller size of the feed materials, as in a compartment mill, shown in Figure 1.19, and a conical ball mill.

2. A vibration mill is driven by eccentric motors to apply a small but frequent impact or shear to the grinding media. Loose bodies or media contained in a shell cause it to vibrate. In contrast to tumbling mills, the media in vibration mills move only a few millimeters through a complex path, shearing as well as impacting the material between them. The apparent amount of media is 75 to 85% by volume, which is about two times as much as filling mills. One of the advantages of vibration mills compared with tumbling mills is the higher grinding rate in the range of fine particles. But it is unsuitable for heat-sensitive materials.

3. A planetary mill consists of a revolving base disk and rotating mill pots, as shown in Figure 1.20. Materials are ground in a large centrifugal field by the force generated during revolution and rotation. The intensity of acceleration can be increased up to 150g on the scale of gravitational acceleration. It is predicted that the grinding mechanism consists of compressive, abrasive, and shear stresses of the balls. Planetary mills are also used in the study of mechanical alloying and composite particles.[31]

A fluid-energy mill is widely noted as a jet mill. In a jet mill, the materials are ground by the collision of a particle with a particle, a wall, or a plate of the grinding vessel (chamber). The collision energy is generated by a high-speed jet flow. Fluid-energy mills may be classified in terms of the mill action. In one type of mill, the energy is generated by high-velocity streams at a section or whole periphery of a grinding and classification vessel. There are the Micronizer and Jet-O-Mizer shown in Figure 1.21 and others of this type. In Majac and other mills, two streams convey particles at high velocity into a vessel where they impact on each other. A fluid-energy mill has no movable mechanical parts, and it has advantages such as dry and continuous operation without a temperature rise, and controlling the particle size by the feed rate of materials and the velocity of jet stream. But the energy efficiency is low, and the power cost is high.

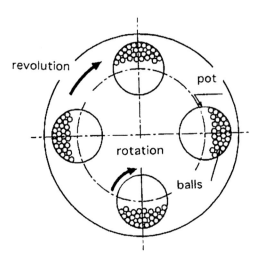

FIGURE 1.20 Planetary mill.

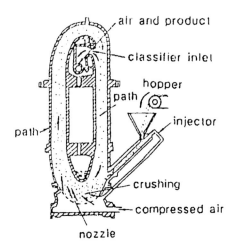

FIGURE 1.21 Jet mill.

Ultrafine Grinding Equipment

Ultrafine grinding equipments produce particles finer than about 1 μm. The medium agitating mills (stirred media mills) are in this category.

A medium agitating mill is regarded as one of the most efficient devices for micronizing materials and has been actively used for preparation of ultrafine particles. In this technique, a large number of small grinding media are agitated by impellers, screws, or disks in a vessel. Breakage occurs mainly by collision of the media. It is classified into three types by agitating mode. Medium agitating mills could produce submicron particles.

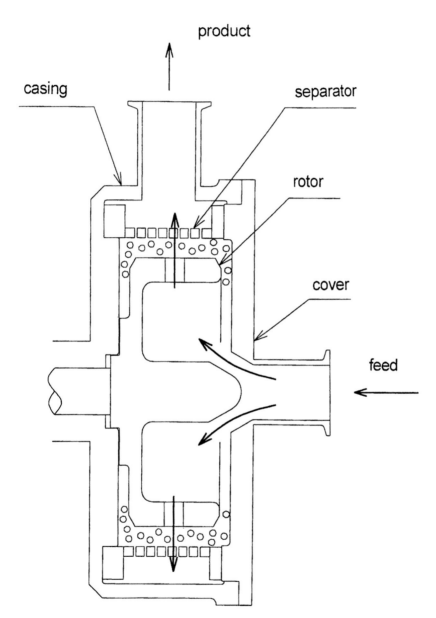

FIGURE 1.22 Stirred mill.

1. An agitating-tank type utilizes vertical agitation of balls and pebbles in a tank. A typical stirred mill is shown in Figure 1.22.
2. In a flow-tank type, disks, pins, or paddles are attached to a shaft to agitate balls or beads and grind particles in a flowing suspension.
3. A tower type is a large standing mill that has a central shaft with a screw to convey balls upward, and the centrifugal force causes them to collide with the wall.

Recently, the demand for ultrafine (submicron) particles has been increasing in many kinds of industries. The research and development to find ultrafine grinding processes has been performed for years for wet grinding, grinding aids, and closed-circuit grinding. The number of particles to be ground varies inversely with the cube of the particle size of the feed material. Then, in ultra-fine grinding equipment, it is necessary to increase the collision probability of the particle and grinding medium. Figure 1.22 shows a stirred mill. With the mill, media is evenly distributed at the outside of the rotor, creating even shear forces. In this mill, the minimum diameter of media is 0.1 mm.

REFERENCES

1. Walker, W. H., Lewis, W. K., McAdams, W. H., and Gilliland, E. R., *Principles of Chemical Engineering*, McGraw-Hill, New York, pp.254, 1937.
2. Bond, E. C., *Trans. AIME*, 193, 484–494, 1952.
3. Holmes, J. A., *Trans. Inst. Chem. Eng.*, 35, 125–140, 1957.
4. Lowrison, G. C., *Crushing and Grinding*, Butterworth, London, p. 54, 1974.
5. Timoshenko, S. and Goodier, J. N., *Theory of Elasticity*, Mc Graw-Hill, New York, p. 372, 1951.
6. Hiramatsu, Y., Oka, T., and Kiyama, H., *Mining Inst. Japan*, 81, 1024–1030, 1965.
7. Kanda, Y., Samo, S., and Yashima, S., *Powder Technol.*, 48, 263–267, 1986.
8. Rumpt, H., *Chem. Ing. Technol*, 31, 323–337, 1959.
9. Griffith, A. A., *Proceedings of 1st Inter Congr.\Appl.\Mech*, pp. 55–63, 1924.
10. Weibull, W., Ing. *Vetenshaps Akad Handle*, 151, 1939.
11. Epstein, B., *J. Appl. Phys.*, 19, 140–147, 1948.
12. Yashima, S., Kanda, Y., and Sano, S., *Powder Technol.*, 51, 277–282, 1987.
13. Yashima, S., Saito, F., and Hashimoto, H., *J. Chem. Eng. Japan*, 20, 257–264, 1987.
14. Bond, F. C., *Trans. AIME*, 217, 139–153, 1960.
15. Bond, F. C., *Brit. Chem. Eng.*, 6, 543–548, 1961.
16. Bond, F. C., *Brit. Chem. Eng.*, 6, 378–385, 1961.
17. Ishihara, T., *J. Min. Metal. Inst. Japan*, 80, 924–928, 1964.
18. Kotake, N., Soji, H., Hasegawa, M., and Kanda, Y., *J. Soc. Powder Technol. Japan*, 31,626–630, 1994.
19. Gilvarry, J. J., *J. Appl. Phys.*, 32, 391–399, 1961.
20. Gaudin, A. M., and Meloy, T. P., *Trans. AIME*, 223, 43–51, 1962.
21. Broadbent, S. R., Callcot, T. G., *J. Inst. Fuel.*, 29, 524–528, 1956.
22. Nakajima, Y. and Tanaka, T., *Ind. Eng. Chem. Process. Des. Dev.*, 12, 23–25, 1973.
23. Chujo, K., *Kagaku Kougaku to Kagakukikai*, 7, 1–83, 1949.
24. Shoji, K., Austin, L. G., Smaila, F., Brame, K., and Luckie, P. T., *Powder Technol.*, 31, 121–126, 1982.
25. Tanaka, T., *J. Soc. Powder Technol. Jpn.* 31, 25–31, 1994
26. Ouchiyama N., Tanaka T., and Nakajima, Y., *Ind. Eng. Chem. Process. Des. Dev.*, 15, 471–473, 1976.
27. Furuya, M., Nakajima, Y., and Tanaka, T., *Ind. Eng. Chem. Process. Des. Dev.*, 10, 449–456, 1971.
28. Furuya, M., Nakajima, Y., and Tanaka, T., *Ind. Eng. Chem. Process. Des. Dev.*, 12, 18–23, 1973.
29. Tanaka, T., *J. Soc. Powder Technol. Jpn.*, 31, 333–337, 1994.
30. Schonert, K., and Lubjuhn, U., 7th European Symposium Comminution Preprints, Part 2, 747–763, 1990.
31. Mizutani, U., Takeuchi, T., Fukunaka, T., Murasaki, S., Kaneko, K., and Mater, J., *Sci. Lett.*, 12, 629–632, 1993.

1.2 Classification

Kuniaki Gotoh
Okayama University, Okayama, Japan

Hiroaki Masuda
Kyoto University, Katsura, Kyoto, Japan

Hideto Yoshida
Hiroshima University, Higashi-Hiroshima , Japan

Jusuke Hidaka
Doshisha University, Kyoto, Japan

1.2.1 BASIS OF CLASSIFICATION

The unit operation for the separation of particulate material depending on its characteristics, such as particle diameter, density, shape, and so forth, is called "classification." However, this term is commonly used to express the size classification that is the separation of particles depending on their diameter. In this section, the basis of size classification is described.

Size classification is meant to separate particles of a certain diameter that is called the "cut size" or the "classification point," and to collect them as fine powder and coarse powder, although separation and collection of dust is aimed to collect all particles suspended in a fluid.

Classification Efficiency

When F [kg] of a powder (= raw material) is classified into A [kg] coarse powder and B [kg] fine powder, the total mass balance of the powder can be written as

$$F = A + B \tag{2.1}$$

If the cumulative oversize fractions of the raw material—coarse powder and fine powder—at the cut size D_{pc} are R_f, R_a, and R_b, respectively, the mass balance of the powder, which is coarser than the cut size D_{pc} is as follows:

$$FR_f = AR_a + BR_b \tag{2.2}$$

By the above equations, recovery, which is one of the representative values of classification efficiency, can be expressed.

Recovery of coarse powder r_a

$$r_a = \frac{AR_a}{FR_f} = \frac{R_a\left(R_f - R_b\right)}{R_f\left(R_a - R_b\right)} \tag{2.3}$$

Recovery of fine powder r_b

$$r_b = \frac{B(1-R_b)}{F(1-R_f)} = \frac{(R_a - R_f)(1-R_b)}{(R_a - R_b)(1-R_f)} \tag{2.4}$$

These recoveries are the mass ratio of the mass collected as coarse or fine powder to that fed to the classifier. In the powder collected as coarse powder, finer powder than cut size D_{pc} may be included, while coarser powder than cut size D_{pc} also may be included in the powder collected as fine powder. These contaminations mean the deterioration of classification performance. Therefore, any unnecessary component in the collected powder should be addressed by concern for separation efficiency. Newton's efficiency η_n defined by the following equation is one aspect of separation efficiency in which the unnecessary component is taken into consideration

$$\eta_n = r_a + r_b - 1 = \frac{(R_f - R_b)(R_a - R_f)}{R_f(1-R_f)(R_a - R_b)} \tag{2.5}$$

Newton's efficiency is defined as the difference between the recovery of product (useful component) and the residuum of the unnecessary component. As in Equation 2.5, Newton's efficiency is the same whether the product is the coarse powder or the fine powder.

The recoveries and Newton's efficiency described above represent the classification efficiency by means of one numerical value. These efficiencies are called "total classification efficiency" and "total separation efficiency." The total classification efficiency is a function of the cut size, because the efficiency is defined by the cumulative oversize fractions at the cut size.

Partial Separation Efficiency

In general, the collection probability E_r of each particle into the coarse side of the classifier depends on the size of particle. When dF [kg] of the particles having a diameter in the range from D_p to $D_p + dD_p$ is included in the powder fed into a classifier and when dA [kg] of particles among them is classified as coarse, the recovery of the particles is expressed as follows:

$$E_r(D_p) = dA/dF \tag{2.6}$$

The recovery is called "partial separation efficiency" or "partial classification efficiency." It is often called simply "classification efficiency."

When the frequency of the raw material, coarse powder, and fine powder designated as $f_F(D_p)$, $f_A(D_p)$, and $f_B(D_p)$ respectively, the mass balance of the particles having a diameter in the range from D_p to $D_p + dD_p$ is expressed as follows:

$$dF = Ff_f(D_p)dD_p \tag{2.7}$$

$$dA = Af_A(D_p)dD_p \tag{2.8}$$

By substituting Equation 2.7 and Equation 2.8 into Equation 2.6, the partial separation efficiency can be rewritten as the following equation:

$$E_r(D_p) = Af_A(D_p)/Ff_F(D_p) \tag{2.9}$$

Typical partial separation efficiency shows a curved line. The curved line is called the partial classification curve, partial separation curve, or Tromp curve.

The particle diameters corresponding to E_r = 0.25, 0.5, and 0.75 are defined as 25% of the cut size D_{p25}, 50% of the cut size D_{p50}, and 75% of the cut size D_{p75}, respectively. The 50% cut size D_{p50} is one of the most important representative values of the characteristics of classification.

Sharpness of Classification

The cut size described above is one of the representative values of classification performance. On the other hand, if the separation efficiency can be expressed by a unit function that jumps up from 0 to 1 at the cut size, the classification is an ideal classification. Thus, whether the separation efficiency increases sharply is another characteristic of performance.

The gradient of the partial separation efficiency is one of the expressions of the sharpness of the classification. However, because the separation efficiency usually shows a curved line, the gradient depends on the diameter. Therefore, the following indexes are used as representative values of the sharpness of the classification.
Index of separation accuracy:

$$\kappa = D_{p75}/D_{p25} \tag{2.10}$$

Terra index:

$$E_p = \frac{D_{p75} - D_{p75}}{2} \tag{2.11}$$

Incomplete index:

$$I = \frac{D_{p75} - D_{p75}}{D_{p75} + D_{p75}} \tag{2.12}$$

As found by the above definitions, index κ for an ideal separation is infinity, while the Terra index E_p and incomplete index I, which are the nondimensional form of the Terra index, are zero for an ideal separation. The reciprocal numbers of these indexes are also used as representative values of the sharpness. Although these indexes are defined by means of the 25% cut size D_{p25} and D_{p75}, describing the 90% cut size or the 100% cut size is preferable for a precise expression of the classification performance, because the gradient of a separation efficiency curve decreases rapidly at the diameter below the 25% cut size and over the 75% cut size.

Outlines of Classifiers

There are two types of classifier, depending on the medium of the particle suspension. One is a "wet classifier" using a liquid as the medium of suspension, and the other is a "dry classifier" using a gas as the medium of suspension. In both classifiers, the classification is achieved by applying a certain force to each particle. The applied force causes different particle trajectories reflecting the particle size. By means of the difference of the trajectories, particles are classified into coarse or fine, depending on their size. The force applied to the particles is gravity, centrifugal force, and so on. As for centrifugal force, it is generated by a semifree vortex or forced vortex. In the case of dry classification, inertia is also utilized for making the difference in the trajectories.

There are advantages and disadvantages for wet and dry classifications:

1. Production rate per unit area of a wet classifier is less than that of a dry classifier. In general, the rate of a wet classifier is 1/50 of that in a dry classifier.

2. Dispersion of the particles, which is a pretreatment for classification, into a liquid is easier than that into a gas. Because of that, the sharpness of the wet classification is better than that of the dry classification.
3. Velocity of particles in a liquid is less than that in a gas, when the same magnitude of separation force is applied. It leads to a narrow size range classification.
4. The collection and transportation of the classified particles is easier in a liquid than in a gas. It leads to a low probability of processing trouble. On the other hand, in case of wet classification, a drying process for classified powder is required after the classification. In other words, wet classification requires additional energy and cost for posttreatment.

As listed above, wet and dry classifications both have advantages and disadvantages. The selection of wet or dry must be made by taking into account the operations before and after the classification.

1.2.2   DRY CLASSIFICATION

Dry classification is widely used in many industrial processes. Compared to wet classification, dry classification does not need drying and slurry treatment. As a result, dry classification is more widely used compared to wet classification. However, in order to classify particles with a cut size of less than about 3 μm, a suitable particle disperser is necessary before using the dry classification apparatus. Theoretical study about the collection efficiency of cyclones was conducted by Leith and Licht.[1] Numerical calculations of fluid flow and particle motions in a cyclone were conducted by Ayers and Boysan,[2] Zhou and Soo,[3] and Yamamoto et al.[4,5] In order to control the cut size, rotational-blade-type classifiers are generally used. However, it is difficult to shift the cut size in the submicron range by use of the forced-vortex-type classifier.

Recent experimental results indicate that it is possible to shift the cut size in the submicron range by use of a flow controlling method applied to the free-vortex-type classifier. Recent industrial interest in dry classification is mainly directed to accurate cut size control in the fine size region. The following are typical air classifiers used in actual industrial processes.

Gravitational Classifiers

Particles are classified by the difference in either their settling velocity or falling position. Horizontal flow, vertical flow, and inclined flow classifiers are all used (see also 1.2.3 for the theory).

Inertial Classifiers

Particles are classified by the difference in their trajectories. Rectilinear (elbow-type), curvilinear (impactor), and inclined (louver) classifiers are commercially available. Figure 2.1 shows various inertial classifiers and Figure 2.2 shows a louver separator with plane or special blades.[6]

Centrifugal Classifiers

Centrifugal force is created either by airflow in cyclone-type classifiers, or by mechanical revolution in air separators. A variety of centrifugal classifiers is available: Micron separator (Figure 2.3), Turbo classifier (Figure 2.4), and Turboplex (Figure 2.5). The characteristics of centrifugal separators are listed in Table 2.1.

Figure 2.6 shows an example of the partial separation efficiency of a special type of cyclone with and without blow-down for the submicron range.[7]

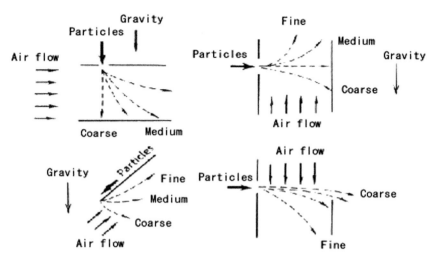

FIGURE 2.1   Principle of classification by traverse flow.

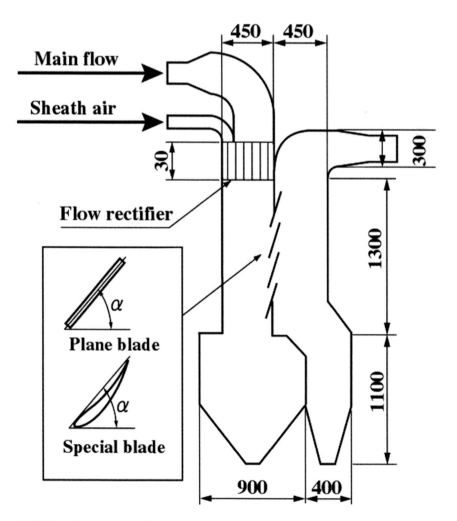

FIGURE 2.2   Structure of louver classifier.

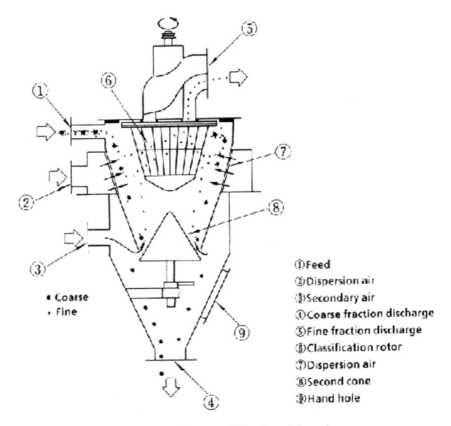

FIGURE 2.3   Micron separator. (Courtesy of Hosokawa Micron.)

①Feed
②Dispersion air
③Secondary air
④Coarse fraction discharge
⑤Fine fraction discharge
⑥Classification rotor
⑦Dispersion air
⑧Second cone
⑨Hand hole

Cut Size Control by Use of Free-Vortex-Type Classifier

In order to control the cut size with a gas-cyclone, the use of a movable slide plate, shown in Figure 2.7, is effective. Figure 2.8 shows the classification performance using a movable slide plate and the blow-down method proposed by Yoshida.[8,9] Both methods are effective in decreasing the 50% cut size. The effect of an apex cone set at the upper part of the dust box is to decrease the cut size and to increase the collection efficiency.

Figure 2.9 shows the simulated fluid vectors with and without the apex cone. Using the apex cone, it is possible to decrease the magnitude of the fluid velocity component in the dust box. As a result, it is possible to reduce re-entrainment of particles from the dust box to the vortex finder. The cut size of cyclones with an apex cone is smaller than that without an apex cone, and Figure 2.10 shows these results. By use of the apex cone, it is possible to shift the cut size to the fine size region.[9]

1.2.3   WET CLASSIFICATION

The principle of wet classification is the same as dry classification. Recently, wet classification has been used for producing ceramic particles and in recycling processes. The following facts should be considered regarding wet classification characteristics.

1. Particle dispersion control is easy compared to dry classification.
2. In order to separate particles from liquid, a drying or dewatering process is necessary.

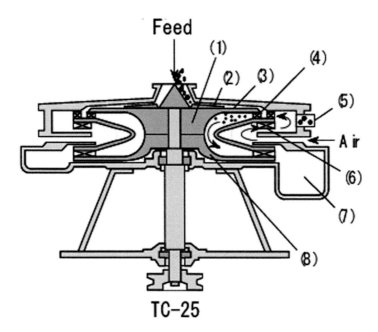

Feed

(1) Classification rotor   (5) Coarse powder outlet
(2) Dispersion blades   (6) Blades
(3) Dispersion disk   (7) Scroll casing
(4) Classification blades   (8) Balance rotor

FIGURE 2.4   Turbo classifier. (Courtesy of Nisshin Engineering.)

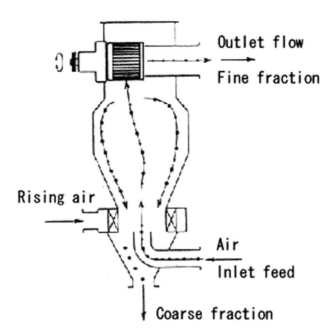

FIGURE 2.5   Turboplex classifier. (Courtesy of Alpine.)

TABLE 2.1   **Characteristics of Centrifugal Air Classifiers**

Type of classifier	Power (kw)	Cut size (μm)	Capacity (kg/h)	Rotor speed (rpm)
Micron separator	5–150	150–12,000	200–2,300	0.75–37
Turbo classifier	0.5–150	150–24,000	700–12,000	1.5–70
Turboplex	2–180	40–35,000	120–22,000	1–45

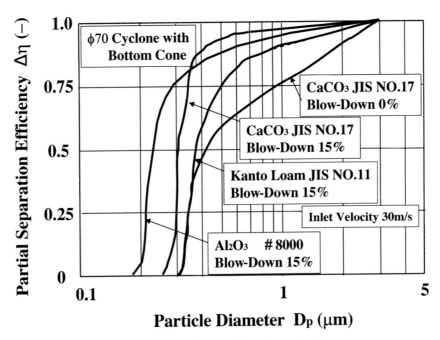

FIGURE 2.6   Particle separation efficiency of 70 cyclone classifiers.

3. The viscosity of the slurry decreases with increasing temperature, and the particle sedimentation velocity increases as the slurry temperature increases.

Classification Theory

Figure 2.11 shows the two types of classifiers, a vertical-flow type and a horizontal-flow type. For the vertical-flow type, the collection efficiency is calculated by Equation 2.13 and Equation 2.14:

$$V_g < u_0 \text{ or } V_g S < Q : E = 0 \tag{2.13}$$

$$V_g > u_0 \text{ or } V_g S > Q : E = 1 \tag{2.14}$$

where V_g is the particle sedimentation velocity, u_0 the upward velocity, S the cross-sectional area, and Q the volume flow rate.

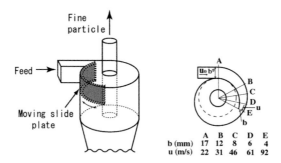

FIGURE 2.7   Cyclone with moving slide plate.

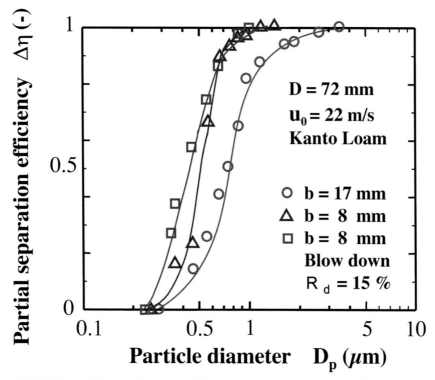

FIGURE 2.8   Effect of slide plate and blow-down on partial separation efficiency-

An ideal classification is realized for this type. However, the performance deteriorates when fluid turbulence is generated in the classifier. The collection efficiency for the horizontal type without fluid mixing is as follows:

$$\frac{V_g S}{Q} \leq 1 \qquad E = \frac{V_g S}{Q} \qquad (2.15)$$

$$\frac{V_g S}{Q} > 1 \qquad E = 1 \qquad (2.16)$$

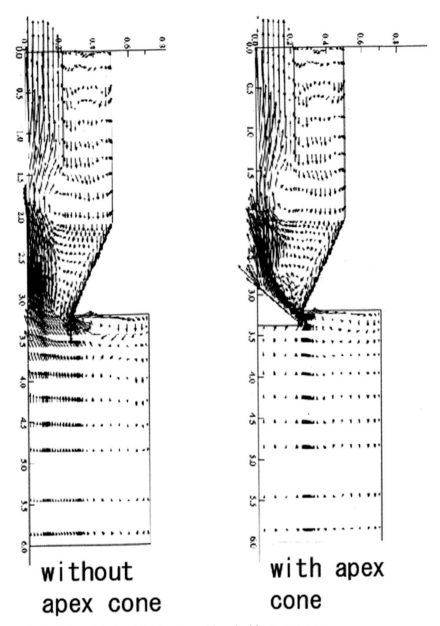

**without
apex cone**

**with apex
cone**

FIGURE 2.9 Calculated fluid vectors with and without apex cone.

For the horizontal type with fluid mixing in the vertical direction, the collection efficiency is calculated by Equation 2.17:

$$E = 1 - \exp(-\frac{V_g S}{Q}) \qquad (2.17)$$

Figure 2.12 shows the relation between the collection efficiency E and the parameter $(V_g S/Q)$ for three cases. Ideal classification is realized for a vertical-flow type. However, classification sharpness decreases for the horizontal-flow type with fluid mixing.

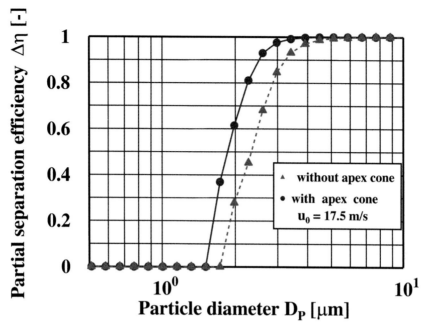

FIGURE 2.10 Partial separation efficiency with and without apex cone.

Sedimentation Velocity

The particle sedimentation velocity in a fluid is important in estimating the separation efficiency. Assuming a steady state, the Stokes sedimentation velocity is as follows:

$$V_s = \frac{(\rho_p - \rho_f)D_p^2 g}{18\mu} \qquad R_{ep} \leq 2 \qquad (2.18)$$

When the particle Reynolds number is greater than about 2, the sedimentation velocities calculated by the following equations are more accurate:

$$R_{ep} = \frac{D_p V_s \rho}{\mu}$$

$$\text{Allen region } V_s = (\frac{4(\rho_p - \rho_f)^2 g^2}{225\mu \rho_f})^{1/3} D_p \quad 2 \leq R_{ep} \leq 500 \qquad (2.19)$$

$$\text{Newton region } V_s = (\frac{3g(\rho_p - \rho_f)D_p}{\rho_f})^{1/2} \qquad 500 \leq R_{ep} \qquad (2.20)$$

Particle sedimentation velocities calculated by Equation 2.18 through Equation 2.20 are shown in Figure 2.13 for various particle densities. When the particle concentration increases, the particle sedimentation velocity calculated by Equation 2.18 through Equation 2.20 should be corrected using the following equation:

$$V_{sc} = \frac{V_s}{F(\varepsilon)} \qquad F(\varepsilon) \geq 1 \qquad (2.21)$$

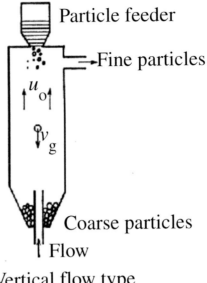

Particle feeder

Fine particles

Coarse particles
Flow

Vertical flow type

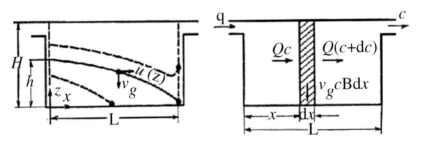

FIGURE 2.11 Vertical- and horizontal-flow-type classifier.

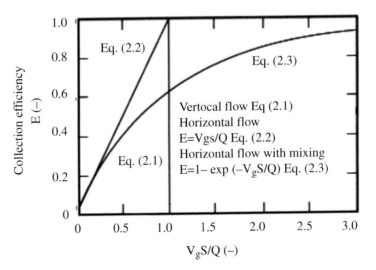

Vertocal flow Eq (2.1)
Horizontal flow
$E = V_g s/Q$ Eq. (2.2)
Horizontal flow with mixing
$E = 1 - \exp(-V_g S/Q)$ Eq. (2.3)

FIGURE 2.12 Relation between collection efficiency and parameter $(V_g S/Q)$.

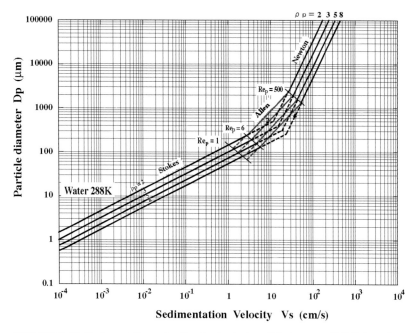

FIGURE 2.13 Sedimentation velocity of spherical particle in water (25°C, 1 atm).

The correction function $F(\varepsilon)$ is experimentally determined

$$F(\varepsilon) = \varepsilon^{-4.65} \tag{2.22}$$

where ε is the volume fraction of the fluid, and the value of $F(\varepsilon)$ increases with decreasing ε.

The sedimentation velocity calculated by Equation 2.21 is referred to as a hindered settling velocity.

Various Types of Classifier

The typical apparatus used in wet classification includes the following:

1. Horizontal-flow type
2. Vertical-flow type
3. Hydrocyclone
4. Centrifugal type with wall or blade rotation

Particle separators of the horizontal-flow type are shown in Figure 2.14. The coarse particles are collected on the bottom part and fine particles are collected to the fluid outlet part. Figure 2.15 shows a spiral classifier. This type is widely used because of its simple design and good performance. The coarse particles in the feed are discharged by the spiral flow, while the fine particles are collected in the overflow side.

Hydrocyclone

Hydrocyclones are widely used in many industrial processes. Experimental studies of the separation efficiency and pressure drop of a hydrocyclone have been reported.[10,11,12] Several flow controlling methods at the outlet pipe of a hydrocyclone have been proposed.[13,14] A prediction of the flow within

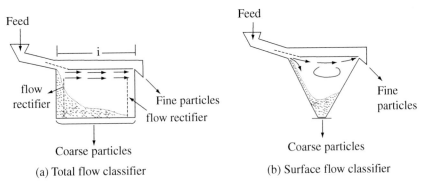

Feed

i

flow
rectifier   Fine particles
                                        flow rectifier

Coarse particles

(a) Total flow classifier

Feed

Fine
particles

Coarse particles

(b) Surface flow classifier

FIGURE 2.14 Particle separators of horizontal-flow type.

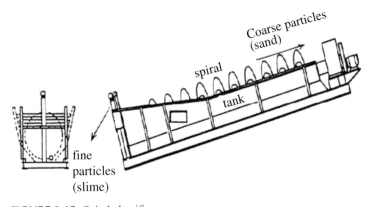

Coarse particles
(sand)

spiral

tank

fine
particles
(slime)

FIGURE 2.15 Spiral classifier.

a hydrocyclone was made using several turbulence models.[15] Performance of the axial-flow-type hydrocyclone was studied by Sineath and DellaValle.[16] Performance between conical and cylindrical hydrocyclones was examined by Chine and Ferrara.[17] The characteristics of hydrocyclones are as follows:

1. The size of a hydrocyclone is small compared to other types of wet classifiers.
2. The cut size decreases with an increase in inlet velocity or with a decrease in the cyclone diameter.
3. The pressure drop is about 0.5 to 5 Kg/cm², depending on the operating conditions.

Figure 2.16 shows the general flow pattern in a hydrocyclone. The coarse and fine particles are collected in the underflow and overflow sides, respectively. The ratio of inlet volume flow rate to underflow volume flow rate is referred to as the underflow ratio. The normal operating conditions for a hydrocyclone are as follows:

1 Cyclone inlet velocity $u_0 = 2 \sim 10$ m/s
2 Cyclone diameter $D_c = 1 \sim 30$ cm
3 Pressure drop $\Delta P = 0.5 \sim 5$ Kg/cm²
4 50% cut size $D_{pc} = 5 \sim 50$ μm
5 Underflow ratio $R_d = 5 \sim 20\%$

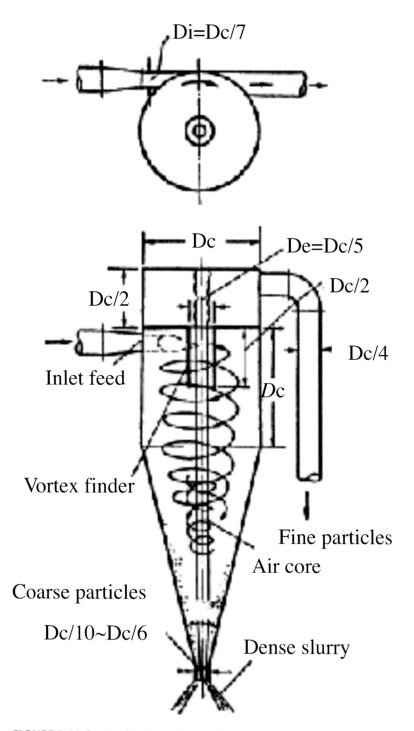

FIGURE 2.16 Standard hydrocyclone and flow pattern.

Figure 2.16 also shows the standard cyclone, and each of the dimensions is indicated using the cyclone diameter D_c.[12] The bottom diameter of the conical section is relatively small compared to a gas-cyclone. As the underflow rate increases, the air core near the axis is observed. In order to improve the sharpness of classification, the air core region should be eliminated. The typical separation efficiency curve obtained using the standard hydrocyclone is shown in Figure 2.17. The partial separation efficiency approaches the underflow ratio R_d as the particle diameter decreases.

In order to normalize experimental data for different underflow ratio conditions, the corrected partial separation efficiency $\Delta \eta_c$ calculated by Equation 2.23 is indicated in the figure.

$$\Delta \eta_c = \frac{\Delta \eta - R_d}{1 - R_d} \tag{2.23}$$

where D_{p50}^* denotes a 50% cut size under conditions where R_d is zero. The corrected collection efficiency indicated in Figure 2.17 can approximately be represented by the following equation:

$$\Delta \eta_c = 1 - \exp(-(\frac{D_p}{D_{p50}^*} - 0.115)^3) \quad 0.02 \le \Delta \eta_c \le 0.98 \tag{2.24}$$

Control of Cut Size

In order to improve the cut size control easily, the modified hydrocyclone shown in Figure 2.18 has been developed.[18,19] The inlet of the hydrocyclone is attached to a movable guide plate. The apex cone at the inlet of the underflow side is also attached. Both underflow and upward flow methods as in gas-cyclone are used to increase classification sharpness in fine size. In order to decrease the 50% cut size, a movable guide plate, shown in Figure 2.19, was used. Under constant feed rate conditions, the inlet velocity u increases with a decrease in inlet width b. The inlet width ratio G, defined by the following equation, is used as a variable parameter.

$$G = \frac{b}{b^*} \tag{2.25}$$

The value of G is equal to one for the standard case. Because of increased centrifugal force and the small radial sedimentation distance, small particles are easily collected for the case of a small inlet

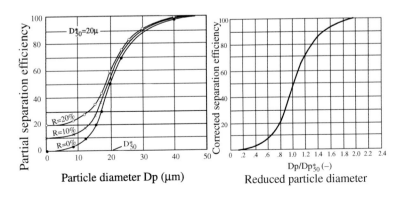

FIGURE 2.17 Particle separation efficiency and corrected partial separation efficiency.

width. Typical experimental results using the guide plate are shown in Figure 2.20. The 50% cut size decreases from 10 μm to 7 μm as the inlet width ratio decreases under a constant inlet flow rate. The partial separation efficiency approaches 0.1 for a particle diameter less than 5 μm. In order to increase classification sharpness in the small particle diameter region, upward flow and underflow methods were used simultaneously. Some experimental results for partial separation efficiency are shown in Figure 2.21. In this case, the upward flow, underflow, and movable guide plate were used.

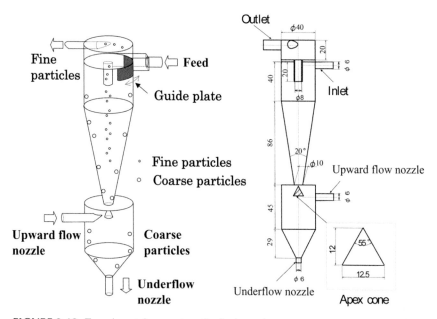

FIGURE 2.18 Experimental apparatus of a hydrocyclone.

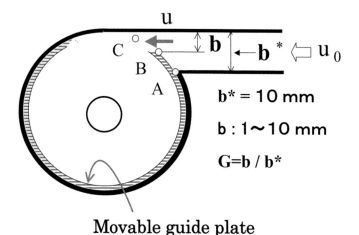

Movable guide plate

G [-]	1.00	0.75	0.50	0.25
u [m/s]	1.4	2.1	2.8	3.5

FIGURE 2.19 Cross section of a cyclone with a guide plate.

The upward flow ratio was 30%, the underflow ratio was 10%, and the inlet width ratio was changed from 1 to 0.1. The 50% cut size changes from 35 μm to 10 μm as the inlet width ratio decreases from 1 to 0.1. The partial separation efficiency in this case approaches zero as the particle diameter decreases.

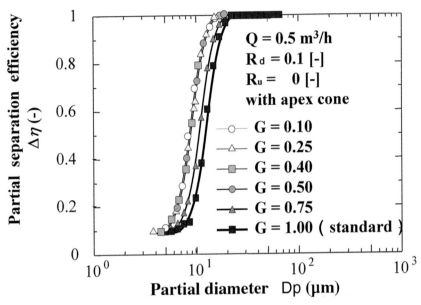

FIGURE 2.20 Classification performance using a guide plate and the underflow method.

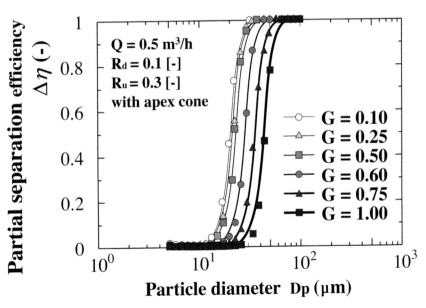

FIGURE 2.21 Classification performance using a guide plate, underflow, and upward flow methods.

Classification Theory of the Centrifugal-Type Separator

Figure 2.22 shows the centrifugal-type separator with a wall rotation.[20] The feed slurry is supplied to the bottom part, and classified coarse and fine particles are collected at the upper part of the rotating cylinder. The limiting particle trajectory satisfies the following conditions:

$$z = 0 : r = r_1, z = L : r = r_2 \tag{2.26}$$

Integrating the equation of particle motion, the partial separation efficiency can be represented as follows:

$$\Delta \eta = \frac{1}{1 - (\frac{r_1}{r_2})^2} (1 - \exp(-\frac{2 V V_g \omega^2}{g Q})) \tag{2.27}$$

The notation V is the volume of the apparatus, and Q is the maximum liquid flow rate defined by the following equation:

$$V = \pi (r_2^2 - r_1^2) L \tag{2.28}$$

$$Q = \frac{V L \omega^2 V_g}{g \ln \frac{r_2}{r_1}} \tag{2.29}$$

where ω is the rotational speed and V_g the terminal velocity of the particle. In order to separate fine particles, it is necessary to increase the length L and the rotational speed ω.

The 50% cut size in this type is small compared to a standard hydrocyclone. However, special care and maintenance are required at the rotor shaft and collection area of coarse particles in the cylindrical wall.

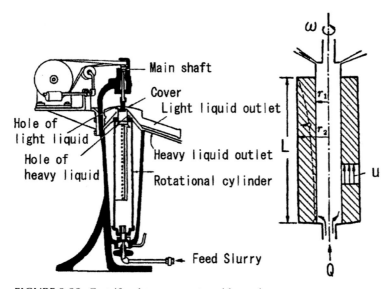

FIGURE 2.22 Centrifugal-type separator with rotation.

1.2.4 SCREENING

Screening is the separation of solids particles having various sizes into different size portions or classes using a screen surface. The screening surface acts as a multiple go/no-go gauge, and the final products consist of more uniform size than those of the original material.

The purposes of screening are (1) to remove fines from a raw material before using grinding equipment, (2) to scalp oversize material or impurities, and (3) to produce or process a commercial-grade product to meet specifications.

Screen Surface

Selection of a proper screen surface is very important; opening size, wire or thread diameter, and open area should all be carefully considered. The screening surface may consist of a perforated or punched plate, grizzly bars, wedge-wire sections, woven-wire cloth, and nylon, polyester, or other bolting cloth. Silk bolting cloth, which was widely used in the past, has largely been replaced by nylon cloth. An electroformed sieve, which has uniform apertures, has also been used for the accurate fractionation of fine powder ranging from 3 to 100 μm.[21,22]

Equipment

Screening machines may be classified into five main categories: grizzlies, revolving screens, vibrating screens, sifters, and air-assisted screening machines. Grizzlies are used primarily for scalping material of 50 mm and coarser size, while revolving screens or trommel screens are generally used for separations of material more than 1 mm in size. The screening machine consists of a cylindrical frame equipped with a wire cloth or perforated plate. The material to be screened is delivered at the upper end, and the oversize is discharged from the lower end. The screens revolve at a relatively low speed of 15 to 20 rpm. Such screens have largely been replaced by vibrating screens, but they are still used for special purposes, for example, in municipal solid waste separation processes.

Vibrating screens, whose screening surface vibrates perpendicular to the screening surface with a frequency greater than 600 rpm, are available for fine powders as well as coarse powders. There are a large variety of vibrating screens on the market, but basically they can be divided into two main types: mechanically vibrated screens and electrically vibrated screens. The most important factors for the selection of vibrating screens are amplitude and frequency. The centrifugal effect K is defined by the formula[23,24]

$$K = \frac{r\omega^2}{g} \tag{2.30}$$

where r = (amplitude in m)/2, $\omega = 2\pi n$, n = (frequency in rpm)/60, and g is the acceleration of gravity. If the centrifugal effect is too small, the near-size particles will wedge into the screen openings. This reduces the open area available for the passage of other fines and creates probably the single greatest limitation on the capacity of the screen. The proper vibrating action not only keeps the feed materials moving on the screen, but it serves primarily to reduce "blinding" to a minimum. Figure 2.23 shows the correlation between the amplitude and frequency for commercially used values of K. Sifters are characterized by low-speed (300 to 400 rpm), large-amplitude (smaller than about 50 mm) oscillation in a plane essentially parallel to the screening surface. Sifters are usually used for material of 30 μm to 400 μm or more in size and have many applications in chemical processes.[25] Most gyratory sifters have an auxiliary vibration caused by balls bouncing against the lower surface of the screen cloth. Figure 2.24 shows the schematics of a typical screening surface and driving mechanisms of vibrating screens and sifters. The air-assisted screening machine, shown in Figure 2.25, has been used for fine powders. The stream of air accelerates the passage of fine particles through the screen and removes the blinding particles in the apertures of screen.

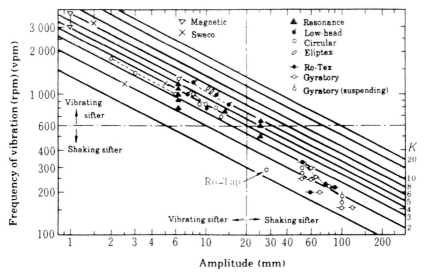

FIGURE 2.23 Centrifugal effect K of screening machines.

Relation between 50% Separation Size and Screening Length

According to Gaudin's probability theory, the passage probability P of particles of size D_p through an opening of size a in the cloth is given by

$$P = \frac{(a-D_P)^2}{a^2} = \left(1 - \frac{D_P}{a}\right)^2 \tag{2.31}$$

Introducing the concept of the number of passage trials i, the oversize fraction of particles after i trails, η_i, which is the partial separation efficiency, is expressed by

$$\eta_i = (1-P)^i = \left[1 - \left(\frac{a-D_P}{a}\right)^2\right]^i \tag{2.32}$$

If i is sufficiently large, Equation 2.32 is written approximately as

$$\ln \eta_i = -i\left(\frac{a-D_P}{a}\right)^2 \tag{2.33}$$

Hence the particle size at 50% separation efficiency, D_{p50}, is derived, substituting $\eta_i = 0.5$ into Equation 2.33:

$$D_{P50} = a - \frac{0.832a}{\sqrt{i}} \tag{2.34}$$

Vibrating screen		Sifter	
①	Inclined	⑦	Reciprocating
②	Low-head	⑧	Reciprocating
③	Hum-mer Rhewum	⑨	Exolon-grader
④	Ty-Rock	⑩	Traversator-sieb Sauer-meyer
⑤	or Gyrex	⑪	Gyro-sifter
⑥	Eliptex	⑫	Rotex screen

(a)

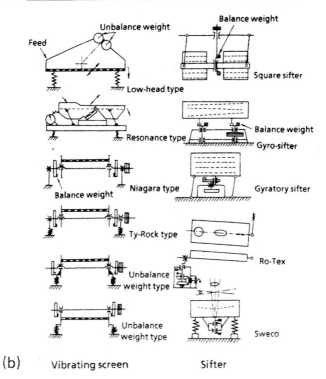

(b)   Vibrating screen   Sifter

FIGURE 2.24 (a) Motion of screen surface of vibrating screens and sifters; (b) driving mechanisms of vibrating screens and sifters.

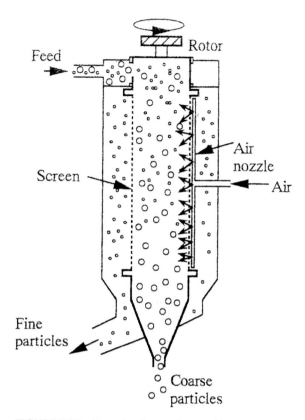

FIGURE 2.25 Air-assisted screening machine.

If $i = \xi l$ where ξl is the trial of particle passage per unit length of screen, the following equation holds approximately:

$$D_{P50} = a - \frac{0.832a}{\sqrt{\xi_1}}$$

(2.35)

REFERENCES

1. Leith, D. and Licht, W., *AIChE Symp. Ser.*, 126, 196–206, 1972.
2. Ayers, W. H. and Boysan, F., *Filtrat. Separat*, 22, 39–43, 1985.
3. Zhou, L. X. and Soo, S. L., *Powder Technol.*, 3, 45–53, 1990.
4. Yamamoto, M., Kitamura, O., and Arakawa, C., *Trans. Jpn. Soc. Mech. Eng.*, 59, 1959–1964, 1993.
5. Yamamoto, M. and Kitamura, O., *Trans. Jpn. Soc. Mech. Eng.*, 60, 4002–4009, 1994.
6. Yoshida, H. et al., *J. Soc. Powder Technol. Jpn.*, 36, 454–461, 1999.
7. Iinoya, K. et al., *KONA Powder Particle*, 11, 223–227, 1993.
8. Yoshida, H. et al., *KONA Powder Particle*, 12, 178–185, 1994.
9. Yoshida, H. et al., *Kagaku Kogaku Ronbunshu*, 27, 574–580, 2001.
10. Bradley, D., *Ind. Chemist*, Sept., 473–485, 1958.
11. Yoshioka, N. and Hotta, Y., *Chem. Eng.*, 19, 632–642, 1955.
12. Yoshioka, N., *Ekitai-Cyclone*, Nikkan Kogyo, 1962, pp. 38–41.
13. Chu, L. Y., Chen, W. M., and Lee, X. Z., *Chem. Eng. Sci.*, 57, 207–212, 2002.

14. Yamamoto, K. and Jiao, X., *Trans. Jpn. Soc. Mech. Eng. Ser. B*, 63, 133–138, 1997.
15. Petty, C. A. and Parks, S. M., *Filtrat. Separat.*, 28, 28–34, 2001.
16. Sineath, H. H. and DellaValle, J. M., *Chem. Eng. Prog.*, 55, 59–69, 1959.
17. Chine, B. and Ferrara, G., *KONA Powder Particle*, 15, 170–179, 1997.
18. Yoshida, H. et al., *J. Soc. Powder Technol. Jpn.*, 34, 690–696, 1997.
19. Yoshida, H. et al., *J. Soc. Powder Technol. Jpn.*, 38, 626–632, 2001.
20. Makino, K. et al., *Kagaku Kougaku Gairon*, Kaitei 11 Han, Sangyo Tosyo, Japan, 1989, pp. 270–272.
21. Hidaka, J. and Miwa, S., *Powder Technol.*, 24, 159–166, 1979.
22. Pierre, B., Jean, C., and Albert, L., *Particle Particle Syst.*, 10, 222–225, 1993.
23. Giunta, J. and Colijn, H., *Powder Handling Process.*, 5, 45–52, 1993.
24. Modrzewski, R. and Wodzinski, P., *Powder Handling Process.*, 10, 167–171, 1998.
25. Sato, Y., Uehara, K., Yasui, A., and Sakata, Y., *Trans. Jpn. Soc. Mech. Eng. Part C*, 59, 2688–2693, 1993.
26. Ishikawa, S., Shimosaka, A., Shirakawa, Y., and Hidaka, J., *J. Chem. Eng. Jpn.*, 36, 623–629, 2003.

1.3 Storage (Silo)

Minoru Sugita
Ohsaki Research Institute, Chiyoda-ku, Tokyo, Japan

1.3.1 GENERAL CHARACTERISTICS OF SILOS

General characteristics of silos used for storing powder and granular materials are as follows:

1. Granular materials can be collected, distributed, and stored in bulk efficiently.
2. Transportation costs, which influence the costs of raw materials and products, can be reduced.
3. Compared with storage on a flat surface such as a floor, a silo's storing capacity is several times greater in the same space.
4. Equipment cost per unit of storage capacity is small.
5. Automatic loading, unloading, and control of storage volume are possible.
6. Operations such as pressurization, heat insulation, moisture proofing, and fumigation are easily accomplished.
7. Quality change, decomposition, breakage, and damage of stored materials by insects and rats can be prevented.
8. A silo can be incorporated easily as a part of an industrial production system and has labor-saving advantages.

1.3.2 CLASSIFICATION OF SILOS

Shallow Bins and Deep Bins

When studying static powder pressure acting on silo walls, silos are classified into shallow bins and deep bins. The classification is based on the following formulas[1,2]:

Deep bins: $h > 1.5d (h > 1.5a)$
Shallow bins: $h \leq 1.5d (h \leq 1.5a)$

where h is the height of the silo (meters), d is the inside diameter of a circular silo (meters), and a is the length of a short side of a rectangular silo (meters).

Single Bins and Group Bins

For a single bin, a circular cross section is frequently used because of some advantages in design and construction. In recent years, coal silos 40–50 m in diameter and as high as about 40 m have been constructed, many of which are independent shallow bins. In addition, large single bins such as cement silos and crinker silos have been constructed. Many steel silos are of the single-bin type.

An example of a group bin is the silo for storing grains. Several to several tens of connected bins in a variety of shapes (e.g., circular, rectangular, and hexagonal in cross section) are used to store various types of powder and granular materials in bulk.

Closed and Open Types

Most bulk silos are of the closed type, equipped with a roof onto which loading equipment is installed. Some silos are airtight to permit the fumigation of imported grains, for example, using poisonous gas to exterminate vermin. Vacuum silos are also used to prevent powder clogging.

51

Open silos are used simply for storage and supply of granular material, the quality of which does not change upon exposure to rain or dust. Examples of these are ore silos (bunkers) and crushed limestone silos.

1.3.3 PLANNING SILOS

Calculation of Silo Capacity

In designing a silo, its capacity should be determined from the total storing weight of the materials, the types of the materials to be stored, and the conditions of use. Silo capacity has two components: total capacity (geometric capacity) and the capacity of loaded stored materials (effective capacity).

Geometric capacity, also called water capacity, is used as a standard value for calculating the fumigation gas to be employed in treating imported grains. Effective capacity is the base for calculating the storing weight and location for taking in materials whose angle of repose should be taken into consideration.

If geometric capacity is represented by V_W (m³) and effective capacity by V_E (m³), the loss volume V_L (m) becomes

$$V_L = V_W - V_E \text{ (m}^3) \tag{3.1}$$

The loss volume of a cylindrical silo, illustrated in Figure 3.1, can be determined from the formula 3:

$$V_L = \frac{4}{3} R^3 \left\{ 3F \int_0^{\pi/2} \cos^2 x \sqrt{(1-F) + F \cos^2 x} \, dx \right.$$
$$\left. + \int_0^{\pi/2} \left[(1-F) + F \cos^2 x \right]^{3/2} dx \right\} \tan \phi_r = c f_L R^3 \tan \phi_r \tag{3.2}$$

where

$$F = \left(1 - \frac{a}{R} \right)^2 \tag{3.3}$$

For a cylindrical container, Figure 3.1 symbols are used to calculate the effective capacity of silos:

$$V_E = \pi R^2 H + \frac{1}{3} \pi R^3 \tan \alpha \tag{3.4}$$

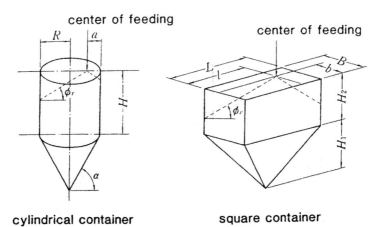

FIGURE 3.1 Symbols used to calculate the effective capacity of silos.

cf_L is the loss coefficient for the cylindrical silo. The effective capacity can be determined from Equation 3.1 through Equation 3.3.

Similarly, the calculation of the capacity of a rectangular silo can be made as follows. The loss capacity of a rectangular silo is

$$V_L = \frac{l^3}{6}\left[2\beta\sqrt{1+\beta^2} + \beta^2 \ln\left(\frac{1+\sqrt{1+\beta^2}}{\beta}\right) + \ln\left(\beta+\sqrt{1+\beta^2}\ \tan\phi_r\right)\right]$$

(3.5)

$$= _s f_L^3 \tan\phi_r$$

$$V_w = LBH_2 + \frac{1}{3}LBH$$

(3.6)

TABLE 3.1 Loss Coefficients

a/R or b/l	$_c f_L$	$_s f_L$	a/R or b/l	$_c f_L$	$_s f_L$
0.00	3.5555		0.50	2.4808	0.2966
0.02	3.5025	0.0100	0.52	2.4509	0.3115
0.04	3.4500	0.0200	0.54	2.4222	0.3267
0.06	3.3982	0.0301	0.56	2.3947	0.3423
0.08	3.3471	0.0403	0.58	2.3683	0.3581
0.10	3.2967	0.0506	0.60	2.3431	0.3742
0.12	3.2471	0.0610	0.62	2.3191	0.3906
0.14	3.1983	0.0715	0.64	2.2962	0.4074
0.16	3.1503	0.0822	0.66	2.2746	0.4245
0.18	3.1031	0.0931	0.68	2.2541	0.4418
0.20	3.0569	0.1041	0.70	2.2349	0.4595
0.22	3.0115	0.1154	0.72	2.2169	0.4776
0.24	2.9670	0.1268	0.74	2.2001	0.4959
0.26	2.9235	0.1384	0.76	2.1845	0.5146
0.28	2.8809	0.1502	0.78	2.1701	0.5336
0.30	2.8394	0.1623	0.80	2.1570	0.5530
0.32	2.7988	0.1746	0.82	2.1451	0.5728
0.34	2.7592	0.1871	0.84	2.1345	0.5927
0.36	2.7206	0.1998	0.86	2.1251	0.6130
0.38	2.6831	0.2129	0.88	2.1169	0.6337
0.40	2.6467	0.2261	0.90	2.1100	0.6547
0.42	2.6113	0.2397	0.92	2.1044	0.6761
0.44	2.5770	0.2535	0.94	2.1000	0.6979
0.46	2.5438	0.2676	0.96	2.0968	0.7199
0.48	2.5117	0.2819	0.98	2.0950	0.7424
—	—	—	1.00	2.0943	0.7651

where $\beta = b/l$ and $_sf_L$ is the loss coefficient for a rectangular silo. Values determined as a function of b/l are listed in Table 3.1. Effective capacity can be determined from Equation 3.1, Equation 3.5, and Equation 3.6.

Unloading Devices

Generally, a gravity discharge system is employed for unloading, but various special devices are used to discharge powder materials of high cohesiveness or materials that are liable to segregate, depending on granular size and composition. Although slight differences exist, many silos used to store cement, aluminum, and flour are equipped with an air slide system to discharge powder by blowing air from the bottom. Vibration is also used to discharge cohesive powders.

Even when the gravity discharge system is employed, the unloading device itself (e.g., rotary valve, screw feeder, or chain feeder) will differ depending on the materials being stored, and the hopper shape differs depending on the discharge system employed. Care should be taken in construction and design because the discharge system generates a large difference in the powder pressure acting on the silo walls when discharge is taking place. Some reports indicate the differences in dynamic pressure while discharging materials depend on the location of the discharge openings.

1.3.4 DESIGN LOAD

Design Recommendation for Storage Tanks and Their Supports, issued by the Architectural Institute of Japan in June 1983, and revised in March 1990,[1] is available for the structural design of silos. The design methods described include, for the first time, Japan's new earthquake-proofing requirements. According to the *Design Recommendation,*[1] the following loads should be considered: (1) dead load, (2) live load, (3) snow load, (4) wind load, (5) earthquake load, and (6) loads appropriate to the containers, such as impact and absorption due to the movement of bulk materials inside the containers.

1.3.5 LOAD DUE TO BULK MATERIALS

In designing bulk silos, a proper understanding of the behavior and pressure of bulk materials inside silos is clearly necessary. However, many unsolved and unanticipated problems remain. Bulk pressure changes in complexity, depending on various properties of the materials stored and the operating conditions of silos, thus relegating silo design to specialists.

According to the ISO 11697, *Bases for Design of Structures: Loads Due to Bulk Materials,*[4] the bulk pressures inside deep bins are discussed for two specified loading conditions. The filling pressures of bulk materials depend mainly on the material properties and the silo geometry. Discharge pressures are also influenced by the flow patterns that arise during the process of emptying. Therefore, an assessment of material flow behavior shall be made for each silo design.

In the assessment of bulk-material flow, it is necessary to distinguish among three main flow patterns:

1. Mass flow (Figure 3.2a): A flow profile in which all the stored particles are mobilized during discharge.
2. Funnel flow (Figure 3.2.b–3.2f): A flow profile in which a channel of flowing material develops within a confined zone above the outlet and the material adjacent to the wall near the outlet remains stationary. The flow channel can intersect the wall of the parallel section or extend to the top surface. In the latter case, the pattern is called "internal flow" (Figure 3.2c–3.2e).
3. Expanded flow (Figure 3.2f): A flow profile in which mass flow develops within a steep-bottom hopper, combining with a stationary in an upper, less steep hopper at the bottom of the parallel section. The mass flow zone then extends up the wall of the parallel section.

Storage (Silo)

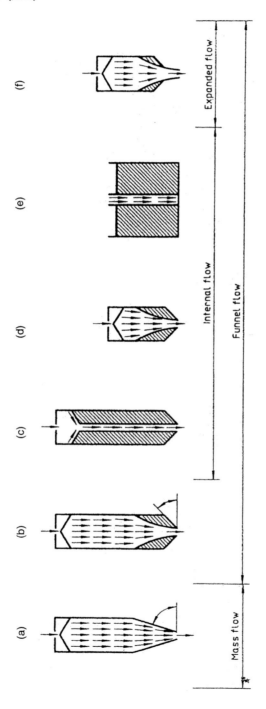

FIGURE 3.2   Flow patterns.

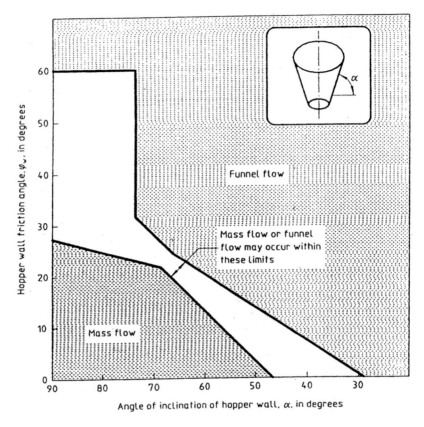

FIGURE 3.3   Limit between mass flow and funnel flow for circular hoppers.

Different pressure distributions are associated with each of the above flow patterns. The conditions necessary for mass flow depend on the inclination of the hopper wall and the wall friction coefficient. They can be estimated using Figure 3.3 for conical and axisymmetrical hoppers, and Figure 3.4 for configurations producing plane flow. The transition regions shown in Figure 3.3 and Figure 3.4 represent conditions in which the flow pattern can change abruptly between mass and funnel flow, thereby producing unsteady flow with pressure oscillations. If such a condition cannot be avoided, the silo shall be designed for both mass flow and funnel flow.

1.3.6   CALCULATION OF STATIC POWDER PRESSURE

In calculating bulk loads, the static powder pressure is the basis to be determined. The Janssen and Reimbert formulas are employed as silo design standards in various countries. The calculation of the static powder pressure in ISO 11967,[4] is based on Janssen's theory. It is derived from a force balance of the powder stored statically in deep bins, taking into consideration the frictional force generated between the powder and the silo walls.

If a silo of cross-sectional area A (m²) and circumference L (m) is filled uniformly with a powder of bulk density γ (tons/m³) (Figure 3.5), the vertical pressure P_v (tons/m²) inside the silo on the horizontal plane at x (m) can be expressed by

$$P_v \frac{\gamma A}{\mu K L}\left[1 - \exp\left(\frac{\mu K L}{A}x\right)\right] \qquad (3.7)$$

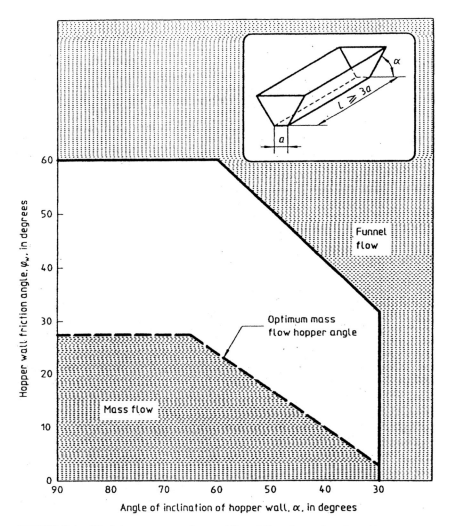

FIGURE 3.4   Limit between mass flow and funnel flow for wedge-shaped hoppers.

and replacing $A/L = R$ (hydraulic radius) by

$$P_v = \frac{\gamma R}{\mu K}\left[1 - \exp\left(-\frac{\mu K}{R}x\right)\right]$$ (3.8)

where $R = D/4$ for cylindrical silos of diameter D. Equation 3.8 is called the Janssen formula.

According to the assumption of Janssen's theory, the horizontal pressure P_h (tons/m²) is proportional to the vertical pressure P_v and can be expressed as follows, where the proportional constant is denoted by K:

$$\frac{P_h}{P_v} = K$$ (3.9)

$$P_h = KP_v = \frac{\gamma R}{\mu}\left[1 - \exp\left(-\frac{\mu K}{R}x\right)\right]$$ (3.10)

Assuming that $x \to \infty$ in Equation 3.8 and Equation 3.10, the maximum static pressure can be obtained as follows:

$$P_{vmax} = \frac{\gamma R}{\mu K}, \qquad P_{hmax} = \frac{\gamma R}{\mu} \qquad (3.11)$$

In calculating the K value, which is called the Janssen coefficient, the Rankine formula, used in soil mechanics, gives the relation to the angle of internal friction ϕ_i:

$$K = \frac{1 - \sin \phi_i}{1 + \sin \phi_i} \qquad (3.12)$$

To use the foregoing formulas in practice, it is necessary to know certain physical properties of the powder. The bulk density γ (tons/m³) is generally measured in a laboratory. Some powders have increased bulk density due to accumulated consolidation. The angle of internal friction of powder ϕ_i is measured by the shear cell method or triaxial compression test. The most difficult measurement is that of the friction angle (coefficient) of the powder against the wall. At present, it is difficult to obtain a correct understanding of the influence of the smoothness of silo walls on powder friction.

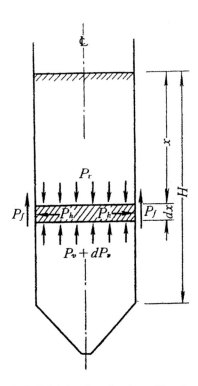

FIGURE 3.5   Powder is uniformly filled.

The static pressure of powder in shallow bins is almost negligible with regard to frictional forces between the silo walls and the powder. According the *Design Recommendation*,[1] the following formulas are specified for shallow bins:

$$P_v = \gamma x \tag{3.13}$$

$$P_h = K\gamma x \tag{3.14}$$

K, called the Rankine constant, can be determined from Equation 3.12 for both shallow and deep bins.

1.3.7   DESIGN PRESSURES

To calculate the design pressure, it is necessary to take into consideration the fact that dynamic pressure can be generated during discharge, and impact pressure can be generated during loading. Figure 3.6 shows distributions of powder pressure measured in actual silos. It is clear that powder pressures measured during discharge are larger than those obtained from the Janssen formula, especially the maximum pressure, which is four to five times higher than the value from Janssen's theory. In the design standards of various countries, the correction factors for dynamic overpressure during discharge and impact pressure during loading are introduced based on practical data.

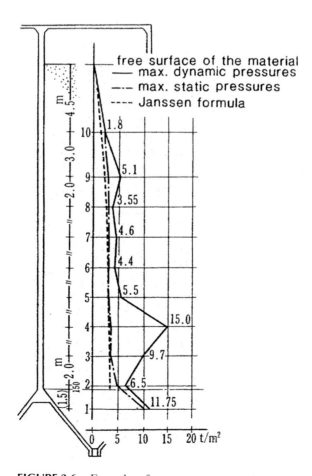

FIGURE 3.6   Examples of pressure measurement.

The correction factor for dynamic pressure is usually taken as 2.0 and that for impact pressure is 1.0–2.0. By multiplying the static pressure in the preceding section by the correction factor, the minimum design powder load is determined.

ISO 11697[4] shows the design wall pressure as follows. In silos where the flow zone intersects the wall (i.e., all flow patterns expect internal flow), the design discharge pressures shall be obtained by multiplying the filling loads by the overpressure coefficient C. The value of C shall be related to the silo aspect ratio h/d:

for $h/d < 1.0$, $C = 1.0$
for $1.0 < h/d < 1.5$, $C = 1.0 + 0.7(h/d - 1.0)$
for $h/d > 1.5$, $C = 1.35$

These values apply only to materials listed in tables that are indicated in this international standard.

REFERENCES

1. Architectural Institute of Japan, *Design Recommendation for Storage Tanks and Their Supports*, Architectural Institute of Japan, 1990.
2. Soviet Code, Ch-302-63, 1965.
3. Vaillant, A., *Chem. Eng.*, 69, 148, 1962.
4. ISO 11697, *Bases for Design of Structures: Loads Due to Bulk Materials*, International Standards Organization, 1995.

1.4 Feeding

Hiroaki Masuda and Shuji Matsusaka
Kyoto University, Katsura, Kyoto, Japan

1.4.1 INTRODUCTION

Powder feeders play an important role in powder-handling processes such as feeding raw materials or discharging product materials from a storage vessel. The powder load or mass to be treated in the processes depends on the accuracy of the feeders. Product quality might depend on the powder load, and low-quality materials will be produced if the load is too high for the process. Feeders should be selected carefully, because some feeders might be unsuited to the process, resulting in a failure of feeding itself. Further, dynamic characteristics of the feeders could also affect the accuracy of feeding. If the feeder cannot respond to a fast change in the manipulating signal, a fluctuation of flow caused by a disturbance cannot be suppressed and could cause a dynamic error in the feeding process.

The static and dynamic characteristics of feeders will depend on powder properties such as particle size, shape, internal friction coefficient, and powder flowabilities. They will also depend on the operating conditions, including temperature, pressure, and moisture content of powder in the process.

Further, feeders are always associated with feed hoppers, and their design might affect the function of feeders. If the feed hopper is not well designed, the flow mode in the hopper will be an unfavorable one called funnel flow, which can cause particle bridging, rat holing, and flushing. Failure of feeding will occur when the selection or maintenance of feeders is inadequate or hopper design is improper.

The power supply should be sufficient so that the feeder will operate even under conditions of a light overload caused by a disturbance. Additional parts, such as a shear pin, are usually incorporated in the feeder design to prevent corapus. However, if the shear pin is too weak, the feeder will often malfunction. Variations in powder properties or operating conditions are also important factors in feeding failure.

Unstable flow is caused by poor feeder selection or inadequate hopper design. Unstable flow with bridging or flushing will also be caused by a variation in powder properties or operating conditions. The wall surface can be coated with ultrahigh-molecular-weight materials so as to prevent these troubles. Suitable vibrators can also be utilized for this purpose. The addition of a small amount (0.2–5 wt%) of fine particles, such as kaolin, diatomaceous earth, cornstarch, silica gel, or magnesium stearate, will decrease both the wall and internal frictions of powders. Granulation, drying, or encapsulation will also decrease the friction so that the powder flow can be changed from the funnel flow mode to the mass flow mode.

The following items are necessary for feeders:

1. They must be suitable for the properties of the powder.
2. The operating range must be sufficiently broad.
3. The static characteristics, such as the relationship between powder feed rate and rotational speed of a drive motor, must be stable, and the repeatability of feed rate should be high.
4. They must have excellent dynamic characteristics.

As mentioned earlier, powder properties affect powder flow. It is necessary to know properties such as particle size, particle shape, cohesiveness, frictional property, and abrasive property so as to

TABLE 4.1   Empirical Rule for the Selection of Feeders[a]

Feeder	Particle Size[b]					Flowability[c]				Abrasiveness, Etc.[d]					
	a	b	c	d	e	f	g	h	i	j	k	l	m	n	o
Gate valve	D	O	O	O	D	X	O	D	D	D	X	D	O	D	D
Rotary	D	O	O	X	X	O	O	X	X	D	X	X	D	X	X
Table	D	O	O	O	D	D	O	D	D	O	D	X	O	X	D
Belt	D	O	O	O	O	D	O	O	D	O	O	O	O	D	O
Screw	O	O	O	D	X	O	O	D	D	D	X	X	D	D	D
Vibrating	O	O	O	O	X	D	O	O	D	O	X	O	O	X	O

[a] O, Applicable; D, difficult; X, no use.

[b] a, less than 100 μm; b, 100 μm–1 mm; c, 1 mm–1 cm; d, 1 cm–10 cm; e, larger than 10 cm.

[c] f, excellent; g, moderate; h, low; i, cohesive.

[d] j, abrasive; k, fragile; l, low bulk density; m, high temperature; n, slurry; o, flaky, fibrous.

TABLE 4.2   Characteristics of Feeders as Final Control Means

Feeder	Operating Method	Control Element	Dynamics	Statics
Electromagnetic	Voltage	SCR	Fast	Nonlinear
Screw	rpm[a]	VS motor	Moderate	Linear
Belt	Gate	Servomotor	Dead time	Nonlinear
	Belt speed	VS motor	Moderate	Linear
Table	rpm[a]	VS motor	Fast	Linear[b]
	Scraper	Servomotor	Fast	Nonlinear
Rotary	rpm[a]	VS motor	Moderate	Linear[b]
Gate valve	Gate	Servomotor	Fast	Nonlinear

[a] rpm, nominal speed.

[b] Restricted to lower rotational speed.

select appropriate feeders. If the feeder is inadequate, it becomes impossible to control the powder feed rate. Table 4.1 shows an empirical rule for the selection of various feeders. The rotary feeder, for example, can be applied to particles between 100 μm and 1 cm in diameter, but it is difficult to utilize for particles below 100 μm. As shown in Table 4.1, no feeder is suitable for use with adhesive powders, and careful attention should be paid to feeder selection.[1,2] Each of the feeders is described in detail in the next section.

The operating range of a feeder should be wide enough to cover the feed rate range required in the process, with a margin of 10%, so that the feeder will work well under conditions of unexpected disturbance. Also, the dynamic response should be as fast as possible. The dynamic characteristics of typical feeders are outlined in Table 4.2.

1.4.2   VARIOUS FEEDERS

Rotary Feeders

A rotary feeder consists of a rotor, a rotor case, and a motor drive. Powder in a hopper flows into the rotor space due to gravity and is discharged through the exit after a half revolution. Figure 4.1 shows

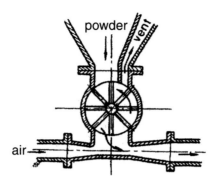

FIGURE 4.1 Rotary feeder.

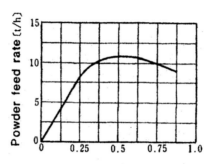

Peripheral velocity of rotor (m/s)

FIGURE 4.2 Powder feed rate of a rotary feeder.

a rotary feeder applied in a pneumatic transport line. There are various types of rotary feeders, and they are sometimes called rotary valves, rotary dischargers, or vane feeders.

The powder feed rate depends on the rotational speed of the rotor, as depicted in Figure 4.2. Corresponding volumetric efficiency decreases with increasing rotational speed. The static characteristics depend on powder properties, depth of rotor space, and inlet area.[3–5] The instantaneous powder feed rate fluctuates periodically because of the rotor configuration. Some modifications are incorporated in the rotor (helical rotor) or inlet configuration in order to suppress fluctuations. Although a rotary feeder can be utilized to feed particles against a pressure difference below 2 atm, the feed rate decreases with increasing pressure difference.[3] The dynamic characteristic of a rotary feeder is modeled as a first-order time delay.

Screw Feeders

A screw feeder consists of a screw, a U-shaped trough or cylindrical casing, and a motor drive. As the screw rotates, particles are forced to move from the hopper to the outlet of the feeder. Figure 4.3 shows a screw feeder with a U-shaped trough. A modified screw (e.g., tapered screw) is utilized so as to obtain a uniform flow pattern in the feed hopper.[6,7] Feeding against a pressure difference is also possible by modifying the screw configuration. For wet powder feeding, a screw feeder consisting of twin screws is suitable. A coil called an auger may also be utilized instead of a normal screw.

The powder feed rate is proportional to the rotational speed of the screw as long as the powder compressibility is negligible. The instantaneous powder feed rate fluctuates periodically as the screw rotates. The fluctuation can be suppressed by using a smaller screw and higher rotational speed, or

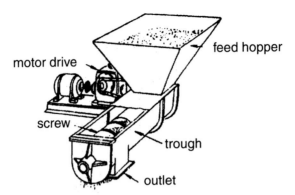

FIGURE 4.3   Screw feeder.

by using a coil screw (an auger feeder). The dynamic characteristic is modeled as a second- or third-order time delay.[8]

Table Feeders

A table feeder consists of a turntable, a scraper, and a motor drive, as shown schematically in Figure 4.4. The powder feed rate can be adjusted by changing either the scraper position or the rotational speed of the turntable. The skirt clearance S in Figure 4.4 is also changed when an extremely wide operational range is required. The powder feed rate is proportional to rotational speed in the practical range of operation, but it is a nonlinear function of the scraper position. The dynamic characteristics are modeled as either a proportional element or a derivative element.[9,10] Further, the instantaneous feed rate can be fairly smooth compared with that of the rotary or screw feeder. The flow pattern in the feed hopper is affected by the hopper inclination and the scraper position. Flow distortion can be controlled by using more than two scrapers.[11]

For fine powders, a special scraper is used to extend the powder uniformly on the turntable before feeding by normal operation.[12] Also, in some types of table feeders, a rotating shell is utilized instead of the turntable. A stationary scraper strips off particles from a gap between the rotating shell and a plate. Table feeders of this type are Auto-feeder, Omega feeder, Bailey feeder, Bin-discharger, and Com-Bin feeder. In these feeders, powder is confined in a vessel.

Belt Feeders

An endless belt pulls out the powder from a feed hopper as shown schematically in Figure 4.5. Belt feeders are easily combined with load cells, and they work as constant feed weighers or belt scales. Powder feed rate is adjusted by changing the belt speed or a gate opening. The relationship between the feed rate and the belt speed is linear, but it is a nonlinear function of the gate opening.[8] Therefore, flow-rate adjustment by changing the belt speed is more preferable than changing the gate opening. Further, the dead time associated with the gate operation causes deterioration of the dynamic response in this feeding system. For fine powders, a deaeration chamber should be attached. The chamber is also effective to prevent flushing of powder.

Vibrating Feeders

Vibrating feeders utilize either electromagnetic or electromechanical drives. Figure 4.6 shows an electromagnetically vibrating feeder. The vibrating trough transports the particles smoothly. The vibration is selected near the resonance frequency.[13] The flow pattern in the feed hopper is affected by the feeder. For fine powders, an appropriate system is necessary to prevent flushing. The dynamic characteristic is modeled as a first-order time delay.[14]

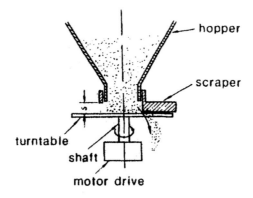

FIGURE 4.4   Table feeder.

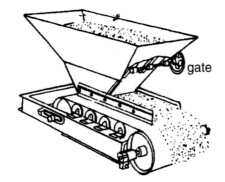

FIGURE 4.5   Belt feeder.

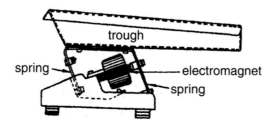

FIGURE 4.6   Vibrating feeder (electromagnetic type).

Also, there is a method using a vibrating capillary tube.[15,16] Even for micron-sized particles, micro-feeding is possible at a constant rate as small as milligrams per second. For adhesive fine powders, ultrasonic vibration is effective in the feeding.

Valves and Dampers

Valves and dampers are utilized in controlling the flow rate of free-flowing particles. Valves in common use in powder-handling industries are cut gates, slide valves, flap valves, vibrating dampers, inclined chutes, lock hoppers, and sleeve valves. These valves are specially modified to prevent particle clogging. There is a kind of valve where the powder flow rate is controlled by blowing off a heap of particles through air injection.[17] A special chute called an air slide is made of porous material through which air is supplied. The powder feed rate through the air slide is controlled by the use of a gate and the airflow rate supplied.[18]

REFERENCES

1. Aoki, R., Ed., *Funryutai no Choso to Kyokyu Sochi (Feeders and Hoppers in Powder Handling Processes)*, Nikkan Kogyo Shimbun, Tokyo, 1963, Chap. 5.
2. McNaughton, K., Ed., *Solids Handling*, McGraw-Hill, New York, 1981, Sec. 2.
3. Jotaki, T. and Tomita, Y., *J. Res. Assoc. Powder Technol. Jpn.*, 7, 534, 1970.
4. Masuda, H., Kameda, T., and Iinoya, K., *Kagaku Kogaku*, 35, 917–924, 1971.
5. Finkbeiner, T., *VDI Forschungsh*, 563, 1974.
6. Johnason, J. R., *Chem. Eng.*, 76 (Oct. 13), 75, 1969.
7. Bates, L., *Trans. ASME*, B91, 295, 1969.
8. Masuda, H., Masuda, T., and Iinoya, K., *J. Res. Assoc. Powder Technol. Jpn.*, 7, 479–484, 1970.
9. Masuda, H., Masuda, T., and Iinoya, K., *Kagaku Kogaku*, 35, 559–565, 1971.
10. Masuda, H., Miura, K., and Iinoya, K., *J. Soc. Mater. Sci. Jpn.*, 21, 577–581, 1972.
11. Masuda, H., Han, Z., Kadowaki, T., and Kawamura, Y., *KONA Powder Sci. Technol. Jpn.*, 2, 16–23, 1984.
12. Masuda, H., Kurahashi, H., Hirota, M., and Iinoya, K., *Kagaku Kogaku Ronbunshu*, 2, 286–290, 1976.
13. Arima, T., *Kagaku Kojo*, 9(2), 34, 1965.
14. Iinoya, K. and Gotoh, K., *Seigyo Kogaku*, 7, 646, 1963.
15. Matsusaka, S., Yamamoto, K., and Masuda, H., *Adv. Powder Technol.*, 7, 141–151, 1996.
16. Matsusaka, S., Urakawa, M., and Masuda, H., *Adv. Powder Technol.*, 6, 283–293, 1995.
17. Bendixen, C. L. and Lohse, G. E., in *Symposium on Solid Handling*, 1969.
18. Mori, Y., Aoki, R., Oya, K., and Ishikawa, H., *Kagaku Kogaku*, 19, 16, 1955.

1.5 Transportation

Yuji Tomita
Kyushu Institute of Technology, Kitakyushu, Fukuoka, Japan

Hiromoto Usui
Kobe University, Nada-ku, Kobe, Japan

1.5.1 TRANSPORTATION IN THE GASEOUS STATE

Introduction

Transportation of powder in the gaseous state is known as pneumatic conveying, of which familiar application is a vacuum cleaner, and uses air forces acting on particles exposed in an air stream in a pipe. The transport distance is 2 km at most, and to select flexible routes is easy due to pipeline transportation. Determination of an optimum pipeline and air source is required in design for a given transport distance and mass flow rate of given powder. Conveying systems consist of an air source, powder feeder, conveying pipe, gas particle separator, and air filter. In the positive systems (Figure 5.1), powder is fed into an air stream through a feeder at above ambient pressure and is continuously discharged into the outside at destinations. Foreign substances are not mixed in transported materials. When the air pressure at the feeding point is high, a continuous feeding is difficult and a batch system using a blow tank is employed (Figure 5.2). The positive systems are favored to deliver the powder to several destinations. In the negative systems, the powder can be continuously fed into an air stream at ambient pressure, but it is difficult to discharge it continuously to the outside at the destinations. While the available pressure for transport is limited, the powder does not leak out of the pipeline. These systems are used to collect powder from several points and deliver it to one destination. Besides, there are positive–negative systems that take advantage of both systems. While most systems are open, using air as a conveying medium, closed systems are used for specific cases such as toxic, explosive, and hygroscopic powders in a controlled environment. Circulating systems are employed for fluidized beds.

Model of Gas Particle Flow in a Pipeline

A working model of gas solid flow in a pipeline is a steady two-phase flow model,

$$\rho_s c v \frac{dv}{ds} = -\frac{d}{ds}(pc) - \rho_s c g \frac{dz}{ds} - \frac{4\tau_{ws}}{d} - F \tag{5.1}$$

$$\rho(1-c)u \frac{du}{ds} = \frac{d}{ds}\{p(1-c)\} - \rho(1-c)g \frac{dz}{ds} - \frac{4\tau_{wa}}{d} - F \tag{5.2}$$

where ρ and ρ_s are the gas and solid material densities, v and u solid and gas velocities, τ_{ws} and τ_{wa} the solid and gas wall shear stresses, p the pressure, c the solid concentration, z the height, g the gravitational acceleration, d the pipe diameter, s the coordinates along the pipe axis, and F the

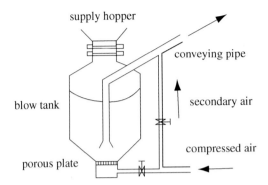

FIGURE 5.1   Positive pressure system.

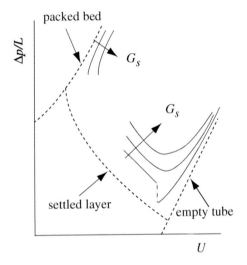

FIGURE 5.2   Blow tank solid conveyor.

interaction force between gas and solid. Based on this model, the pressure drop along the pipeline is given by

$$(p_2 - p_1) = \frac{4G_s}{\pi} \int_{v_1}^{v_2} \frac{dv}{d^2} + \frac{4G}{\pi} \int_{u_1}^{u_2} \frac{du}{d^2} + g \int_{z_1}^{z_2} \{\rho_s c + \rho(1-c)\} dz$$
$$+ 4 \int_{s_1}^{s_2} \left(\frac{\tau_{ws} + \tau_{wa}}{d} \right) ds + \sum_i \Delta p_{si} + \Delta p_{ai} \tag{5.3}$$

where G and G_s are the gas and solid mass flow rates, and Δp_{si} and Δp_{ai} are the local pressure losses in pipeline due to solid and gas.

Flow Patterns of Gas Solid Flow in Pipes

Since $\rho_s/\rho \sim 10^3$, the influence of gravity on the flow is strong and particle flow patterns depend on pipeline configurations. Figure 5.3 schematically shows a phase diagram for horizontal flow,[1] where $\Delta p/L$ is the pressure drop per unit length and U is the superficial air velocity defined by $u(1 - c)$.

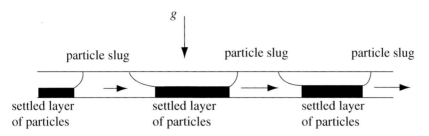

FIGURE 5.3 Phase diagram for horizontal flow.

Transportation of particles is possible in a region enclosed by lines of empty tube, packed bed, and settled layer. Air velocity at which $\Delta p/L$ becomes minimum for a given G_s is called the saltation velocity U_s,[2] which is important in the design of conveying system and above which the particles are conveyed suspended in air flow. Matsumoto et al.[3] gives the following equation for U_s based on their measurements for particles of 0.29 mm $\leq D \leq$ 2.6 mm and pipe of $d = 26$ and 49mm

$$m = 0.448 \left(\frac{\rho_s}{\rho} \right)^{0.50} \left(\frac{u_g}{10\sqrt{gD}} \right)^{-1.75} \left(\frac{U_s}{10\sqrt{gd}} \right)^{3.0} \tag{5.4}$$

where $m = G_s/G$ is the solid loading ratio, D the mean size of particle, and u_g the terminal settling velocity of the single particle. Below U_s, the flow becomes heterogeneous and takes various patterns depending on the pipe size and particle properties. Low-velocity transport is advantageous to avoid pipeline erosion and degradation of particles. However, there is danger of pipeline blockage if the air velocity is too low. For coarse particles, the extrusion flow is observed near the packed bed line in a short pipeline, where particles are fully filled throughout the pipeline. When decreasing U below U_s, the flow becomes intermittent and unstable. Further reduction in U, a stable slug flow (Figure 5.4), appears where there remain particle settled layers along the pipeline. Particles in a slug are always replaced with those in the settled layer that are transported by a definite distance between every slug passing. A plug flow appears when the settled layer disappears. When particle flow is heterogeneous and is not continuous, it is difficult to locate such a flow pattern in the phase diagram, and the relation between $\Delta p/L$ and U depends on conditions such as feeding method and length of pipeline. For fine powder, the flow pattern changes smoothly to fluidized dense phase flow from suspension flow. Figure 5.5 shows a phase diagram for vertical flow, where A is the point of incipient fluidization and B is the terminal settling velocity of a single particle. Vertically upward transportation is possible in a region enclosed by lines of empty tube, packed bed, and fluidization. Fluidization is a counterpart of the settled layer in horizontal flow in a sense that there is no net particle transport while there is particle circulation suspended by the air stream. Corresponding to U_s, there is a choking velocity $U_c = u_c \epsilon_c$ in the vertical flow for which Yang[4] gives the following empirical relation:

$$G_s = \frac{\pi d^2}{4} \rho_s (1 - \epsilon_c)(u_c - u_g), \quad \frac{2gd(\epsilon_c^{-4.7} - 1)}{(u_c - u_g)^2} = 6.81 \times 10^5 \left(\frac{\rho}{\rho_s} \right)^{22} \tag{5.5}$$

Above U_c the particle flow is suspension. Below U_c various flow patterns are observed and some of them are similar to those in horizontal flow. Figure 5.6 shows a slug flow in a vertical pipe where the particle slugs propagate upward in substantial particle holdups.

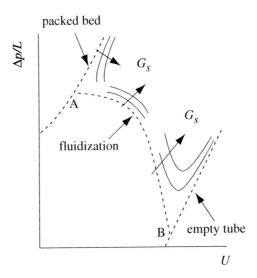

FIGURE 5.4   Slug flow in a horizontal pipe.

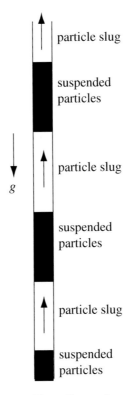

FIGURE 5.5   Phase diagram for vertical flow.

Pressure Loss Calculation for Suspension Flow

We can use Equation 5.3 to calculate a total pressure loss in pipeline for suspension flow. We need v for calculating the first term on the right-hand side of Equation 5.3, and c for the third term. When it is difficult to find them, we can use the following approximation:

$$v \cong U - u_g \quad \rho_s c + \rho(1-c) \cong \rho(m+1) \tag{5.6}$$

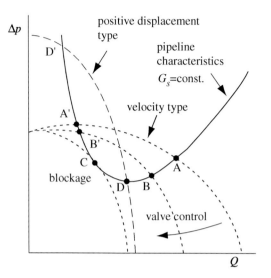

FIGURE 5.6 Slug flow in a vertical pipe.

There are many types of correlation for the friction term and the following correlation is recommended:

$$\left(\lambda + m\lambda_z\right)\frac{\rho U^2}{8} = \tau_{wa} + \tau_{ws} \tag{5.7}$$

where λ is the pipe friction coefficient for gas and λ_z is the additional pressure loss coefficient due to powder. Although there are numerous references for various materials, it is best to use the measurement for a given powder, since this term is important. Weber[5] gives the following empirical equation for horizontal flow of several granules:

$$\lambda_z = 4.56 m^{0.6} \left(U/\sqrt{gd}\right)^{-1.970} \tag{5.8}$$

It is a practice for vertical flow to regard the gravitational term as a loss and include it in the additional pressure loss together with the friction term as follows:

$$m\lambda_z \frac{1}{d}\frac{\rho U^2}{2} = g\rho_s c + \frac{4\tau_{ws}}{d} \tag{5.9}$$

We can approximately assume λ_z of this definition for the vertical flow as twice that for the horizontal flow. A serious local loss is due to 90° bends, Δp_{ab}, Δp_{sb}, which depends on the bend configurations, and for the rough estimate we can use[6]

$$\xi_b (m+1)\frac{\rho U^2}{2} = \Delta p_{ab} + \Delta p_{sb} \tag{5.10}$$

where, $\xi_b = 1.5, 0.75$ and 0.5 for $R_b/d = 2, 4$ and more than 6, respectively, R_b being the radius of the bend. It becomes accurate to consider the change of gas density along the pipeline when calculating the pressure loss. For this purpose we divide the pipeline into many sections, and we assume that the gas density is constant in each section. Furthermore, we assume that the gas flow is isothermal, and we use $\rho = p/RT$, where R is the gas constant and T the absolute temperature of the gas. At first, we guess the total pressure loss in a section and calculate the mean pressure in the section. Then we estimate the mean gas density and mean gas velocity U and revise the total pressure loss based on Equation 5.3. If this is close to the first guess, we stop calculation. Otherwise, we repeat the process until a given convergence is obtained. In the positive system, the pressure at a pipe exit is known, and the calculation is from the pipe exit toward the pipe inlet, while in the negative system it is *vice versa*. The pressure loss is in proportion to U^2 and $U = 4G/\rho\pi d^2$, then it is effective to keep U constant by increasing the pipe diameter in stepwise fashion toward the downstream in order to reduce power consumption, which is approximately estimated by $p_oQ_o \ln\{1 + (\Delta p/p_o)\}$ where Δp is the total pressure loss along the pipeline, p_o the ambient pressure, and Q_o the volumetric flow rate of free air. Furthermore, it is important to keep U always larger than the design velocity, in particular, at the pipe inlet where U is minimum.

Matching Air Source and Pipeline Characteristics

Operating points of air source are crossings between the air source and pipeline characteristics, as shown in Figure 5.7, where Q is the volumetric flow rate of air and Δp is the delivery pressure of air source or the total pressure loss for a given pipeline. The stable operating points are those at higher flow rate, that is, points A, B, C, and D. The points, A', B', and C' are unstable operating points. When the air source is the velocity type and we want to choose point B as an operating point, we can change the operating points by the valve or speed control. In the figure, the exit valve changes the characteristics of the velocity type of the air source. In this case, there is a limit point C that is called blockage velocity, below which the pipeline is blocked. The blockage velocity depends on the characteristics of the conveying system. A positive-displacement type of air source is used for low-velocity transport.

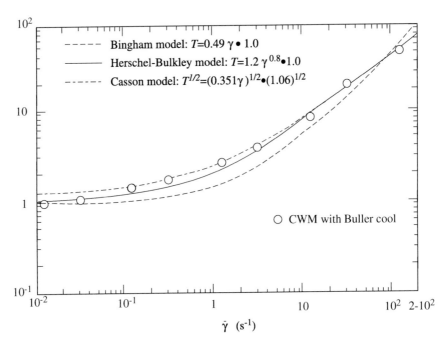

FIGURE 5.7   Operating points of conveying system.

1.5.2 TRANSPORTATION IN THE LIQUID STATE

Hydrotransport of solid particles is now an acknowledged alternative to conventional forms of transport, such as pneumatic or mechanical conveying. In the coal industry, for example, the use of highly loaded coal–water mixtures with reduced viscosity and the direct combustion of fines with a high water content have been proposed. In the transport of minerals and the disposal of waste materials, hydrotransport techniques have been employed for the reduction of manpower and cost. Conveying distances up to 400 km have been utilized. The capacity of a slurry pipeline is usually less than 12 million tons per year, and a pipe diameter is selected in the range 60–500 mm.

The following classification of slurries is useful to discuss the hydrotransport technique:

1. Homogeneous or nonsettling
2. Nonhomogeneous or settling

Homogeneous Slurries

The design of a slurry-handling system usually involves two steps: selection of pipe diameter and determination of frictional head loss. A homogeneous slurry can be tested in a Couette viscometer without significant settling. Some examples of the shear stress–shear rate relationship are shown in Figure 5.8. Three purely viscous non-Newtonian fluid models[7] (i.e., Bingham model, Hershel–Bulkley model, and Casson model) are compared in this diagram. Also, the power-law model, which shows a straight line in a log-log diagram such as shown in Figure 5.8, can be used to fit the narrow range of rheological data. A suitable rheological model should be selected by taking into account the shear rate range of the practical problem.

Laminar flow in cylindrical tubes has been analytically solved,[8] and the relationships between flow rate Q and pressure drop ΔP for some non-Newtonian fluid models are summarized as follows:

Power-law model: $\tau = m\gamma^n$

where
$$Q = \left(\frac{\pi R^3}{\left(\frac{1}{n} \right) + 3} \right) \left(\frac{\tau_R}{m} \right)^{\frac{1}{n}}, \text{ where } \tau_R = \frac{\Delta P R}{2L} \tag{5.11}$$

Bingham model: $\rho = \eta\gamma + \tau_y$

where
$$Q = \left(\frac{\pi R^3 t_R}{4n} \right) \left[1 + \left(\frac{3}{4} \right) \left(\frac{\tau_y}{\tau_R} \right) + \left(\frac{1}{3} \right) \left(\frac{\tau_y}{\tau_R} \right)^4 \right], \text{where } \tau_R = \frac{\Delta P R}{2L} \tag{5.12}$$

In the case of turbulent flows, prediction methods are proposed by Kemblowski and Kolodziejski[9] for the power-law model and by Wilson and Thomas[10] for the Bingham model.

Nonhomogeneous Slurries

Nonhomogeneous slurries cause the settling of solid particles. Thus, the flow situations are significantly different between horizontal and vertical positions. However, the long-distance transportation

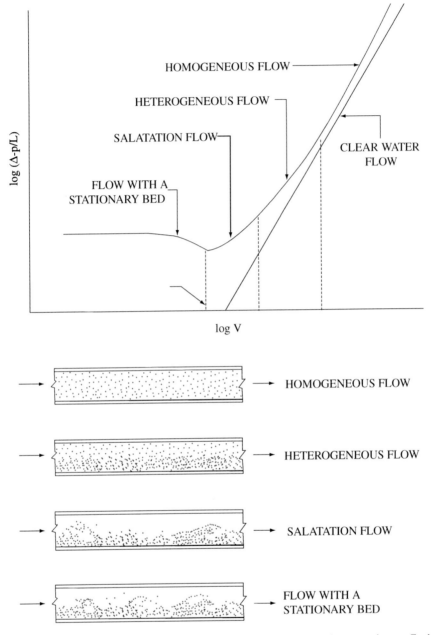

FIGURE 5.8 Shear stress–shear rate relationship obtained for a coal–water mixture. Coal concentration is 68.2 wt%. Slurry temperature and pH are 298 K and 8.6, respectively.

of slurries is mainly concerned with the horizontal pipeline. Horizontal transportation is discussed in this section. Flow regimes of slurries can be classified into four regions. The friction factor defined by $f = (-\Delta P/L)(D/2\rho v^2)$ is given for each flow regime as follows[11]:

Flow with a stationary bed:

$$f = f_w + 0.4036C^{0.7389} f_w^{0.7717} C_D^{-0.4054} \left(\frac{v^2}{Dg(s-1)} \right)^{-1.096} \tag{5.13}$$

Saltation flow:

$$f = f_w + 0.9857 C^{1.018} f_w^{1.046} C_D^{-0.4218} \left(\frac{v^2}{Dg(s-1)} \right)^{-1.354} \tag{5.14}$$

Heterogeneous flow:

$$f = f_w + 0.5518 C^{0.8687} f_w^{1.200} C_D^{-0.1677} \left(\frac{v^2}{Dg(s-1)} \right)^{-0.6938} \tag{5.15}$$

The deposition velocity, or limit-deposit velocity, is the usual lower limit of the slurry transportation velocity. Below this limit, a stationary deposit of particles forms on the bottom of the pipe. This makes stable pipeline operation very difficult. The correlation of deposition velocity was reported by Durand,[12] Oroskar and Turian,[13] and Gillies and Shook.[14] Gillies et al.[15] have proposed a method for predicting the pressure drop in the horizontal slurry pipeline flow based on the two-layer model.

The hydraulic capsule pipeline is another candidate to transport solid materials. A capsule pipeline system for limestone transportation has been in commercial use since 1983, with a capacity of 2 million tons per year, a transportation distance of 3.2 km, and a pipe diameter of 1.0 m.[16] In this case, a pneumatic capsule transport technique is used. A new capsule transportation technique, the coal log pipeline, has been proposed by Liu and Marrero[17] and Liu[18] and is utilized by the hydraulic driving force. The coal, extruded and compressed into a log shape, is used as the capsule. Consequently, the coal log pipeline is simpler to handle than the container type (hydraulic capsule).

Notation

C Solid volume fraction
C_D [$= (4/3)gd(s-1)/v\infty$] drag coefficient for a free-falling sphere
D Pipe diameter
d Diameter of solid particle
f Friction factor
f_w Friction factor for a liquid flow
g Gravity acceleration
L Pipe length
m Consistency index of the power-law model
n Power-law index
ΔP Pressure drop
Q Flow rate
R Pipe radius
s ($= \rho_s/\rho$) relative density
v Cross-sectional averaged velocity
$v\infty$ Terminal velocity of particle settling in an unbound fluid
γ Shear rate
η Viscosity
ρ Density of liquid
ρ_s Density of solid

τ Shear stress

τ_R Wall shear stress

τ_y Yield stress

REFERENCES

1. Welschof, G., *VDI-Forsch,* 492, 1962.
2. Zenz, F. A., *Ind. Eng. Chem. Fundam.,* 3, 65–75, 1964.
3. Matsumoto, S., Hara, M., Saito, S., and Maeda, S., *J. Chem. Eng. Jpn.,* 7, 425–430, 1974.
4. Yang, W. C., *Powder Technol.,* 35, 143–150, 1983.
5. Weber, M., *Bulk Solids Handling,* 11, 99–102, 1991.
6. Engineering Equipment Users Association, *Pneumatic Handling of Powdered Materials,* Constable, London, 1963. pp. 52–64.
7. Bird, R. B., Stewart, W. E., and Lightfoot, E. N., *Transport Phenomena,* Wiley, New York, 1960, p. 10.
8. Skelland, A. H. P., *Non-Newtonian Flow and Heat Transfer,* Wiley, New York, 1967, pp. 82 and 110.
9. Kemblowski, Z. and Kolodziejski, J., *Int. Chem. Eng.,* 13, 1973, 265.
10. Wilson, K. C. and Thomas, A. D. (1985). *Can. J. Chem. Eng., 63:*593.
11. Turian, R. M. and Yuan, T.-F., *AIChE J.,* 23, 232, 1977.
12. Durand, R., In *Proceedings of the Minnesota International Conference on Hydraulic Conveying,* 1953, p. 89.
13. Oroskar, A. R. and Turian, R. M., *AIChE J.,* 26, 550, 1980.
14. Gillies, R. G. and Shook, C. A., *Can. J. Chem. Eng.,* 69, 1225, 1991.
15. Gillies, R. G., Shook, C. A., and Wilson, K. C., *Can. J. Chem. Eng.,* 69, 173, 1991.
16. Kosuge, S., in *Proceedings of the Seventh International Symposium on Freight Pipelines,* Vol. 1, 1992, p. 13.
17. Liu, H. and Marrero, T. R., U.S. Patent No. 4,946, 317, 1990.
18. Liu, H., *Freight Pipelines,* Elsevier, New York, 1993, p. 215.

1.6 Mixing

Kei Miyanami
Osaka Prefecture University, Sakai, Osaka, Japan

1.6.1 INTRODUCTION

Powder mixing is an operation to make two or more powder ingredients homogeneous with, if necessary, some amount of liquid. In some industrial fields, it is called blending. The size of the powders to be mixed ranges widely, and the states of moisture range from dry to pendular. Because solid particles are subjected to various interactive forces and are not self-diffusive, they cannot be set in motion without any external force such as mechanical agitation. Different external forces can be applied to the powders to be mixed, and many types of mixers have been developed for a variety of applications.

1.6.2 POWDER MIXERS

Table 6.1 shows a classification of various powder mixers, based on the manner by which the powders are set in motion. This table also lists rough ranges of powder properties appropriate to each type of mixer. Typical structures of powder mixers are depicted in Figure 6.1. Although mixer performance should be evaluated on the basis of the powder properties being handled, operating conditions, and the application purpose, the general features of each mixer are as described below.

Rotary Vessel Type

The rate of mixing is rather low in a rotary vessel, but a good final degree of mixedness can be expected. The powders to be mixed are charged up to 30–50% of the vessel volume. The rotational speed is set at 50–80% of the critical rotational speed, N_{cr}, given as

$$N_{cr} = \frac{0.498}{\sqrt{R_{max}}} \; \left(s^{-1}\right) \tag{6.1}$$

where R_{max} (m) is the maximum radius of rotation of the mixer. For mixing powders with poor flowability and large differences in particle densities and diameters, various types of guide plates or internals are installed in the rotating vessel, as shown in Figure 6.1a and 6.1b. The addition of mixing aids or an operation at a rotational speed close to N_{cr} may be useful in some cases.

Stationary Vessel Type

With Mechanical Agitation

In stationary vessels using mechanical agitation, large amounts of powders can be handled in a small space. Specialized atmospheres as well as normal temperatures and pressures are accessible for multipurpose operations. Some types can be used in both batch and continuous modes.

Powder-Handling Operation

TABLE 6.1   Classification of Powder Mixers and Range of Their Services[a]

Classification		Mixer	Operation		Range of Particle Diameter (mm)				Flowability Angle of Repose (deg)				Differences in Powder Properties		Abrasive	Water Content		Symbol in Fig. 9.1
			Batchwise	Continuous	Over 1.0	1.0–0.1	0.1–0.01	Under 0.01	Under 35, H	35–45, M	Over 45, L	Cohesive	Small	Large		Dry	Wet	
Rotary vessel	Horisontal axis of rotation	Horizontal cylinder	○	○	○	○	●		○	●			○		○	○		
		Inclined cylinder	○		○	○	●		○	●				●	○	○		
		V-type	○		○	○	●		○	●			○		○	○		
		Double cones	○		○	○	●		○	●				●	○	○		
		Cubic	○		○	○	●		○	●			○		○	○		
		S-type	○		○	○	●		○	●			○		○	○		
		Continuous V-type		○	○	○	●		○	●			○		○	○		(c)

Mixing

			(d)			(e)	(f)	(g)		(h)	(i)	(a)	(b)		(j)
Stationary vessel	Horizontal axis of rotaion	Ribbon	●	○	○	●	○	○				○	○	○	○
		Screw	○	○	○	○	○	○	○	○	○	○	○	○	○
		Rod or pin	○	○					○	○	○	○	○	○
		Double-axle paddles	●	○	○	●	○		○	○	●	○	○	●
	Vertical axis of rotation	Ribbon	○	○					○	○	●	○	○	●
		Screw	○	○		○					○	○	○	○
		Screw in cone	●	○	○	●	●		●	●	○	●	○	○
		High speed	●	○										○
		Rotating disk	○	○	●	●	●	○						
		Muller	○	○		○		○	○	○	○	○	○	○
	Vibration	Vibratory mill	○	○	●	●	●	○	○	○	●	●	○	○
		Sieve			○	○		○	○	○	○	○	○
	Gas flow	Moving or fluidized bed	○						○	○				○
	Gravity	Motionless	○	○	○				○	○	○	○	○
	Internals in rotat-ing vessel	Horizontal cylinder	●	○	○	●	●	○	○	○	○	○	○
Complex		V-type	●	○	○	●	●	○	○	○	○	○	○
		Double cones	●	○	○	●	○	○	●	○	○	○	○
		Gas flow and mechanical agitation	○	○	○	○	○	○	○	○	○	○	○
		Vibration and mechanical agitation	○	○	○	●	○	○	○	○	○	○	○

[a]○, suitable; ●, usable.

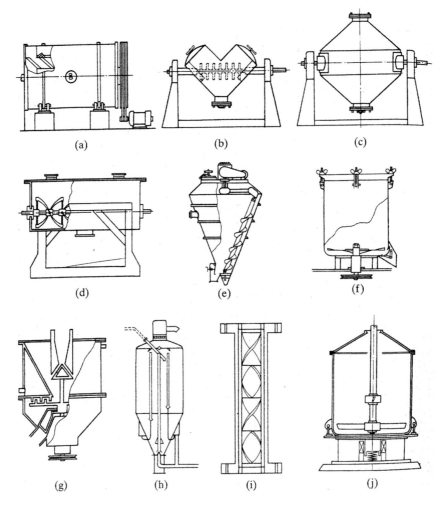

FIGURE 6.1   Typical examples of powder mixers.

With Gas-Flow Agitation

Stationary vessels using gas-flow agitation are used primarily for batch mode mixing. The powders to be mixed can be charged to more than 70% of the vessel volume. The vessel also serves as a storage container. Additional equipment, including blowers, dust collectors, and pressure regulators, is necessary so the system as a whole usually becomes rather large.

Free Falling Due to Gravity (Motionless Type)

In motionless stationary vessels, the gravitational flow of powders is repeatedly divided and united to promote mixing, and the degree of mixedness can be adjusted by the number of repetitions.

Complex Type and Others

Impellers (sometimes called intensifiers) or other internal parts can be installed inside a rotary vessel. Mechanical agitation can be carried out in addition to agitation by gas flow. Vibratory motion can also be added to stationary vessel mixers. All these efforts are aimed at making the powder motion in mixers as free as possible from the force of gravity, enhancing the rate of mixing and the degree of mixedness, and extending the applicability of a mixer to powders having fairly large differences in physical properties.

Detailed descriptions of powder mixers are available elsewhere (*Mixing Technology for Particulate Materials,*[1] p. 57).

1.6.3 MIXING MECHANISMS

Degree of Mixedness and Its Final Value

In conventional powder-mixing operations, a perfectly homogeneous mixture is defined such that the powder component under investigation becomes uniform throughout the mixture.

Statistics are used widely to define the degree of mixedness (the degree of homogeneity) for a powder mixture. Let $x_i (i = 1, 2, \ldots, N)$ be the composition of the key component in the ith sample of N spot samples taken randomly from a binary powder mixture. The sample mean $\bar{x}_s$ is given by

$$\bar{x}_s = \sum_{i=1}^{N} \frac{x_i}{N} \tag{6.2}$$

If the charged composition $\bar{x}_c$ is known, the sampling procedure can be examined by comparing $\bar{x}_s$ to $\bar{x}_c$. The variance of the samples σ^2 is defined in the following ways:

$$\sigma_p^2 = \sum_{i=1}^{N} \frac{(x_i - \bar{x}_c)^2}{N} \tag{6.3}$$

or

$$\sigma_s^2 = \sum_{i=1}^{N} \frac{(x_i - \bar{x}_s)^2}{N-1} \tag{6.4}$$

σ_s^2 is an unbiased estimation of the population variance with the degree of freedom, $v = N$, and the following relation holds with the reliability of 90%:

$$\bar{x}_s - \frac{1.64\sigma_s}{\sqrt{N}} \le \bar{x}_c \le \bar{x}_s + \frac{1.64\sigma_s}{\sqrt{N}} \tag{6.5}$$

Therefore, the degree of homogeneity for the mixture can be estimated by evaluating the magnitude of the sample variance σ_s^2. In other words, σ_s^2 can be a measure of the degree of mixedness and is useful in practical applications. However, σ_s^2 is influenced by various measuring conditions and cannot be the universal measure. Table 6.2 lists a variety of dimensionless or normalized expressions proposed so far for the degree of mixedness.[2] In this table, σ_r^2 and σ_0^2 are, respectively, the sample variances in a perfectly random mixture and in a completely segregated mixture, namely,

$$\sigma_r^2 = \frac{\bar{x}_c (1 - \bar{x}_c)}{n} \tag{6.6}$$

TABLE 6.2   Typical Expressions for Degree of Mixedness

Classification			Expression for Degree of Mixedness, M	Completely Segregated M_0	Perfectly Mixed, M_r
Expression for mixedness	I	1	$(\sigma_0^2 - \sigma^2)/(\sigma_0^2 - \sigma_r^2)$	0	1
		2	$1 - \sigma/\sigma_0$	0	1
	II		$(\sigma_0^2 - \sigma_r^2)/(\sigma^2 - \sigma_r^2)$	1	∞
	III		σ_r/σ	σ_r/σ_0	1
Expression for unmixedness	IV	1	σ/σ_0	1	σ_r/σ_0
		2	$(\sigma^2 - \sigma_r^2)/(\sigma_0^2 - \sigma_r^2)$	1	0
	V		$\sigma^2 - \sigma_r^2$	$\sigma_0^2 - \sigma_r^2$	0
	VI	1	σ^2	σ_0^2	σ_r^2
		2	σ	σ_0	σ_r

and

$$\sigma_0^2 = \bar{x}_c \left(1 - \bar{x}_c\right) \tag{6.7}$$

where n is the size of sample or the number of powder particles contained in a spot sample. Although some of the degree of mixedness shown in Table 6.2 is less dependent on measuring conditions as well as the operating conditions of mixers, there is no rationale for the expressions; they are simply conventional and intuitive. The expressions in Table 6.2 are simply transformations from the interval $[\sigma_0^2, \sigma_r^2]$ to the fixed interval $[0, 1]$, a region of which is magnified in each definition.

For a multicomponent mixture, the degree of mixedness can be evaluated by a covariance matrix in the same way as in the binary mixture, but its measurement and calculation procedure are complicated. Therefore, the multicomponent system is regarded as a mixture of the single most important component (the key component) and the others, and then is treated as a binary mixture.

In cases of batch mode mixing, only spatial variation in the composition of the key component is a matter of concern. In cases of continuous mixing, however, the time change in the composition at the outlet becomes important in addition to the spatial variation. The degree of mixedness in continuous mixing can be expressed by the magnitude of the time change in the key composition coming out from the outlet of the mixer in a steady operation. Let be x_i the key composition in the ith spot sample of N spot samples taken randomly or periodically at a constant time interval from the mixture flow at the outlet of the mixer. The variance σ_c^2 of the time change in x_i is obtained in a manner similar to that for binary mixing:

$$\sigma_c^2 = \sum_{i=1}^{N} \frac{\left[\left(x_i - \bar{x}\right)/x_0\right]^2}{N}, \quad \bar{x} = \sum_{i=1}^{N} \frac{x_i}{N} \tag{6.8}$$

where x_0 is the inlet composition of the key component (feed or charge ratio). At a steady state (i. e., $\bar{x} = x_0$), the degree of mixedness σ_c^2 defined by Equation 6.8 is independent of the sampling interval. Also, it has been confirmed that σ_c^2 is practically unaffected by the measuring and operating conditions.[3]

Detailed practical methods of evaluating the degree of mixedness, including sampling methods, have been described elsewhere (*Mixing Technology for Particulate Materials*,[1] p. 19).

Mechanism of Powder Mixing

Roughly speaking, mixing of powders progresses with the following three types of particle motion.

Convective Mixing

A circulating flow of powders is usually caused by the rotational motion of a mixer vessel, an agitating impeller such as a ribbon or paddle, or gas flow. This circulating flow gives rise to convective mixing and contributes mainly to a macroscopic mixing of bulk powder mixtures. Although the rate of mixing by this mechanism is rather high, its contribution to the microscopic mixing is unexpected. Convective mixing is beneficial for batch mode operations but gives unfavorable effects to continuous mode mixing.

Shear Mixing

Shear mixing is induced by the momentum exchange between the powder particles having different velocities (velocity distribution). The velocity distribution develops around the agitating impeller and the vessel walls due to compression and extension of bulk powders. It is also developed in the powder layer in rotary vessel mixers and at blowing ports in gas-flow mixers. Shear mixing can enhance semimicroscopic mixing and be beneficial in both batch and continuous operations.

Diffusive Mixing

Diffusive mixing is caused by the random motion of powder particles—the so-called random walk phenomenon—and is essential for microscopic homogenization. The rate of mixing by this mechanism, however, is low, when compared with convective and shear mixing.

Characteristic Curve of Mixing

Powder mixing proceeds in a mixer where the three mechanisms described above take place simultaneously. The characteristic curve of mixing is the plot of the degree of mixedness M (on a logarithmic scale) against the mixing time t (on a linear scale). The mixing time is the time measured from the start of mixing in a batch mode operation, whereas it corresponds to the mean residence time (the powder volume in a mixer divided by its volumetric flow rate) in a continuous mode operation. The characteristic curve of mixing is useful for the performance evaluation of mixers. Figure 6.2 shows a schematic example of the curve, where the standard deviation is plotted on a logarithmic scale. Generally speaking, convective mixing is dominant in the initial stage (I) and the mixing proceeds steadily by both convective and shear mechanisms in the intermediate stage (II). In the final stage (III), the effect of diffusive mixing appears, and a dynamic equilibrium between mixing and segregation is reached. The degree of mixedness M_∞ at this stage is called the final degree of mixedness, M_∞. Various powder mixers exhibit a variety of patterns in the characteristic curve of mixing (*Mixing Technology for Particulate Materials*,[1] p. 33). The value of M_∞ is also influenced appreciably by operating conditions and powder properties.

Rate of Mixing

The logarithm of the standard deviation changes linearly with time at the initial period I in almost all powder mixers, as shown in Figure 6.2. This can be expressed by the following rate equation of the first order:

$$\frac{d\sigma}{dt} = -K_1\sigma, \quad \sigma = \sigma_0 \quad \text{at } t = 0 \tag{6.9}$$

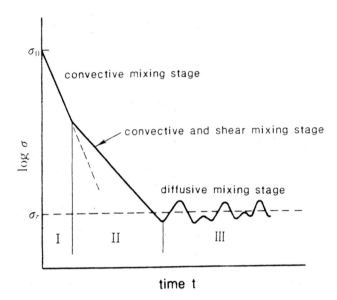

FIGURE 6.2   Characteristic curve of mixing process (schematic).

and solving Equation 6.9, one obtains

$$\sigma = \sigma_0 \exp(-k_1 t) \tag{6.10}$$

If M is expressed by the classification IV-2 in Table 6.2, the rate equation becomes

$$\frac{d\sigma^2}{dt} = k_2 \left(\sigma^2 - \sigma_r^2\right), \quad \sigma^2 = \sigma_r^2 \quad at \; t = 0 \tag{6.11}$$

and hence,

$$M = \exp(-k_2 t) \tag{6.12}$$

These equations are nothing more than phenomenological expressions of the rate of mixing. The coefficients k_1 and k_2 in Equation 6.9 or Equation 6.11, which denote the slope of the straight line at the stage I in Figure 6.2, are called the rate constant of mixing process (the dimension is s^{-1}). The rate constant of the mixing process is affected appreciably by operating conditions and powder properties.

1.6.4   POWER REQUIREMENT FOR MIXING

The following factors should be taken into account in estimating the power requirement for a powder mixer in steady-state operation: (1) the net energy to keep powders in the mixer in steady motion, (2) the net energy to keep the mixer itself or the agitating impellers in steady motion, and (3) the compensatory energy for friction losses in the driving system of the mixer. The net energy is influenced by the mechanical structure of the mixer, the powder properties, and the operating atmosphere and conditions. Let T (Nm) be the axial torque for rotating the mixer vessel or impeller at a rotational speed N_s (rps). The power P is given by

$$P = 2\pi N_s T \tag{6.13}$$

Horizontal Cylinder Mixer

The torque acting on the rotational axis of a vessel consists of (1) the torque against the gravity force acting on the center of gravity of the powder bed in the vessel, (2) the torque from the force required to drive the powder into steady rotational motion, and (3) the torque from the frictional force between the vessel wall and the powder pressed onto the wall by centrifugal force. After lengthy consideration of these factors, Equation 6.14 has been developed[4,5]:

$$\frac{T}{R^3 L \rho_b g} = A + B \frac{N_s^2 R}{g}$$
(6.14)

The term on the left-hand side of Equation 6.14 is called the Newton number, and the second term on the right-hand side is called the Froude number. Coefficients A and B, both dimensionless, can be obtained from Figure 6.3 as functions of the charge ratio f, the angle of repose of powder ø, and the friction coefficient of powder at the vessel wall μ_w.

V-Type Mixer

In this case, the axial torque changes periodically at a period of π, exhibiting a maximum torque T_{max} and a minimum torque T_{min}, because the rotational motion displaces powders from one end to the other in the vessel.

Let R_{max} be the maximum radius of rotation, then T_{max} and T_{min} are given by

$$\frac{T_j}{R_{max}^3 \rho_b g} = A_j + B_j \frac{N_s^2 R_{max}}{g}, \quad j = \text{max or min}$$
(6.15)

The coefficients A_j and B_j can be evaluated from Figure 6.4.

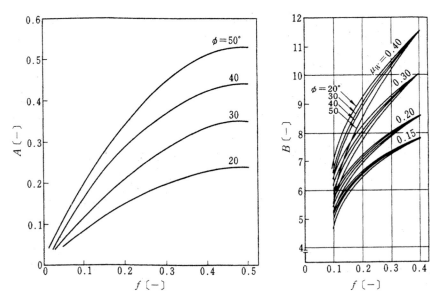

FIGURE 6.3 Coefficients A and B in Equation 6.14 for horizontal cylinder mixer.

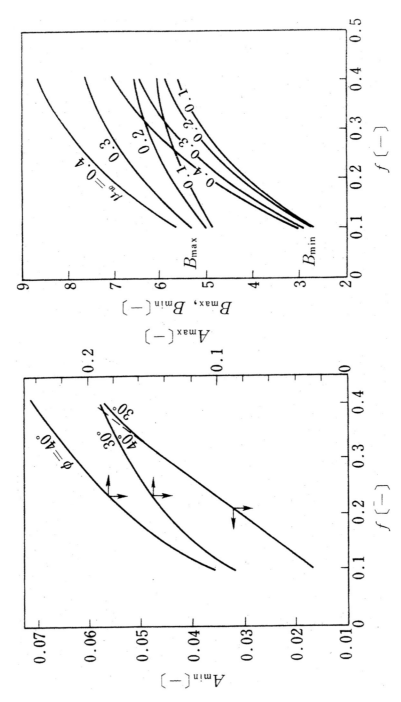

FIGURE 6.4   Coefficients A and B in Equation 6.15 for V-type mixer. Angle of apex = $90°$; ratio of cylinder radius to maximum radius of rotation = 0.5.

TABLE 6.3 Exponents in Equation 6.16

Type of Mixers	α1	α2	α3	α4	α5	α6	α7	α8
				Exponent in Eq. (6.16)				
Vertical ribbon	0	1.0	0.7	1.5	3.0	1.2	−0.8	—
Horizontal ribbon	0	1.0	1.2	1.0	3.3	−0.3	0.7	1.2
Paddle	0	1.0	1.0	1.0	2.0	—	1.0	—

Stationary Vessel Mixer

An empirical formula for estimating the axial torque for a ribbon or paddle impeller in powders having a high flowability has been proposed[6–8]:

$$T = K D_p^{\alpha 1} \rho_b^{\alpha 2} \mu_s^{\alpha 3} Z^{\alpha 4} d^{\alpha 5} \left(\frac{S}{d}\right)^{\alpha 6} b^{\alpha 7} f^{\alpha 8} \quad (\text{Nm}) \tag{6.16}$$

where
  D_p = particle diameter (m)
  ρ_b = bulk density (kg/m³)
  μ_s = internal friction factor
  S = pitch of ribbon impeller (m)
  b = width of impeller (m)
  d = diameter of impeller (m)
  f = charge ratio (–)
  Z = height of powder bed (m)

Table 6.3 gives the exponents $\alpha 1 \sim \alpha 8$ in Equation 6.16, which have been determined experimentally. A marginal torque for setting impellers in rotational motion in a quiescent powder bed should also be considered in the mechanical design.

1.6.5 SELECTION OF MIXERS

How to select a mixer depends on the degree of homogeneity required for the product (final degree of mixedness), the rate of mixing, and the power requirement. The type of operations (batch or continuous modes) and the strength of the powder particles against mechanical agitation should also be examined.

When the mixing data with a bench-scale mixer are available to design a large system, the following factors should be kept in mind:

1. Whether the desired final degree of mixedness is attainable
2. How to determine the operating conditions
3. Amount of power and cost necessary

For rotary vessel mixers in which a circulating flow of powders exists, the ratio of centrifugal force to gravity force (i.e., the Froude number, $N_s^2 R / g$) can be kept constant in scaling up. Although the Froude number can be evaluated for stationary vessel mixers by taking one half of the outer impeller diameter as the representative radius, R, the flow patterns of the powders in a prototype mixer might not become geometrically similar to that in a small-scale mixer. For example, if the clearances between the impeller tips and the vessel walls are geometrically scaled up, a dead zone might be formed in the powder flow in the large-scale mixer. If the scale-up is based on the tip speed of the impeller, a dead zone where agitation is insufficient for mixing may be formed around the impeller axis, necessitating an additional design for the mixer structure in some cases.

To select the pertinent mixer, not only must the performance and characteristics of various mixers themselves be well understood, but also the mixing operation in the entire process. The purposes of the mixing operation, the powder properties, the capacity of production, the desired final degree of mixedness, maintenance problems, unit and running costs, and related matters must also be well defined. The priority of each of these factors depends on individual cases. Unfortunately, no general organized argument is available for the methodology of selecting the target mixer.

Notation

A, B	Coefficients in Equation 6.14
A_j, B_j	Coefficients in Equation 6.15, j = max or min
b	Width of impeller
D_p	Particle diameter of powder
d	Diameter of impeller
f	Charge ratio; volume of powder charged divided by volume of mixer vessel
g	Acceleration due to gravity (m/s^2)
K	Coefficient in Equation 6.16
k_i	Rate constant of mixing process, defined by Equation 6.9 for $i = 1$, by Equation 6.11 for $i = 2$, (s^{-1})
L	Length of horizontal mixer vessel (m)
M	Degree of mixedness
M_∞	Final degree of mixedness
N	Number of spot samples
N_s	Rotational speed of mixer vessel or impeller (s^{-1})
N_{cr}	Critical speed of rotation, defined by Equation 6.1 (s^{-1})
n	Sample size, number of powder particles in a spot sample
P	Power for driving mixer, defined by Equation 6.13 (W)
R	Radius of horizontal mixer vessel (m)
R_{max}	Maximum radius of rotation of mixer (V-type) (m)
S	Pitch of ribbon impeller (m)
T	Axial torque for driving mixer (Nm)
T_j	Maximum or minimum axial torque for driving V-type mixer (Nm)
t	Mixing time (batch operation) or mean residence time (continuous operation) (s)
x_0	Inlet composition of key component for continuous mixing
$\bar{x}_c$	Charged composition of key component for batch mixing
x_i	Composition of key component in the ith spot sample
$\bar{x}_s$	Sample mean of , defined by Equation 6.2
Z	Height of powder bed in mixer vessel (m)
$\alpha1 \sim \alpha8$	Exponents in Equation 6.16
μ_s	Internal friction factor
μ_w	Wall friction factor
v	Degree of freedom in estimation
ρ_b	Bulk density of powder (kg/m^3)
σ^2	Variance of compositions of key component in samples

σ_0^2	Variance of compositions of key component in completely segregated mixture, defined by Equation 6.7
σ_c^2	Variance of compositions of key component at outlet of continuous mixer, defined by Equation 6.8
σ_P^2	Variance of compositions of key component, defined by Equation 6.3
σ_r^2	Variance of compositions of key component in perfectly random mixture, defined by Equation 6.6
σ_s^2	Variance of compositions of key component, defined by Equation 6.4
ϕ	Angle of repose (rad)

REFERENCES

1. Association of Powder Process Industry and Engineering (APPIE), Ed., *Mixing Technology for Particulate Materials*, Nikkan Kogyo Press, Tokyo, 2001.
2. Yano, T. and Sano, Y., *J. Soc. Chem. Eng. Jpn.*, 29, 214, 1965.
3. Yano, T., Sato, M., and Mineshita, Y., *J. Soc. Powder Technol. Jpn.*, 9, 244, 1972.
4. Sato, M., Yoshikawa, K., Okuyama, N., and Yano, T., *J. Soc. Powder Technol. Jpn.*, 14, 699, 1977.
5. Sato, M., Miyanami, K., and Yano, T., *J. Soc. Powder Technol. Jpn.*, 16, 3, 1979.
6. Novosad, J., *Collect. Czech. Chem. Communi.*, 29, 2681, 1964.
7. Makarov, Yu. I., *Apparatus for Mixing of Particulate Materials*, Mashinostroenie, Moscow, 1973, p. 118.
8. Sato, M., Abe, Y., Ishii, K., and Yano, T., *J. Soc. Powder Technol. Jpn.*, 14, 411, 1977.

1.7 Slurry Conditioning

JunIchiro Tsubaki
Nagoya University, Nagoya, Japan

Makio Naito
Osaka University, Ibaraki, Osaka, Japan

1.7.1 SLURRY CHARACTERIZATION

Slurry is usually not a final product but an intermediate one. Slurry as intermediate product is shaped and then dried before final use. Since in the shaping process slurry must have fluidity, measurement of viscosity is very important to characterize the slurry. After shaping, fluidity is not important because the slurry does not need to deform or flow; on the contrary, characterization of the thickening behavior of the slurry becomes important. A settling test is widely done to characterize thickening behavior, and it is reported that the morphology of spray-dried granules can be estimated from this settling test. Slurry composed of dense sediment makes hollow granules, whereas slurry composed of less dense sediment makes solid granules.[1,2] The disadvantage of the settling test is that it takes a long time; for example, in the case of submicron particles it takes weeks or months under gravity, and even in a centrifuge it takes days. Consequently the settling test is hard to use for on-site optimization or control of production process in factories.

To shorten the measurement time, the following characterizing methods have been proposed.

Particles that make a loose sediment coagulate each other and easily network from the bottom in the early settling stage. The bottom of a container supports the mass of the networked particles. Particles that make dense sediment repulse each other and settle freely. In this case, viscous drag supports the particle mass. If we measure the hydrostatic pressure at the bottom of a container during a batch settling test, the pressure changes from $P_{max}(=\rho_s gh)$ at $t = 0$ to $P_{min}(\rho gh)$ up to infinitely. Where ρ_s is the density of the slurry, ρ is the density of the dispersion medium, and h is the slurry depth. $P_{max} - P_{min}$, $P_{max} - P$ and $P - P_{min}$ stand for the whole particles, the deposited particles on the bottom, and the freely settling particles, respectively.

Figure 7.1[3] shows settling test results of the aqueous slurry of alumina abrasive particles (mean diameter is 3μm) adjusted by pH value. As shown in Figure 7.2,[3] the hydrostatic pressure at the bottom can distinguish the coagulation situation in slurry at an early stage.

The mass settling flux from the interface between supernatant and slurry zone is calculated by Kynch's theory, and the depositing mass flux to the bottom can be calculated from the pressure decrease rate dp/dt. Taking the flux ratio as shown in Figure 7.3,[4] the settling behavior can be analyzed more informatively than in the traditional settling test.

Constant pressure filtration is also utilized for characterization of thickening behavior.[5] As the pressure changes, the mechanical properties of the cake are analyzed. If the packing fraction of the cake Φ is constant during filtration, the pressure drop ΔP is described by the following Kozeny–Carman's equation.

$$\Delta P = \mu \left\{ R_m + 5 S_v^2 \frac{\Phi^2}{(1-\Phi)^3} L \right\} \frac{dv}{dt} \tag{7.1}$$

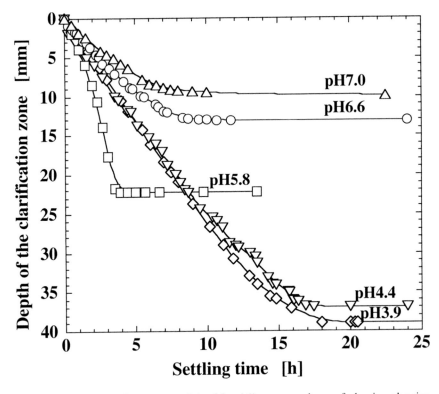

FIGURE 7.1   The settling curves of the 35 vol.% aqueous slurry of alumina abrasive particles. The mean diameter is 3 μm and the slurry is adjusted by pH value.

Where L is the cake thickness, v is the filtrate volume per unit area, t is the filtration time, S_v is the specific surface area of the particles, μ is the viscosity of the filtrate, and R_m is the resistance of the filter medium. Equation 7.1 is rewritten as follows:

$$\frac{\Delta P}{5\mu S_v^2}\frac{dt}{dv} = \frac{\Phi}{(1-\Phi)^3}\left(\Phi L + R_m'\right) \qquad (7.2)$$

The packing fraction of the cake can be read from the plots of the left side vs. ΦL, standing for the cake height as shown in Figure 7.4. Since the cake height is defined as the cake volume per unit area, ΦL can be calculated from the following mass balance equation:

$$\Phi L = \phi(v + L) \qquad (7.3)$$

where ϕ is the slurry volume concentration.

If the plots are on a strait line, the cake has no packing fraction distribution. On the contrary, if the plots are on the curve (a) shown in Figure 7.5, it is suggested that the packing fraction decreases with the cake height. And the curve (b) suggests that the packing fraction increases with the cake height. Moreover, the curve (c) suggests that the primary particles coagulate each other and the settling velocity of the agglomerates is not negligibly small.

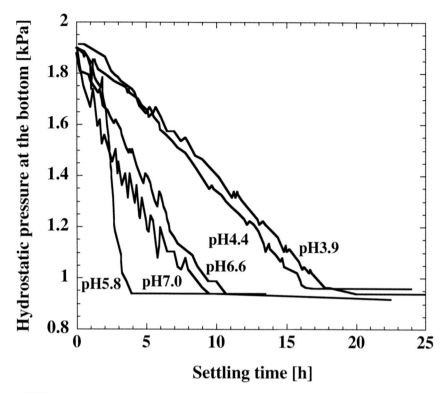

FIGURE 7.2   The hydrostatic pressure at the container bottom. The depth of the sample slurry is 90 mm.

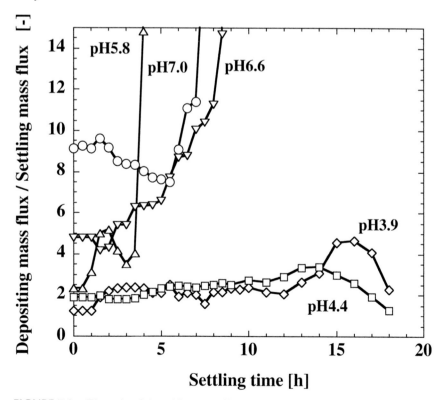

FIGURE 7.3   The ratio of depositing to settling mass flux.

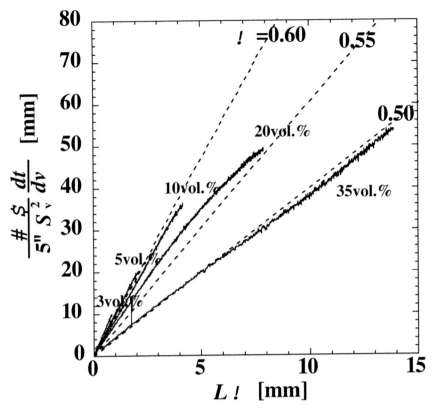

FIGURE 7.4 The constant pressure filtration results of aqueous alumina slurry expressed by Equation 7.2. The particle size is 0.48 μm, and polyacrylic ammonium is added to be the same absorbed.

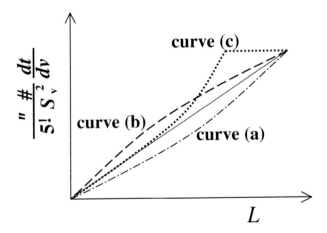

FIGURE 7.5 Filtration pattern of constant pressure filtration.

1.7.2 SLURRY PREPARATION

Slurries are widely used in industry as an intermediate material during the fabrication process of final products, as well as being products themselves. The purposes of slurry usage cover a variety of areas. As intermediate materials, slurry is used for preparing spray-dried granules or for slip-casting green bodies to make ceramics parts. As a product, slurry is used as a material for chemical–mechanical polishing and is a key tool for high-tech polishing techniques.

Therefore, the preparation of a slurry is very important, and its preparation method should be decided by considering various factors related to the dispersion process. They mainly concern the wetting behavior between the suspension medium and the particle surface, the dispersion mechanism of the particles in the liquid, and the stability of dispersion state. In this section, these factors, being considered important for achieving a better slurry preparation, are explained. Practical examples of slurry preparation and its influences in the manufacturing process on the improvement of ceramic products are given to provide understanding of the importance of slurry preparation technique.

Wetting between Suspension Medium and Particle Surface

It is important to evaluate the wetting behavior between the suspension medium and the particle surface to understand the dispersion behavior of particles in the liquid. The direct measurement of the wetting ability between the surface of particle and the liquid is very difficult. Therefore, various kinds of indirect methods have been proposed to determine the wetting ability of a powder. One of the well-known methods is based on the evaluation of the penetration phenomena of a liquid into the powder bed. By measuring the penetration rate of liquid into the powder bed, the contact angle is calculated by using the Washburn equation.[7] To improve the wetting ability, a combination of liquid and particles should be considered, and a low contact angle corresponds to a better dispersion of the particles. Modification of the particle surface for changing the wetting ability is also effective to improve the dispersion of particles.

Dispersion of Particles into a Liquid

After the combination of liquid and solid particles is decided, a good wetting of liquid onto the surface of suspended particles should be achieved. Dispersed particles and the dispersion should be stable in the liquid for a long time. In this regard, the dispersion mechanism between particles and suspension liquid should be taken into account. In an aqueous system, the basic mechanism is based on the DLVO theory. The zeta potential is an important factor to evaluate the stability of particles in liquid. It can be changed by controlling several factors. The interaction between particles can be also controlled by adsorbing dispersants as well as polymers on the surface of particles. By using these methods, optimum dispersion conditions should be selected.

Mechanical dispersion techniques are also crucial to achieve better dispersion of particles. When the powder has powder aggregates, they must be disintegrated into primary particles. This can be achieved by the application of ultrasound or by milling. For example, agitated ball milling is very effective to reduce agglomerates and disperse them into primary particles. There are different kinds of mechanical dispersion methods. Depending on the strength of aggregates, a suitable method should be selected. Mechanical dispersion methods also depend on the ranges of shear rate of the slurry.[8]

Stability of Suspension

After the particles are dispersed in the liquid, the dispersion should be stable until it is actually used. To prevent the sedimentation of powder particles, a continuous stirring of the slurry may also be effective. For example, a ferrite powder has magnetic properties, and the true density is relatively high, therefore, continuous stirring is crucial to maintain the stability of the suspension.

Slurry Conditioning for Better Ceramics

Ceramic manufacturing processes are based on various powder-processing techniques. They are classified according to the shaping process used. These include the granule compaction process, slip-casting, and extrusion and injection molding. The slurry properties actually influence the properties of the spray-dried granules and green bodies, thus leading to a change of the quality of the resultant ceramics.

The granule compaction process of alumina is introduced to explain the effect of slurry preparation conditions on the properties of the resultant ceramics. Table 7.1 shows the data of slurry preparation conditions and its apparent viscosity.[9,10] The density, fracture toughness, average fracture strength, and Weibull modulus are also listed in the same table.[9,10] The table clearly shows that the properties of sintered ceramics change with slurry preparation conditions. Although a low soda alumina (AL160-SG4, Showadenko, Japan) was used, the strength of the sintered ceramics changes from 363 to 486 Mpa, depending on the slurry preparation conditions.

The correlation between slurry preparation conditions and the properties of sintered ceramics is explained by the property change of spray-dried granules. Table 7.2 shows the structure and strength of spray-dried granules according to each set of slurry preparation conditions.[9,10] The internal structure of the granules was observed by the liquid immersion method.[11] The compressive strength was measured by a compressive strength tester for single granules,[12] and the average strength and Weibull modulus were determined subsequently.

As shown in Table 7.2, granules prepared from the slurry with higher apparent viscosity (#4) have a close-packed structure. They have a low average compressive strength compared to that obtained under conditions #1–3, but the Weibull modulus is relatively low. Therefore, the granules made by the conditions of #4 lead to a uniform green body by pressing, which leads to a higher average strength of the corresponding sintered body. However, the low Weibull modulus of the granules allows the green body to contain a low amount of extremely high compressive strength granules, working as a fracture origin in the sintered ceramics. As a result, this leads to higher average strength and a lower Weibull modulus of ceramics.

On the other hand, the low-viscosity slurry leads to a dimple structure of the granules during spray-drying,[10] and a high Weibull modulus of granules compared to that obtained under condition #4. The average compressive strength of the granules increases with the amount of dispersant (ammonium polyacrylate). As higher strength granules are harder to fracture during the compaction process,

TABLE 7.1   Slurry Preparation Conditions and Properties of the Sintered Ceramics

No.	pH	Dispersant amount [mass%]	Viscosity [mPa.s]	Density [kg/m³]	Fracture toughness [MPam^{1/2}]	Average strength [MPa]	Weibull modulus
#1	10	0.2	43	3.91×10^3	3.7	486	20
#2	9.1	0.5	22	3.94×10^3	3.8	430	16
#3	8.1	2.0	54	3.89×10^3	3.8	363	13
#4	9.0	0.2	125	3.94×10^3	3.8	480	8

TABLE 7.2   Structure and Strength of Spray-Dried Granules

No.	Internal structure	Average compressive strength [MPa]	Weibull modulus
#1	Dimple structure	0.31	10
#2	Dimple structure	0.91	6
#3	Dimple structure	5.4	10
#4	Closed packed structure	0.32	3

it leads to lower fracture strength of the sintered ceramics. With a low-viscosity slurry, a higher Weibull modulus of ceramics is obtained because all granules have a dimple structure, leading to uniformly distributed pores in the sintered ceramics. As a result, the ceramics made under condition #1 reached the highest fracture strength and Weibull modulus of the four slurry conditions. When granules are made by spray-drying, the dispersant will work as solid bridges to increase the compressive strength of granules, as shown in Table 7.2.[9] Therefore, slurry conditioning is a key issue to achieve a higher quality of products, and slurry should be processed according to its purposes.

REFERENCES

1. Tsubaki, J., Yamakawa, H., Mmori, T., and Mori, H., *J. Ceram. Soc. Jpn.*, 110, 894–898, 2002.
2. Mahdjoub, H., Roy, P., Filiatre, C., Betrand, G., and Coddet, C., *J. Euro. Ceram. Soc.*, 23, 1637–1648, 2003.
3. Tsubaki, J., Kuno, K., Inamine, I., and Miyazawa, M., *J. Soc. Powder Technol.*, 40, 432–437.
4. Kuno, K., *Analysis of the Settling and Depositing Process in Dense Slurry,* Master's thesis, Nagoya University, Nagoya, Japan, 2002.
5. Tsubaki, J., Kim, H., Mori, T., Sugimoto, T., Mori, H., and Sasaki, N., *J. Soc. Powder Technol.*, 40, 438–443[2].
6. Ato, K., *Analysis of Cake Forming Process During Constant Pressure Filtration,* Bachelor's thesis, Nagoya University, Nagoya, Japan, 2004.
7. Washburn, E. W., *Phys. Rev.,* 17, 273–278, 1921.
8. Reed, J. S., Ed.; *Principles of Ceramics Processing,* John Wiley, New York, 1995, pp. 282–288.
9. Abe, H., Hotta, T., Kuroyama, T., Yasutomi, Y., Naito, M., Kamiya, H., and Uematsu, K., *Ceram. Trans.,* 112, 809–814, 2001.
10. Abe, H., Naito, M., Hotta, T., Kamiya, H., and Uematsu, K., *Powder Technol.,* 134, 58–64, 2003.
11. Uematsu, K., *Powder Technol.,* 88, 291–298, 1996.
12. Naito, M., Nakahira, K., Hotta, T., Ito, A., Yokoyama, T., and Kamiya, H., *Powder Technol.,* 95, 214–219, 1998.

1.8 Granulation

Isao Sekiguchi
Chuo University, Bunkyo-ku, Tokyo, Japan

1.8.1 GRANULATION MECHANISMS

Outline of Granulation Techniques

The heading "granulation" in this section is used as a main title to describe all forms of granules, small compacts, and small grains or prills. The granulation techniques in a variety of industrial fields are adopted to gain the advantages of solids handling, as listed in Table 8.1. There are two principal modes of granulation: granule growth modes and machine-made product modes. In the former, granulation is normally achieved by tumbling, agitating, or fluidizing raw materials in the presence of liquid binders, and the final products (i.e., agglomerates) have a wide distribution in granule sizes. The latter includes granulation processes in which materials are forced to flow in a plastic or sticky condition through dies or screens or in molds or other devices. Forming droplets from a fused material or suspension such as spray prilling or drying also belongs in the machine-made product mode.

Main Mechanisms of Bonding

The important mechanisms of bonding between solid particles can be classified into four principal groups.

Deformation and Breakage of Solid Particles

In the compaction of powders, adjacent particles are pressed together so that if the particles have a plasticity, the compacting action leads to permanent bonding at the contact points.[1] If the failure of brittle solid particles or dry agglomerates within a bed occurs during compaction,[2,3,4] the powder compact is deformable and easy to handle.

Heating of Solid Particles[5]

Hot powders heated to an appropriate high temperature just before the compaction bring about various active bonding forces among particles owing to the improved contact surface of particles due to alterable surface structure, desorption, chemical activation, gentle sintering, and so on. At further elevated temperature, immediately below the melting point of the constituent components, sintering or nodulizing causes the formation of hard agglomerated materials in a rotary kiln or hot fluidized bed.

Addition of Binding Liquids

The lower the activation energy to form droplet nuclei on the surface of fine particles, the gentler agglomeration formed by moistening the powder with a water vapor or other solvent occurs. Some amount of liquid binder is necessary to achieve some degree of plasticity of powders. The choice of liquid binder affects the properties of powders to be agglomerated. Especially the interparticle force among the particles created by the presence of nonviscous liquids shows a dependence on the value of $\sigma \cos \theta$, where σ is the surface tension of the liquid and θ is the wetting contact angle on the particle surface.[6-8] The interparticle force caused by highly viscous binding materials such as liquid

TABLE 8.1   Beneficial Effects of Granulation or Compaction

To prepare particulate materials for more convenient solids handling or processing

Nonsegregating blends: to achieve more uniform composition of mixture of solids
and to prevent segregation

Compaction: to ensure uniform filling and pressing in a die or mold

Densification: to densify particulates for packing or feeding without dust losses or
hazard

Flowability: to reduce the tendency of particulates, generally hygroscopic, to form
lumps, and also to improve the flow properties

To make a product suitable for chemical reaction, heat, and/or mass transfer

Solubility: to prevent the formation of undissolved lumps of fine powders in fluid
phases

Permeability: to achieve a more stable flow of fluids through fixed beds of
particulates

Release: to obtain controlled release of chemicals in granular materials

Immobilization: to create a so-called carrier in which the activity of catalyst or
biocatalyst may remain

paraffin, pitch, or asphalt[9-12] enhances the viscous and plastic deformation of the binding material that precedes the fracture.

Addition of Foreign Powders

A very small amount of solid lubricants such as talc, stearate, magnesium oxide, or paraffin wax is added to the compacting material to improve the strength of compact. The addition of special substances such as clay, bentonite, or other plasticizers to damp powders makes it possible to produce stronger agglomerates in a tumbling or agitating granulator.

Testing Methods for Evaluating the Agglomerative Granulation of Damp Powders

Flocculent masses of damp powders prepared by mixing dry powder with water or other liquid binders have a marked tendency for agglomerative granulation under the tumbling, vibrating, or agitating action in the various devices. A new testing device to examine how the generation and growth of well-defined agglomerates are made from the damp powders has been developed.[13] Figure 8.1 shows a schematic diagram of a tapping-type gyratory agglomerator. This testing device consists basically of a shaking pan with a rubber bottom (disk diameter = 200 mm) tapped at the central part through a conic swing of the pendulum. The tapping action is induced by an eccentric circular motion of the rubber disk and by the subsequent circular swing of the pendulum rod when the lower front of the pendulum strikes the cylindrical obstacles located at the cyclic positions, as shown in Figure 8.2.

Damp powders on the rubber disk accompanied by both the gyratory motion and the subsequent tapping action, as mentioned above, change from a flocculent mass to harder and denser agglomerates in a visible state. The agglomeration behavior of damp powder is subjected to the method of preparing the feed and is often brought into a powdery or pasty state. The quality of the agglomerate at various liquid contents C_d can be evaluated by an agglomeration index defined as

l_g = (Liquid binder content for preparing a damp powder)/(Liquid binder content at a plastic limit)

$$= \frac{C_d}{[\varepsilon_0 / (1 - \varepsilon_0)](\rho_l / \rho_s)} \tag{8.1}$$

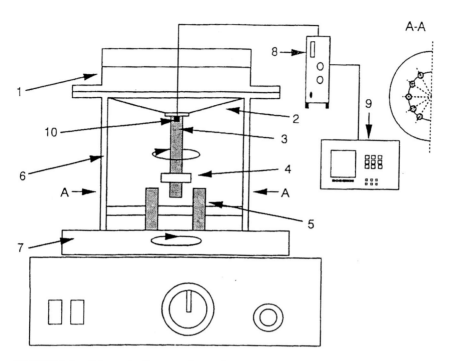

FIGURE 8.1   Schematic diagram of a tapping-type gyratory agglomerator.

where ε_0 is the void fraction in a close-packed bed of dry powder made by a tapping action. In addition, the plastic limit of the damp powder in Equation 8.1 can be determined by an agitating torque method in which the torque exhibits a distinct maximum of a liquid binder content corresponding to the plastic limit.[14] An acceptable value of I_g for establishing stable agglomerates favorable to a tumbling or agitating granulation generally lies in the range of about 0.50 to 0.75.

As an example, Figure 8.3 shows a relationship between agglomerate diameter $X_{v\infty}$ and vibration intensity G for calcium carbonate (volume mean particle diameter = 10.2 μm, ε_0 = 0.476) methanol system. The agglomerate diameter in the figure is the final equilibrium value in the agglomerate growth. The growth of equal-sized agglomerates becomes excellent in both ranges $0.55 \leqq I_g \leqq 0.65$ and $0 < G \leqq 38.4$. It is apparent that the agglomerates in such a region become progressively larger in lower values of I_g with increasing vibration intensity. In this region, an important characteristic of agglomerate growth is such that the agglomerates suddenly grow to larger ones in the course of the agglomeration process under sufficient vibration intensity, as shown in Figure 8.4. This tendency is remarkable in lower values of I_g because of the combined action of breakage and coalescence. However, smaller agglomerates formed near the critical value of I_g = 0.65 experience simple random coalescence in the nuclei region, yielding extremely different agglomerate sizes.

Granulation Due to Granule Growth Modes

A drum granulator in batch operation is important in characterizing the agglomeration behavior of damp powders. The growth mechanism of granules in this operation is divided into the following three stages: (1) nucleation, nuclear growth, and coalescence of nuclei or small granules (nuclear growth region), (2) rapid coalescence of small granules (transition region), and (3) layering of fragments onto large granules, and compaction by tumbling (balling growth region).

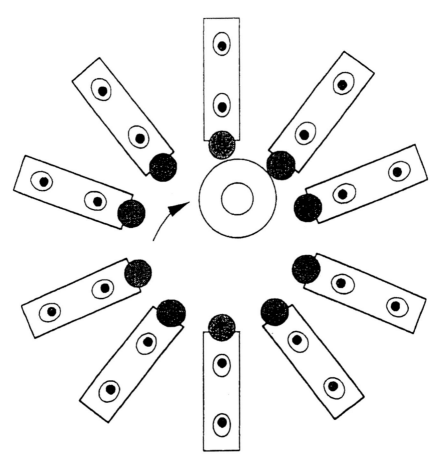

FIGURE 8.2 A circular swing of the pendulum rod striking the cylindrical obstacles, which are located at the cyclic positions.

The growth rate of granules in the nuclear growth region is expressed in most cases as follows[15]:

$$\frac{dX_a}{dt} = \frac{k_n}{3} X_a \qquad (8.2)$$

where X_a is the arithmetic mean diameter of granules on a mass basis and k_n is the time-invariant rate constant of granule growth by coalescence. If a nuclear or smaller agglomerate assembly in the floclike mass of damp powders has a relatively narrow size distribution, a rate equation of granule growth in the first stage is derived from a random disappearance–coalescence model, to which Equation 8.2 is applicable. The size distribution of granule originating from a floclike mass of damp powders, as mentioned above, is formulated with the aid of statistical mechanics to become[16]

$$R_m(X) = \exp(-BX^3) \qquad (8.3)$$

where $R_m(X)$ is the cumulative mass fraction of granules coarser than diameter X, and B is a constant. In the case of a damp powder having a relatively narrow-sized floclike mass prepared by screening or kneading, the granule size data obey Equation 8.2 and Equation 8.3. The careful preparation of feed material is a prerequisite for a successful granulation in the early stage of tumbling granulation.

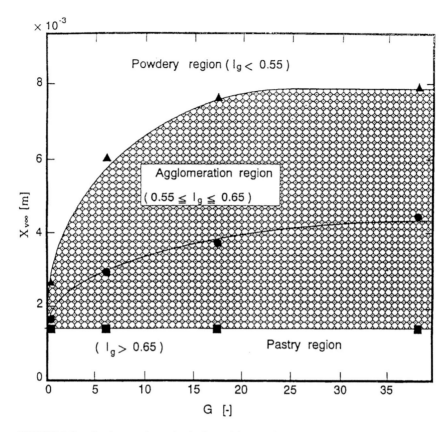

FIGURE 8.3   Agglomeration criteria for calcium carbonate powder agglomerated with methanol.

The granule growth period in the transition region is very short, and the rate of granule growth obtained from the plot of dX_a/dt versus t has a momentary maximum.[17] Formation of large granules in the balling growth region is probable because the layering mechanism of fragments onto the growing granules is extremely dependent on the breakage of granules due to impaction, compaction, or abrasion. The rates of granule growth in the above three regions are given by the following[18]:

1. Nucleii growth region: $dX_a/dt \propto X_a$
2. Transitional growth region: $dX_a/dt \propto 1/X_a$
3. Ball growth region: $dX_a/dt \propto 1/X_a^3$

For an unstable granulation period, some generalized macroscopic-population balance models have been proposed. A majority of these models describe the self-preserving size spectra of granulated materials in batch tumbling granulation systems.[17–24]

The point of granulation in a hot fluidized bed with spray is that a liquid binder, suspension, or melt is atomized into the fluidized bed of the particles to be granulated.[25–30] Typical growth of granules formed by a batch fluidized-bed-spraying system is shown in Figure 8.5. The formation of granules at a time interval between A and B is very stable, but the subsequent granulation step forms larger granules or crumbs, so that the fluidized bed becomes extremely bubbly until defluidization occurs. Normal operation of the bed is carried out for the period A to B, in which it is a matter of course that the heat and mass balance must be satisfied. Further treatments of granulation in the granule growth mode, including pan-type granulators, are summarized in Table 8.2. Spouted beds

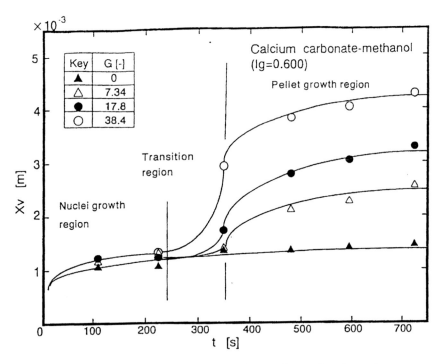

FIGURE 8.4   Effect of vibration intensities on the growth rate of agglomerates.

developed as the improved type of fluidized bed are often employed to produce granules or coating particles. Spray liquid is introduced as a spray into the conical base of the bed together with the hot spouting gas. This granulation method is usually suitable for the coating or layering particle process because of the rapid evaporation of spray liquid in the spouting gas stream at high temperature.

Machine-Made Product Modes

The compaction of solid particles or small granules in a confined space can be carried out by punch, piston, roll, or other means of applying pressure. The degree of consolidation achieved during compacting press depends mainly on the stress–strain relation[2] and on the physical and chemical properties of solid particles. The compaction mechanism of solid particles is divided into four stages as follows: (1) increase in the resistance due to the mutual friction of movable particles, (2) breakage of the interlocking among adjacent particles, (3) deformation and fracture of constituent particles, and (4) work-hardening of the particle bed. Such behavior seems to occur in the compaction of narrow solid particles. For a wide variety of solid particles, many empirical equations are proposed for the relationship between compacting pressure and void fraction.[31]

As many different machines are used for making the desired compacts, several important subjects should be described briefly here. The forming mechanism of a thick disk-shaped compact in a cylindrical die[32,33] is expressed by

$$\frac{F_U}{F_L} = \exp\left(\frac{4\mu_w kL}{D}\right) \tag{8.4}$$

where F_U and F_L are respectively the applied and transmitted forces, k is the ratio of radial to axial stress, μ_w is the coefficient of wall friction, and D and L are the diameter and thickness of a compact.

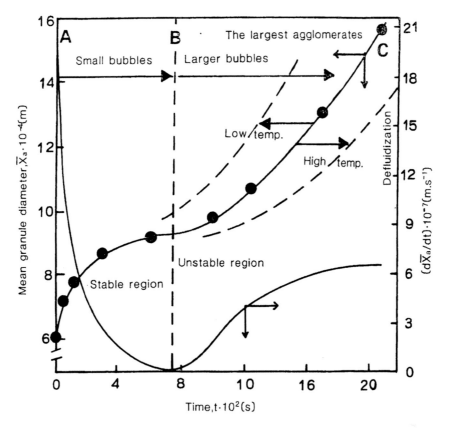

FIGURE 8.5 Rate of granule growth in a hot fluidized-bed spray granulator. Seed particles, sodium chloride(X_a = 0.000606 m); binder, gelatine solution(20.0 wt%). W = 6 kg; F_p = 0.00063 kg/s ; u = 0.703 m/s; air temperature = 353 K.

Further analysis reveals that Equation 8.4 has some serious drawbacks. Nevertheless, the expression is straightforward and simple to use.

In roll compactions, except briquetting machines, particulate material with or without binders is fed into the compacting zone between two smooth rolls, as shown in Figure 8.6, and the compacted sheet can be cut to the desired granule size. The relationship between the void fraction of compacted sheet and the angle of nip θ_n is given by

$$\cos\theta_n = 1 + \frac{D_S}{D_B}\left(1 - \frac{1-\varepsilon_p}{1-\varepsilon_n}\right)$$

(8.5)

where ε_p is the void fraction of compacted sheet and ε_n is the void fraction of the material between two rolls at the nip angle θ_n. The horizontal force F_H exerted on the roll-bearing blocks is given by

$$F_H = \frac{\pi D_R B}{360}\int_0^{u_n} p\cos d\theta$$

(8.6)

where B is the length of the active roll surface and p is the pressure exerted by the material undergoing compaction on a certain portion of the roll surface. To integrate Equation 8.6, it is necessary to measure the compaction pressure distribution and the void fraction along the circumference of the roll.[34] F_H in Equation 8.6 is derived from the theory of solids conveyance between two smooth rolls.[35]

TABLE 8.2 Typical Operating Data for Pan or Drum Granulators and Fluidized-Bed Spray Granulators

Granulator type	Feed Conditions	Operating Conditions	Residence-time Distribution
Pan	1. Powder and spray droplets 2. Continuous operation 3. Ig=0.50–0.75	Nopt = (0.40–0.75)Nc Nc=2538(sin/D)1/2 H = (0.10–0.25)D	
Drum	1. Damp powders, or powder and spray droplets 2. Batch or continuous operation 3. Ig = 0.50–0.75	Nopt = (0.45–0.85)Nc h = 0–8	1. Backmixing at a ring dam 2. Ideal pistol flow
Fluidized-bed spray	1. Powder and spray droplets 2. Batch or continuous operation	umf < u0 < 0.5uf	Ideal perfect mixing flow
Pan	W/Fp = K(Fp/PpNdD3) x (Ha1/2r/DeNd) (Ohabayashi, 1972)	dV/dt= K1(Ve – V) (Tohata et al., 1966). Rate of granule growth due to an elliptical orbit model on the bottom of pan (Macavi, 1965)	Larger granules due to size-classifying properties of inclined pan can be separated in part at the discharge rim
Drum	W/FP = (La1/2r/DhNd) x (FP/ρPNd) a(H/D)b (Sekiguchi et al., 1970)	Xa/Xa,0 = exp(knZd/3)* Xa= Xa,0 + ksZd + Granule population balancer, (Ref. 18, 20)	Equation (8.2) Ref. 16 Granule size distribution taken both sizes of upper and lower limits into consideration (Capes and Danckwerts, 1965; Ref. 17)
Fluidized-bed spray	W/FP = AfLf(1 - ef)FP Nv = f(u0/uf)	Xa/Xa,0 = exp(Ct)‡ Xa/Xa,0 = 1/(1 – Ct)§ (Ref, 40)	Mass balance in a batch‡ or continuous operation§

*Granule growth by coalescence.
†Granule growth by layering (Umeya, K. and Sekiguchi, I., *J. Soc. Mater. Sci. Jpn.*, 24, 664–668, 1975).
‡Batch operation (Harada, K., Fujita, J., and Yoshimura, M., *Kagaku Kogaku*, 33, 793–799, 1969).
§Continuous operation (Harada, K., Fujita, J., and Yoshimura, M., *Kagaku Kogaku,* 33, 793–799, 1969).

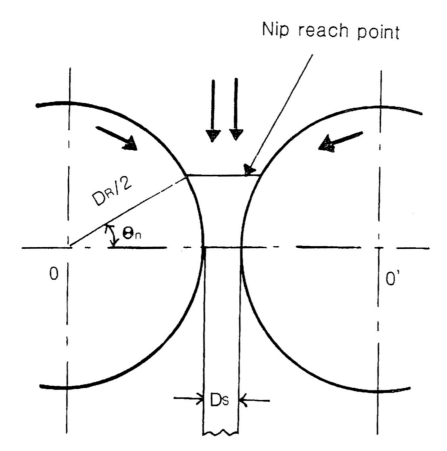

FIGURE 8.6　Schematic diagram of a roll compaction. [Kurtz, B. E. and Barduhn, A. J., *Chem. Eng. Prog.*, 56, 67–72, 1960. With permission.]

In the application of screw-type extruders, the delivery rate of a screw rotating within the barrel should be exactly equal to the flow rate of material through the die. The delivery mechanism of the screw has some difficulties, but there is a theory of the screw conveyance of solids in clay extrusion.[36] On the other hand, in connection with extruding flow characteristics of carbon paste[37] or other solid–liquid mixtures,[38] the pressure P required to extrude the material through a nozzle or multihole die is expressed as

$$(V - V_0) = K(P - P_0)$$

(8.7)

where V is the velocity of piston movement, P_0 and V_0 are the ultimate limit of both P and V, and K is a constant related to the fluidity of the material through the dies and related to the structure of a nozzle or multihole die. Further useful granulation methods are outlined in the following section.

1.8.2 GRANULATORS

Figure 8.7 shows the basic designs of various granulation equipments.

Tumbling Granulation Methods

Typical inclined pan granulators consist mainly of an inclined disk with a rim and a granule size separating ability. Raw materials are fed continuously from above onto the nearly central part of the

1. GRANULE GROWTH MODES FOR COALESCENCE, BREAKAGE,
   AND LAYERING

1.1 Tumbling agglomeration methods:

(1) Inclined pan   (2) Drum   (3) Conical drum

1.2 Fluidized-bed spray methods:

(1) Fluidized bed   (2) Fluidized bed   (3) Spouted bed
                              with a conical
                              distributor

1.3 Agitation methods (mixer agglomeration):

(1) Pug mill   (2) High-speed flow-   (3) Horizontal pan
                              type mixer   mixer

2. MACHINE-MADE PRODUCT MODES THE MACHINES
   MAINLY PERFORM

2.1 Forced screening methods:

(1) Vertical knives   (2) Horizontal   (3) Rotor bars
                              knives

FIGURE 8.7   Typical granulation methods and equipment.

2.2   Compaction methods:

Roll compaction   (2) Roll briquetting   (3) Tableting

2.3   Extrusion methods:

(1) Screw extruder   (2) Ring die   (3) Multi-armed
                                                   spreader

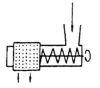

2.4   Fusion methods:

(1) Prilling tower   (2) Spouted bed   (3) Steel belt

      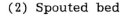

FIGURE 8.7   (*continued*)

disk, as shown in Figure 8.8, and larger granules formed are discharged beyond the rim. Binding liquid is sprayed onto incoming fine particles and smaller granules at a spraying zone on the disk. To form narrow granules, the I_g value defined by Equation 8.1 is adjusted in the range of about 0.5 to 0.75. Other operating conditions are given Table 8.2. A batch drum granulator is often available in laboratories experimenting with damp powders. In industrial practice, this type is used for the continuous granulation of damp powders within a slightly inclined rotating cylinder. If the equipment is made suitable for an ideal piston flow, experimental data for the batch drum granulator are applicable as is. Of course, it is necessary to have a good tumbling action of growing granules in the drum. Therefore, the inside wall of the drum is covered with a rugged rubber sheet or wire net and equipped with an adjustable or reciprocating scraper. When the granule product discharged has a wide size distribution, more than two rotating drums are used in a closed-circuit system. A rotating conical drum granulator has a granule size separating ability, and its best example is found in the study of simultaneous granulation and separation in a horizontal rotating vessel.[39]

Fluidized-Bed Spray Methods

The basic method for making granules in a fluidized bed (i.e., the fluidized-bed spray granulator) can be obtained by the combination of a hot gaseous fluidized bed and a two-fluid spray nozzle. A binding liquid or solution is sprayed onto a bed surface of fluidized particles or is sprayed directly into the bed. It tends to show that compared to seed particles, larger spray droplets generally result in the formation of agglomerates, which is influenced by a wide variation of operating conditions. In relation to granule growth of simultaneous coating and agglomeration in a hot fluidized bed, the size distribution and growth rate of granules under continuous operation can be derived from the material balance of the bed particles.[40–43] The general principal in the spouted bed spray granulator is to spray solution,[44–47] melt,[48] or slurry[49,50] onto seed particles that circulate in a bed spouted by hot gas. This type of granulator is used to granulate several materials of industrial interest.

Agitation Methods

The mixing or blending of solid–liquid mixtures not only results in the dispersion of liquid in powders but is also suitable for the formation of small granules or agglomerates. A pugmill is generally used in the pretreatment for further granulating operation, but its application is limited to special fields such as fertilizer formulation or clay preparation. Other mixers, operated by means of pins, pegs, or blades, are utilized in a similar manner as the above. Granular products formed by those mixers tend to show a wide size distribution. The high-speed flow-type mixer is of growing importance in the formation of very small granules and is often used to coat coarse particles with a he-particle layer rather than coalesce with each other. The horizontal pan mixer consists mainly of a rotating mixing pan, a counterrotating blade, and a stationary scraper for cleaning the interior surface of the pan. Dry powders fed into the pan are wetted by adding a liquid binder so that granules are formed depending, on the moisture content and the rotational speed of the pan or the blades.

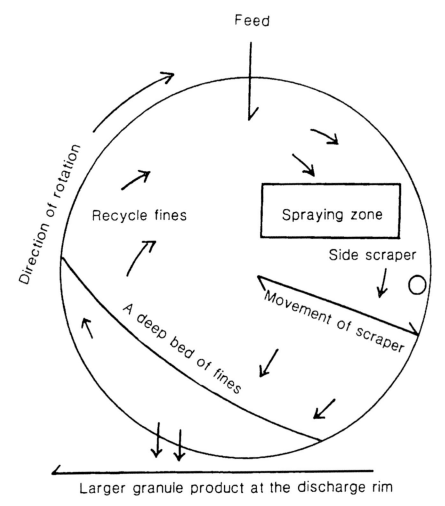

FIGURE 8.8   Flow of material in an inclined pan granulator.

Forced Screening Methods

A comminuting granulator consists of rotating vertical knives and a chamber, the lower part of which holds interchangeable screens. This machine is advantageous for the disintegration of dry clumps or cakes, or larger agglomerates. In contrast to this, the tornado mill, a type of screen granulator, is well suited for the disintegration of sticky agglomerates, in which moist material passes through a perforated screen by means of the high-speed rotor with a row of cutting knives. The oscillating granulator is used for processing large granulated materials, and the disintegration mechanism depends on a rubbing action created by a combination of a rotor bar and a wire screen with a suitable aperture.

Compaction Methods

The roll compaction machine consists of two opposing counterrotating rolls and a feeding arrangement such as gravity feeder or tapered screw feeder. The product, compacted in sheet form, is granulated by a simple disintegrator or mill. The principal feature of a roll briquetting machine is similar to roll compaction, and the feed material, with or without a binding agent, is compacted to egg- or pillow-shaped briquettes, according to the pockets on the two rolls. The formation of a tablet on a single-punch machine is usually as follows. Granular material or powder is fed from a hopper in a die when the lower punch is at the descent position. The excess filling from hopper is scraped off by rotating the die table. Subsequently, the upper punch is lowered to press the powder, and then the compacted tablet is ejected when both punches are raised. The production capacity of this machine is about 100 tablets per minute. High-speed rotary tabulating machines can produce up to about 8000 tablets per minute.

Extrusion Methods

The screw-type extruder used in the formation of small rodlike pellets consists of two main parts: a screw and a cylindrical or flat multihole die placed at the forward position of the barrel. Moist powder containing a binding liquid is introduced to the barrel through a hopper and kneaded sufficiently to form a plastic mass that is extruded through the multihole die. The die roll extrusion machine known as a pellet mill is also used extensively. In a typical pellet mill, the material fed into the interior of a large ring-shaped die roll is pressed by means of one or two smaller press rollers arranged eccentrically inside the die roller. At that time, adjustable knives on the exterior of the die roller cut the rodlike extrudates into pellets. Another machine of extruding granulation extrudes plastic or sticky material through a vertical perforated cylinder attached to the top hopper. The extruding force introduced by a multiarmed spreader decreases with the increased resistance of material flow through the holes of a perforated cylinder, so that the preparation of feed material becomes important.

Fusion Methods

The prilling method yields a granular product that is essentially spherical and uniform in size. The principle is that of solidifying droplets of melt in air, water, or oil. In air prilling, for example, molten material is dispersed by spraying it from the top of the prilling tower, and the falling droplets solidify in the cooling air stream.[51,52] Other fusion methods include a spouted bed and a steel belt or drum flaker. It is necessary to have more detailed knowledge of granulation techniques in practice because so many different machines have been used to obtain the desired granules in a variety of industries. For more detailed information, readers are referred to *Handbook of Granulation,* published in 1991.[53]

Notation

F_L Force transmitted through powder mass in cylindrical die (Pa)
F_p Mass flow rate (kg/s)
F_u Force applied to powder in cylindrical die (Pa)

H	Depth of pan, height of ring dam in drum (m)
K	Constant concerning the fluidity of solid–liquid mixture through a nozzle (m^3/N s)
k	Ratio of radial to axial stress
k_n	Rate constant of granule growth by coalescence (s^{-1})
k_s	Rate constant of granule growth by layering (kg/m^2 s)
L	Thickness of powder mass in cylindrical die (m)
L_c	Height of fixed bed (m)
L_f	Height of fluidized bed (m)
N_c	Critical rotating speed of drum or pan (s^{-1})
N_d	Rotating speed of drum or pan (s^{-1})
N_{opt}	Optimum rotating speed of drum or pan (s^{-1})
N_v	Expansion ratio ($= L/L_c$)
P	Pressure exerted by the material undergoing compaction certain portion of roll surface (Pa)
p	Compacting or extruding pressure (Pa)
P_o	Ultimate limit of extruding pressure (Pa)
R_m	Cumulative mass fraction of granules coarser than a stated size
t	Time (s) _
u_o	Superficial gas velocity in a fluidized bed (m/s)
u_t	Terminal velocity of a particle in upward direction (m/s)
u_{mf}	Minimum fluidization velocity (m/s)
V	Volume of a granule (m^3)
V	Velocity of piston movement (m/s)
V_o	Extruding piston velocity at the ultimate pressure (m/s)
V_e	Ultimate volume of a granule (m^3)
W	Holdup (kg)
X	Diameter of granule (m)
X_a	Arithmetic mean diameter of granules given by mass fraction in a size distribution (m)
X_p	Diameter of solid particle (m)
Z_d	Cumulative drum revolution
α_r	Angle of repose (deg)
ε	Void fraction
ε_f	Void fraction of fluidized bed
ε_n	Void fraction of the material between the rolls at the angle of nip
ε_p	Void fraction of the product when there is no plastic deformation during the compaction process
η	Angle of inclination to the horizontal (deg)
θ	Angle of inclined pan (deg)
θ_n	Angle of nip (deg)
μ_w	Coefficient of wall friction
ρ_p	Density of particle (kg/m^3)

REFERENCES

1. Dahneke, B. J., *Colloid Interface Sci.*, 40, 1–13, 1972.
2. Umeya, K. and Sekiguchi, I., *J. Soc. Mater. Sci. Jpn.*, 24, 608–612, 1975.
3. Takahashi, M., Kobayashi, T., and Suzuki, S., *J. Soc. Mater. Sci. Jpn.*, 32, 953–957, 1983.
4. Mort, P. R., Sabia, R., Niesz, D. E., and Riman, R. E., *Powder Technol.*, 79, 111–119, 1994.
5. Sekiguchi, I., *Kagaku Kogaku*, 32, 745–747, 1968.
6. Newitt, D. M. and Conway-Jones, J. M., *Trans. Inst. Chem. Eng.*, 36, 422–441, 1958.
7. Rumpf, H., in *Agglomeration*, Knepper, W. A., Ed., John Wiley, New York, 1962, pp. 379–418.
8. Pietsch, W. E., Hoffman, E., and Rumpf, H., *Ind. Eng. Chem. Prod. Res.*, 8, 58–62, 1969.

9. Umeya, K. and Sekiguchi, I., *Kagaku Kogaku,* 37, 704–712, 1973.
10. Sekiguchi, I. and Umeya, K., *Kagaku Kogaku,* 37, 744–747, 1973.
11. Mazzone, D. N., Tardos, G. I., and Pfeffer, R., *J. Colloid Interface Sci.,* 113, 544–556, 1986.
12. Mazzone, D. N., Tardos, G. I., and Pfeffer, R., *Powder Technol.,* 51, 71–83, 1987.
13. Sekiguchi, I. and Ohta, Y., *J. Powder Technol. Jpn.,* 28, 232–238, 1991.
14. Michaels, A. S. and Puzinaskas, V., *Chem. Eng. Prog.,* 59, 604–614, 1954.
15. Umeya, K. and Sekiguchi, I., *J. Soc. Mater. Sci. Jpn.,* 24, 664–668, 1975.
16. Sekiguchi, I. and Tohata, H., *Kagaku Kogaku,* 32, 1012–1020, 1968.
17. Kapur, P. C. and Fuerstenau, D. M., *Ind. Eng. Chem. Process. Des. Dev.,* 8, 56–62, 1966.
18. Ouchiyama, N. and Tanaka, T., *Ind. Eng. Chem. Process. Des. Dev.,* 13, 383–389, 1974.
19. Kapur, P. C. and Fuerstenau, D. M., *Trans. AIME,* 229, 348–355, 1964.
20. Kapur, P. C., *Chem. Eng. Sci.,* 27, 1863–1869, 1972.
21. Ouchiyama, N. and Tanaka, T., *Ind. Eng. Chem. Process. Des. Dev.,* 14, 286–289, 1975.
22. Sastry, K. V. S. and Fuerstenau, D. M., *Trans. AIME,* 250, 64–67, 1971.
23. Sastry, K. V. S. and Fuerstenau, D. M., *Trans. AIME,* 252, 254–258, 1972.
24. Sastry, K. V. S., *Int. J. Mining Process.,* 2, 187–203, 1975.
25. Dencs, B. and Ormos, Z., *Powder Technol.,* 31, 85–91, 1982.
26. Dencs, B. and Ormos, Z. *Powder Technol.,* 31, 93–99, 1982.
27. Smith, P. G. and Nienow, A. W., *Chem. Eng. Sci.,* 38, 1223–1231, 1983.
28. Smith, P. G. and Nienow, A. W., *Chem. Eng. Sci.,* 38, 1233–1240, 1983.
29. Huang, C. C. and Kono, H. O., *Powder Technol.,* 55, 35–49, 1988.
30. Ennis, B. J., Tardos, G., and Pfeffer, R., *Powder Technol.,* 65, 257–272, 1991.
31. Kawakita, K. and Ludde, K. H., *Powder Technol.,* 4, 57–60, 1971.
32. Spencer, R. S. and Gilmore, G. D., *J. Appl. Phys.,* 21, 527–531, 1950.
33. Train, D. and Twis, D., *J. Trams. Inst. Chem. Eng.,* 35, 258–266, 1957.
34. Kurtz, B. E. and Barduhn, A. J., *Chem. Eng. Prog.,* 56, 67–72, 1960.
35. Murakami, K., *Kobunshi Kagaku,* 17, 571–576, 1960.
36. Parks, J. R. and Hill, M. J., *J. Am. Ceram. Soc.,* 42, 1–6, 1959.
37. Jimbo, G., Iwamoto, F., Hosoya, M., Sugiyama, Y., and Moro, Y., *Kagaku Kogaku,* 36, 654–660, 1972.
38. Jimbo, G. and Kambe, T., *J. Soc. Powder Technol. Jpn.,* 9, 451–457, 1972.
39. Sugimoto, M. and Kawakami, T., *Kagaku Kogaku Ronbunshu,* 8, 530–532, 1982.
40. Harada, K. and Fujita, J., *Kagaku Kogaku,* 31, 790–794, 1967.
41. Harada, K., Fujita, J., and Yoshimura, M., *Kagaku Kogaku,* 33, 793–799, 1969.
42. Harada, K., Fujita, J., and Yoshimura, M., *Kagaku Kogaku,* 34, 102–104, 1970.
43. Harada, K., *Kagaku Kogaku,* 36, 1237–1243, 1972.
44. Uemaki, O. and Mathur, K. B., *Ind. Eng. Chem. Process. Des. Dev.,* 15, 504–508, 1976.
45. Singiser, R. E., Heiser, A. L., and Prillig, E. B., *Chem. Eng. Prog.,* 62, 107–111, 1966.
46. Robinson, T. and Waldie, B., *Trans. Inst. Chem. Eng.,* 57, 121–127, 1979.
47. Kurcharski, J. and Kmiec, A., *Can. J. Chem. Eng.,* 61, 435–439, 1983.
48. Weiss, P. J. and Meisen, A., *Can. J. Chem. Eng.,* 61, 440–447, 1983.
49. Pavarini, P. J. and Coury, J. R., *Powder Technol.,* 53, 97–103, 1987.
50. Liu, L. X. and Litster, J. D., *Powder Technol.,* 74, 215–230, 1993.
51. Tohata, H. and Sekiguchi, I., *Kagaku Kogaku,* 26, 818–825, 1962.
52. Tohata, H., Sekiguchi, I., and Suzuki, H., *Kagaku Kogaku,* 32, 5–55, 1968.
53. Association of Powder Process Industry and Engineering, Zouryuu Handbook, Ohmsha, Tokyo, 1991. [In Japanese.]

1.9 Kneading and Plastic Forming

Minoru Takahashi
Nagoya Institute of Technology, Tajimi, Aichi, Japan

The term "kneading" is sometimes distinguished from "mixing" to emphasize compounding of powder with liquid having a high viscosity, or coating of a powder surface with liquid. In industrial applications, kneading is usually combined with the plastic forming processes. In this chapter, the fundamentals of kneading and plastic forming processes are described.

1.9.1 KNEADING

Powder–Gas–Liquid Dispersion System

The packing conditions of powder–gas–liquid mixture systems are ideally classified in Table 9.1.[1] The systems are distinguished by the differences in the existence and continuity of each phase. Kneading can be defined as the mixing operation in the mud system. Kneading of either film type or matrix type is employed.[2] Film-type kneading is the thin coating of the surface of each particle with liquid in the funicular state, whereas matrix-type kneading is the coating of the surface of each particle with a large amount of liquid or the dispersing of particles into the liquid in the capillary state.

Critical Powder Volume Concentration

Compounds for plastic forming in a narrow sense should be prepared to a capillary state where the voids among particles are saturated with liquid. The powder concentration corresponding to a capillary state is called a critical powder volume concentration (CPVC). CPVC is very dependent on the size, size distribution, and shape of particles and can be estimated experimentally from the mixing torque versus liquid content curve using a torque rheometer.[3] With the addition of liquid, torque gradually increases, reaches a maximum, and then decreases sharply, as shown in Figure 9.1.[1,4] CPVC is determined by the powder concentration at the maximum torque. The figure also shows that the condition of the powder–gas–liquid system shifts as follows: powder → (powder + granules) → granules → (granules + paste) → paste → slurry.

Mechanisms and Machines

Kneading is attained by the mechanisms of convection, shear, and diffusion, as is dry or slurry mixing. However, the convection and shear mechanisms are relatively important in the kneading process because the mixture systems, having a quite high viscosity, are dealt. Therefore, kneading machines capable of producing high shear force must be used to coat the particles with liquid. Also, kneading machines with high shear speed will be required to shorten kneading time. Representative kneaders and their applicabilities are listed in Table 9.2.[2] Dispersion of particles into the matrix becomes more critical with decreasing particle size and increasing powder concentration. Special kneading with the assistance of high temperature or vacuum will be useful for spreading the liquid on the particle surface or removal of trapped air. Another route for attaining sufficient dispersion

TABLE 9.1　Packing Characteristics of a Powder–Air–Water System

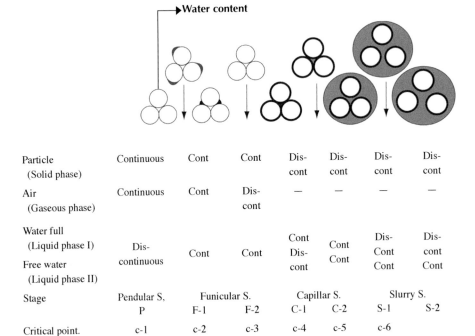

Particle (Solid phase)	Continuous	Cont	Cont	Dis-cont	Dis-cont	Dis-cont	Dis-cont
Air (Gaseous phase)	Continuous	Cont	Dis-cont	—	—	—	—
Water full (Liquid phase I)	Dis-continuous	Cont	Cont	Cont Dis-cont	Cont Cont	Dis-Cont	Dis-cont
Free water (Liquid phase II)						Cont	Cont
Stage	Pendular S, P	Funicular S. F-1　　F-2		Capillar S. C-1　　C-2		Slurry S. S-1　　S-2	
Critical point.	c-1	c-2　　c-3		c-4　　c-5		c-6	
		P.L. (Plastic Limit)		L. L. (Liquids Limit)			
Rheological System	Power System		Mud System			Slurry System	
Mix	Dry mix	Semi-Dry (or Semi-Wet) mix		Wet mix		Slurry mix	

Source: Umeya, K., *J. Soc. Rheol. Jpn.,* 13, 145–166, 1985.

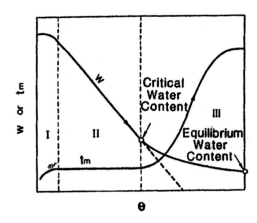

FIGURE 9.1　Pelletizing effect in powder–air–water system. [From Umeya, K., *J. Soc. Rheol. Jpn.,* 13, 145–166, 1985. With permission.]

TABLE 9.2 Types and Factors of Kneaders

Type	Name	Sketch	Material Factor			Kneader Factor		
			Particle size [a]	Binder [b] content	Viscosity [c]	Power (kw)	Kneading Force [d]	Kneading Speed [d]
Wheel	Simpson		M–C	FI–C	H	0.75–150	4	2
	Wetpan		M–C	FI–FII	H	5.8–15	4	2
	Counter Flow		M–C	FI–FII	M–H	7.5–11	3	2
	Speed muller	Wheel	M–C	FI–FII	M–H	15–75	4	4
Ball	Ball mill	Ball	F–C	FI–S	L–H	3.7–37	5	1–2
	Sand Grinder	Beads Blade	F	S	L–M	0.2–37	5	1–2
Blade	Paddie	Blade	M–C	P–FII	M–C	2.2–37	1	3
	Kneader		F–M	FI–C	H	0.25–1500	5	2
	Auger	vac	F	FII–C	H	0.75–19	5	2
	Turbulizer	Blade	M–F	P–FI	L	3.7–95	2	5
	Henschal	Propeller	M–F	P–FI	L	1.8–200	2	5
	Vertical	Blade	F–C	FI–C	M–H	0.2–15	2	2
Roll	Roll Mill	Roller	F	C–S	H	0.4–200	5	2
	Taper Roll		F	C–S	H	2.2–200	5	2
	Banbury		F	FII–C	H	5.6–1900	5	2
	Continuous	Roller Screw	F	FII–C	M–H	0.75–300	5	2
Other	Pestle	Pestle	F–C	P–C	L–H	0.1–2.3	3	2

[a] F, Fine <50 μm; M, middle 50–500 μm; C, course >500 μm.
[b] Refer to Table 8.1.
[c] L, Low <10 mPa·s; M, middle 10^2–10^3 mPa·s; H, high >10^4 mPa·s.
[d] Estimated value, 5–1; highest kneading force and speed, f.

Source: Hashimoto, K., in *Powder Technology Handbook,* Iinoya, K., Gotoh, K., and Higashitani, K., Eds., Marcel Dekker, New York, 1990, p. 653.

is surface modification of the particles, which improves the wetting between the particles and the organic medium and promotes penetration of the viscous fluid into the particle agglomerates.

1.9.2 PLASTIC FORMING

Plastic forming methods among a variety of forming methods include wet pressing, jiggering, extrusion, and injection molding. Plastic bodies prepared by kneading processes are treated in these forming methods.

Wet Pressing

Wet pressing is similar to dry pressing and is often called semidry pressing, or dust pressing. The forming cycle consists of three stages: (a) filling raw powders into the die cavity, (b) compaction, and (c) ejection of the pressed body, as shown in Figure 9.2. Usually, feed material containing 5–10% water is pressed in dies made of steel, cast iron, or tungsten carbide. The packing state of the feed mixture is in the range of funicular I to funicular II in Table 9.1. The feed is occasionally prepared by crushing the filtered cake of slurry without the use of a kneader. Compared with the dry feed, the wet feed deforms plastically during compression. Therefore, it is possible to fabricate parts with complex contours by this method. However, dimensional tolerances of wet pressing are only held to ±2%, lower than those of dry pressing.[5] The pressed body after ejection from the die must be treated carefully because it will deform easily from the force of gravity.

Jiggering

Jiggering is one of the oldest plastic forming techniques, where a skilled hand-worker can make any hollowware from plastic clay supported on a rotating wheel. Presently, highly automated jiggering systems are used for the industrial fabrication of hollowware.[6] This creates a schematic of a jiggering machine, or roller machine. The jiggering process consists of three stages: (1) putting a slice of plastic feed onto a spinning plaster mold, (2) pressing the slice by a hot performance die, and (3) jiggering. A slice of de-aired extrudate is commonly used for feed material. Sticking of the plastic feed to the die is a serious problem. To prevent this, lubrication is produced by heating metal roller tools in order to generate steam and facilitate release, or a lubricant is sprayed onto a tool to minimize sticking.[5]

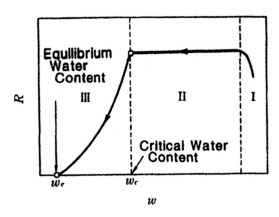

FIGURE 9.2   Sequence of wet pressing for a ceramic component.

Extrusion

A compound adjusted to a capillary state will show a plastic flow necessary for extrusion. If the powder concentration is too low, a squeezing of liquid from the mixture will occur during extrusion at high pressure. On the other hand, too high a concentration will bring about the breakage of the extruding machine because of insufficient plasticity. Granulated feed will be fully consolidated in a compression zone of the extruding machine. In the nonclay system, organic additives are added to promote plasticity.

Extrusion is usually carried out by an auger having mixing equipment. In the mixing before extrusion, feed material is shredded and de-aired under vacuum. Major stages of extrusion are (1) feeding, (2) compression in a barrel, (3) flow through a die, and (4) ejection. The exit, with a uniform section, is supported on a suitable surface to prevent distortion and is cut to proper lengths. The very complex parts such as honeycomb structures are fabricated by the use of a specially designed die.[7] Sometimes, the columnar exit can be used as blanks for further machining.

The plastic mixture after compression can be characterized as a Bingham body, which shows two flow modes in a tube: one is laminar flow and the other is plug flow. The laminar flow occurs in the outside portion of the tube, where the shear stress is greater than the yield stress of the body; the laminar flow occurs in the inside portion. The laminar flow causes orientation of particles along its axis if platelike particles such as clay particles are involved in the body. However, the inside portion has the same random orientation as the original feed.[8] Inevitably, this nonuniform packing structure of the exit results in a differential shrinkage during drying and sintering. The cracks arising from the differential shrinkage often become serious problems in extrusion forming.

Injection Molding

Raw powders must be mixed with an organic vehicle consisting of a binder, lubricant, plasticizer, and dispersing or wetting agent. Thermoplastic or thermosetting resins are commonly used as the binders. The thermoplastic resins are preferable for recycling of a large amount of scrap left in the sprue, runner, and gate of a molding die set. High powder concentration in the compound is generally desired in the case of ceramic injection moldings because the organic vehicle should be removed before sintering. However, too high a concentration over the CPVC results in inadequate fluidity of the compound during mixing and molding. A twin screw extruder and a kneader with pressurizing equipment have been verified to be effective for mixing heavily loaded fine fillers and viscous organic systems.[9] It should be noted that thermal or mechanical degradation of thermoplastic resins will sometimes occur during mixing.[10]

The molding processes consist of four stages: (1) preheating of feed materials within an injection machine, (2) injection of the plasticized feed into the cavity, (3) solidification of the cast body by heating and/or cooling, and (4) ejection of the body. Parts accompanied by thermoplastic or thermoset resins can be made by a screw injection machine. The injected body should be free of defects such as weld lines, pores, and cracks.

REFERENCES

1. Umeya, K., *J. Soc. Rheol. Jpn.,* 13, 145–166, 1985.
2. Hashimoto, K., in *Powder Technology Handbook,* Iinoya, K., Gotoh, K., and Higashitani, K., Eds., Marcel Dekker, New York, 1990, pp. 653.
3. Markhoff, C. J., Mutsuddy, B. C., and Lennon, J. W., in *Forming of Ceramics,* Mangels, J. A. and Messing, G. L., Eds., American Ceramic Society, Ohio, 1984, pp. 246–250.
4. Michales, A. S. and Puzinauskas, V., *Chem. Eng. Prog.,* 50, 604–614, 1954.
5. Reed, R. S., *Introduction to the Principles of Ceramic Processing.* John Wiley, New York, 1988, pp. 355–379.
6. Gould, R. I. and Lux, J., in *Ceramic Fabrication Processes,* Kingery, W. D., Ed., MIT Press, Cambridge, 1958, pp. 98–107.
7. Richerson, D. W., in *Modern Ceramic Engineering,* Marcel Dekker, New York, 1982, pp. 178–216.
8. Norton, F. H., in *Fine Ceramics,* McGraw-Hill, New York, 1970, pp. 130–156.
9. Edirisinghe, M. J. and Evans, J. R. G., *Mater. Sci. Eng.,* A109, 17–26, 1989.
10. Takahashi, M., Suzuki, S., Nitanda, H., and Arai, E., *J. Am. Ceram. Soc.,* 71, 1093–1099, 1988.

1.10 Drying

Hironobu Imakoma
Kobe University, Nada-ku, Kobe, Japan

Morio Okazaki*
Kyoto University, Kyoto, Japan

1.10.1 DRYING CHARACTERISTICS OF WET PARTICULATE AND POWDERED MATERIALS

Water Content and Drying Characteristic Curve

Drying is the removal of relatively small amounts of water (including all states of water molecules, e.g., liquid, adsorbed, bounded, and gaseous) from solids by a hot gas stream or airstream, or other heating sources. To express the amount of water contained, one uses two types of (mean) water content: dry-basis water content w (kg water per kg dry material) and wet-basis water content w' (kg water per kg wet material). The former is much more convenient for calculation than the latter.

$$w = \frac{w'}{1-w'} \tag{10.1}$$

$$w' = \frac{w}{1+w} \tag{10.2}$$

Suppose that a wet nonhygroscopic porous material is being set in a hot airstream of constant temperature, humidity, and velocity. The water content w and the material temperature t_m change with time during drying, as shown in Figure 10.1. The drying process consists of the following three periods: (I) preheating period, (II) constant drying-rate period, and (III) decreasing drying-rate period. Figure 10.2 shows the drying rate R in relation to the water content w. This curve is called the drying characteristic curve.

During period II, free water exists on the surface of the material, and the material temperature t_m remains constant and is equal to the wet-bulb temperature of the hot air, t_w. All of the heat supplied to the material is consumed in water evaporation. Hence, the drying rate becomes constant. This period continues until the water content becomes the critical value w_c, which decreases with reduction of the material size.

During period III, there is no free water on the surface. Because the evaporation surface proceeds to the inside of the material and diffusion of water vapor in the material takes place, the drying rate decreases with time until finally drying ends at the equilibrium water content w_e. Hot-air heat is consumed for heating up the material temperature as well as for water evaporation, and the temperature rises from t_w until it finally reaches the hot-air temperature.

In the case of other materials, except for a nonhygroscopic porous one, one can also observe more or less similar drying characteristic curves. However, shapes of curves in period III are quite different from one another and also from the shape of the nonhygroscopic case.

*Retired.

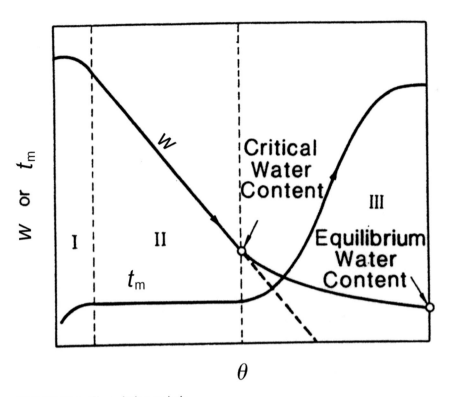

FIGURE 10.1 Three drying periods.

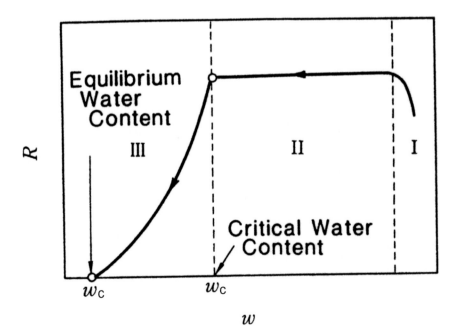

FIGURE 10.2 Drying characteristic curve.

Drying Rate

Because water evaporates on the material surface during the constant drying-rate period, the drying rate R_c is apparently equal to the evaporation rate from the free water surface:

$$R_c = -\frac{W}{A}\frac{dw}{d\theta} = k(H_m - H) \approx \frac{h(H_m - H)}{C_H}$$

(10.3)

When the heat is supplied from the hot air only, the material temperature t_m is equal to the wet-bulb temperature of the air t_w. Hence,

$$R_c = \frac{h(t - t_w)}{r_w}$$

(10.4)

On the other hand, when the heat is supplied by convection, conduction, and/or radiation, the constant drying rate can be estimated by considering that the heat input to the material surface is consumed only for the evaporation of water. The material temperature sometimes deviates from the wet-bulb temperature in the case of other liquids, although a constant material temperature and drying rate are observed.[1-3]

The decreasing drying rate R_d is strongly affected by the material properties and drying method.[3-5] This rate, R_d, is given as a function of water content w. The drying time θ_d during the decreasing drying rate period can be estimated by

$$\theta_d = \int_{w_1}^{w_2} \frac{dw}{R_d} = \int_{w_1}^{w_2} \left(-\frac{d\theta}{dw}\right) dw$$

(10.5)

1.10.2   DRYER SELECTION AND DESIGN

Many types of dryers have been developed and operated, as listed in Table 10.1.[4] First, dryers should be selected based on shape, size, water content of initial and final target wet material, the production rate, and the mode of drying operation. Then, the properties of the material (e.g., stickiness) are taken into consideration, and the selection of dryer type is made. Finally, the volume of the dryer selected should be roughly estimated.

In the selection of dryer type, it is very important to estimate dryer volume. For dryers receiving convective heat from hot air, the rate of heat transfer q is given approximately by

Batch operation:

$$q = h_a V(t - t_m)$$

(10.6)

Continuous operation:

$$q = h_a V(t - t_m)_{lm}$$

(10.7)

For those of conductive-heating type:

$$q = U_k A_k (t_k - t_m)$$

(10.8)

Here, $(t - t_m)_{lm}$ is the logarithmic mean of the temperature differences between the hot air and the material at the material inlet and outlet. Although the material temperature t_m in a dryer of

TABLE 10.1 Rough Estimation of Dryer Capacity[4]

(a) Convective-heating dryer (Batch mode)

Type of dryer	h_a [kJ/(h.K.m³-dryer volume)]	Critical water content, [%]	$(t - t_m)$ [K]	Inlet temperature of hot air[°C]
Box (parallel-flow)	800-1300 (h= 80-130)	>20	30-100[a]	100-150
Box (through-flow)	13000-33000 (granules; layer thickness 0.1- 0.15m)	>20	50[a,b]	100-150
	4000-13000 (mudlike, extruded pieces; layer thickness 0.1- 0.15m)			

(b) Convective-heating dryer (Continuous mode)

Type of dryer	ha[kJ/(h.K.m³-dryer volume)]	Critical water content, [%]	$(t - t_m)l_m$ [K]	Inlet temperature of hot air [°C]
Rotary	400-800	2-3	Countercurrent 80-150	200-600
			Cocurrent 100-180	300-600
Rotary (through-flow)	1300-6000	2-3	80-100	200-350
Pneumatic conveying	13000-25000	1-2	100- 180	400-600
Fluidized bed	8000-25000	2-3	Horizontal (single and multicompartment) 50-150	100-600
			Multistage (countercurrent) 80-100	200-350

Type of dryer		Critical water content [%]		
Spray	80(large particle) ~420(fine powder)	30-50	Countercurrent 80-90 Cocurrent 70-170	200-300 200-450
Through-flow vertical	20000-54000	2-3	Countercurrent 100-150	200-300
Tunnel (parallel-flow)	800-1300 (h= 80-130)	<20	Countercurrent 30-60[a,c] Cocurrent 50-70[a,c]	100-200 100-200
Band (parallel-flow)	170-330	<20	Same as tunnel(parallel-flow)[a,c]	100-200
Band (through-flow)	3000-8000	2-3	40-60[a,b]	100-200

(c) Conductive-heating dryer

Type of dryer	U_k [kJ/(h.K.m²-area contacting with material)]	Critical water content [%]	$t_k - t_m$ [K]
Agitated cylinder or rotary with steam-heated tubes	250-540[d] (small for very sticky materials)	2-5	50-100[a]

[a] Values for the surface evaporation periods, and maximum drying rates will be obtained.

[b] Mean temperature difference between the hot air passing through the layer and the material temperature during the surface evaporation period.

[c] Mean values at both ends of dryers for case a.

[d] No change in the case of vacuum.

Source: Toei, R., *Kanso Souchi*, Nikkan Kogyo Shinbun, Tokyo, 1966, pp. 9–11.

conductive-heating type increases as drying proceeds, the average of the initial and final values is available to roughly estimate the dryer volume.

One can increase the volumetric heat transfer coefficient and prolong the surface evaporation period by lowering the critical water content in order to minimize dryer volume. To do this, agitation or dispersion of particulate materials in hot air is very effective.

The method of making a rough estimation of dryer volume introduced above can be useful only for the selection of dryer type suitable for the target material. Therefore, one should make a more accurate estimation considering static and transport properties within the material after a proper dryer selection by using the rough method. Some books can assist in accurate estimations.[3,5,6] Several problems are solved for difficult materials (e.g., poor fluidizability) in fluidized-bed drying.[7] A guideline for powder manufacture by spray-drying is proposed.[8] Recently, some of other types of dryer (not listed in Table 10.1) are being used, for example, spouted bed, vibrated fluidized bed, and so on.[5,6,9,10] The spouted-bed dryer is essentially a modified fluidized one for larger particles.

Whereas a single type of dryer is usually selected for a target material, the application of combinations of different types of dryer for the material will be more popular. The combination of a pneumatic conveying dryer and a succeeding fluidized-bed dryer allows a remarkable volume reduction of the fluidized-bed dryer in comparison with an independent use of another fluidized-bed dryer, because first removing the surface water of the particles in the pneumatic dryer prevents them from sticking to one another or to the dryer wall, and subsequent recycling of dried material becomes unnecessary. It is noted that one should apply the appropriate combination of dryers for the target material.[11]

Superheated steam is sometimes utilized as a heating medium instead of heated air and may have potential in the next decade.[9,10] Advantages of superheated steam drying are reduced net energy consumption, no oxidative or combustion reactions, and higher drying rates, accompanied by the limitations of a more complex system. Highly moist slurries are sometimes predesaturated by electro-osmotic means for the purpose of reducing energy for the subsequent drying.[9,10,12]

Notation

A	Drying area of material [m^2]
A_k	Heating area of material by conduction [m^2]
C_H	Humid heat of air [kJ/(kg-dry air K)]
H	Humidity [kg-water/kg-dry air]
H_m	Saturated humidity at t_m [kg-water/kg-dry air]
h	Film heat transfer coefficient [kJ/(m^2·h·K)]
h_a	Volumetric heat transfer coefficient [kJ/(m^3·h·K)]
k	Film mass transfer coefficient [kg/(m^2·h·ΔH)]
q	Rate of heat transfer [kJ/h]
R	Drying rate [kg-water/(h·m^2)]
r_w	Heat of evaporation of water at t_w [kJ/kg]
t	Hot-air temperature [°C]
t_k	Temperature of heat source [°C]
t_m	Material temperature [°C]
t_w	Wet-bulb temperature [°C]
U_k	Overall heat transfer coefficient [kJ/(h·K·m^2-heating area by conduction)]
V	Volume of dryer [m^3]
W	Mass of dry material [kg]
w	Dry-basis water content [kg-water/kg-dry material]
w'	Wet-basis water content [kg-water/kg-wet material]
θ	Drying time [h]

REFERENCES

1. Keey, R. B. and Suzuki, M., *Int. J. Heat Mass Transfer,* 17, 1455–1464, 1974.
2. Suzuki, M., in *Drying '80,* Vol. 1, Mujumdar, A. S., Ed., Hemisphere, New York, 1980, pp. 116–127.
3. Suzuki, M., Imakoma, H., and Kawai, S., in *Kagaku Kogaku Binran,* 6th Ed., Maruzen, Tokyo, 1999, pp. 735–787.
4. Toei, R., *Kanso Souchi,* in *Nikkan Kogyo Shinbun,* Tokyo, 1966, pp. 9–11.
5. Keey, R. B., *Drying of Loose and Particulate Materials,* Hemisphere, New York, 1992.
6. Mujumdar, A. S., Ed., *Handbook of Industrial Drying,* 2nd Ed., Marcel Dekker, New York, 1995.
7. Filka, P. and Filkova, I., in *Proceedings of the 11th International Drying Symposium,* Halkidiki, 1998, pp. 1–10.
8. Masters, K., in *Proceedings of the 13th International Drying Symposium,* Beijing, 2002, pp. 19–27.
9. Mujumdar, A. S., in *Proceedings of the 13th International Drying Symposium,* Beijing, 2002, pp. 1–18.
10. Kudra, T. and Mujumdar, A. S., *Advanced Drying Technology,* Marcel Dekker, New York, 2002.
11. Mujumdar, A. S., *Drying Technol.,* 9, 325–347, 1991.
12. Yoshida, H. and Yukawa, H., in *Advances in Drying,* Vol. 5, Mujumdar, A. S., Ed., Hemisphere, Washington, DC, 1992, pp. 301–323.

1.11 Combustion

Hisao Makino and Hirofumi Tsuji

Central Research Institute of Electric Power Industry,
Yokosuka, Kanagawa, Japan

1.11.1 INTRODUCTION

As shown in Figure 11.1,[1] combustion equipment for solid fuels is classified into the fixed-bed type, the fluidized-bed type, and the entrained-bed type, according to slip velocity between gas and particles. The fixed bed uses extremely large solid fuels of more than 10 mm. The actual combustion equipment for particles is mainly the fluidized-bed type and the entrained-bed type.

The bubbling bed, in which the gas velocity is 1–2 m/s, and the circulating bed, in which the gas velocity (4–8 m/s) is faster than that in the bubbling bed, were developed for the fluidized-bed type. In fluidized-bed combustion, the thermal NO_x, which is generated from N_2 and O_2 in the combustion air, becomes lower because the combustion temperature is low.

Pulverized coal combustion is one of the typical entrained-bed combustion materials. In this system, the pulverized coal particles of the order of 40 µm mean diameter are blown from the burner into the furnace to make the flame. Pulverized coal combustion is the most common method, because the combustion efficiency is higher than in the other systems and it is easy to scale up.

In this section, the pulverized coal combustion system is reviewed as an example of combustion of solid fuels.

1.11.2 CONTROL OF THE COMBUSTION PROCESS

In order to utilize the system effectively, it is very important to achieve high combustion efficiency in the operation of combustion systems. In many cases, as the combustion efficiency is increased, concentration of fuel NO_x increases, which is generated from the nitrogen in the coal and the O_2 in the combustion air. It is, therefore, important to develop combustion technology, in which the formation of NO_x is suppressed while improving combustion efficiency. Figure 11.2[2] shows the flame structure for this combustion technology. Figure 11.3[3] represents combustion profiles of the conventional method and the advanced method for low NO_x combustion. In the conventional method, the combustion reaction near the burner is suppressed. In the advanced method, the combustion reaction at the low air ratio region near the burner is accelerated, and the NO_x reduction region is formed effectively downstream of this combustion acceleration region. It is shown that the NO_x emissions from the advanced method is lower than that from the conventional method.

In order to achieve the further improvement of combustion efficiency and to make the NO_x reduction flame wider, the new technology, in which the coal particles are pulverized into fine particles in the order of 10 µm mean diameter, was developed.

1.11.3 COMBUSTION BURNER

Figure 11.4[4] shows the structure of the advanced low-NO_x burner for pulverized coal combustion. This burner was developed for the advanced method shown in the previous section. The

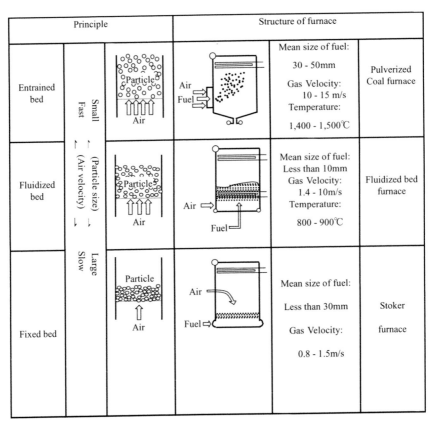

FIGURE 11.1 Combustion equipment for solid fuels. [From Miyasaka, T., *Thermal Nucl. Power Jpn.*, 32, 1056–1060, 1981. With permission.]

pulverized coal particles and the primary air are blown into the furnace from the primary air port. The amount of the primary air is about one fifth of the amount of the total air, which is approximately 1.2 times that of the theoretical air. The secondary air and the tertiary air are provided into the furnace from the outer region of the primary air port. A part of the total air is also provided from the furnace as two-stage combustion air, to suppress the formation of NO_x near the burner. To accelerate the combustion reaction and to make a better NO_x reduction flame, it is necessary that the residence time of the pulverized coal particles near the burner be lengthened by the recirculation flow. To achieve this concept, the flow rate and the swirl force of the secondary air and the tertiary air were optimized. Figure 11.5[4] shows the relationship between NO_x emission and swirl vane angle of the secondary air when using this advanced burner and a conventional burner. For these two burners, NO_x emission changes greatly with the swirl vane angle. NO_x emission from the advanced burner is lower than from that of the conventional burner.

Figure 11.6[4] shows relationship between NO_x emission and the ratio of the secondary air flow rate to the sum of the secondary and the tertiary air flow rates. When decreasing the ratio of the secondary air flow rate, NO_x emission becomes lower.

It is well known that the flame stability of conventional burners decreases when lowering the coal feed rate to reduce the load of the furnace, because the coal particle concentration in the primary air decreases at low-load conditions. In order to improve the flame stability at low-load

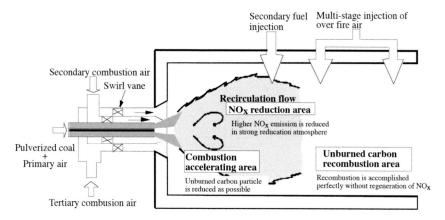

FIGURE 11.2 Concept of advanced low NO$_x$ technology. [From Makino, H., Tsuji, H., Kimoto, M., Hoshino, T., Kiga, T., and Otake, Y., in *Ninth Australian Coal Science Conference*, distributed on CD-ROM, 2000. With permission.]

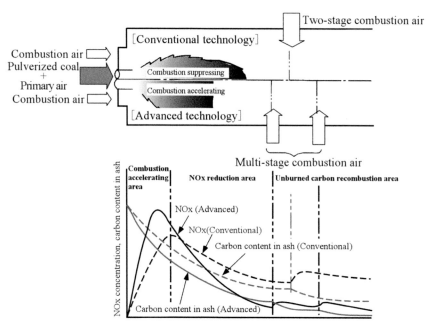

FIGURE 11.3 Combustion profiles of conventional technology and advanced technology. [From Makino, H., Kimoto, M., Kiga, T., and Endo, Y., *Thermal Nucl. Power Jpn.*, 48, 64–72, 1997. With permission.]

conditions, the new burner was developed.[5,6] In the outer region of the primary air port of this burner, the coal particles are concentrated by use of the concentrated ring. Figure 11.7[6] shows the streamlined concentrated ring, which has been developed for the primary air flow without swirl force. At low-load conditions, the ring is moved to the burner exit to improve flame stability. When the load is increased, the ring is moved from the burner exit to unify the whole particle concentration in the burner.

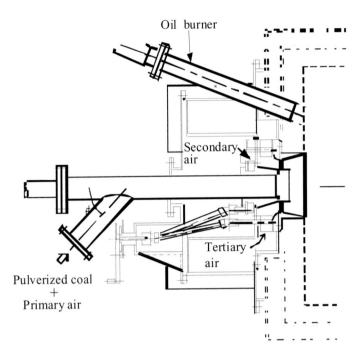

FIGURE 11.4 Structure of the advanced low NO$_x$ burner. [From Makino, H., Kimoto, M., *Kagaku Kogaku Ronbunshu*, 20, 747–757, 1994. With permission.]

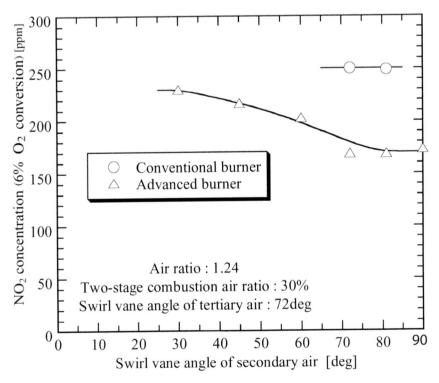

FIGURE 11.5 Relationship between the swirl vane angle of the secondary air and NO$_x$ emission. [From Makino, H., Kimoto, M., *Kagaku Kogaku Ronbunshu*, 20, 747–757, 1994. With permission.]

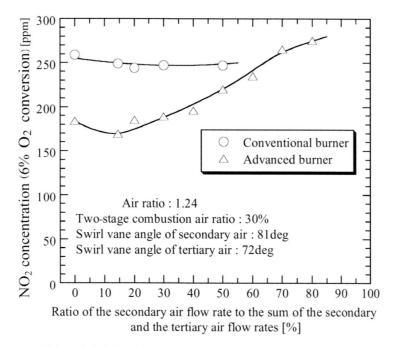

FIGURE 11.6 Relationship between NO$_x$ emission and the ratio of the secondary air flow rate to the sum of the secondary and the tertiary air flow rates. [From Makino, H., Kimoto, M., *Kagaku Kogaku Ronbunshu*, 20, 747–757, 1994. With permission.]

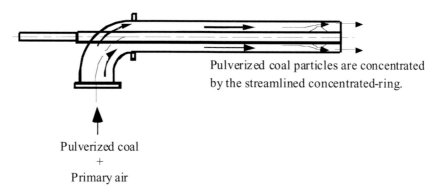

FIGURE 11.7 Structure of the streamlined concentrated ring. [From Makino, H., Kimoto, M., and Ikeda, M., *J. Soc. Powder Technol. Jpn.*, 37, 457–463, 2000. With permission.]

1.11.4 FURNACE AND KILN

Furnace

In the utility boiler, about 10–40 burners are set up in a furnace. Figure 11.8 shows the schematic diagram of a pulverized coal-fired thermal power plant. Several burners are arranged in a row, and each burner makes a flame. Two or three rows are arranged in a furnace. The furnace is classified into the front-firing type, the opposed-firing type, and the corner-firing type. In the front-firing type, burners are

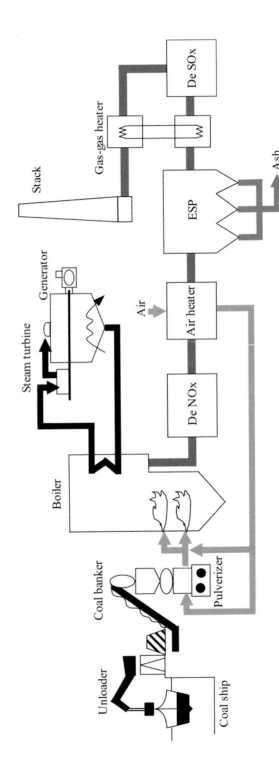

FIGURE 11.8 Schematic diagram of a pulverized coal-fired thermal power plant.

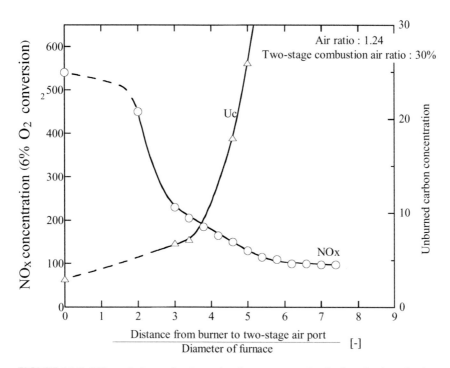

FIGURE 11.9 NO_x emission and unburned carbon concentration in fly-ash when altering position of two-stage air port. [From Makino, H., *Nensho Kenkyu,* 111, 55–68, 1998. With permission.]

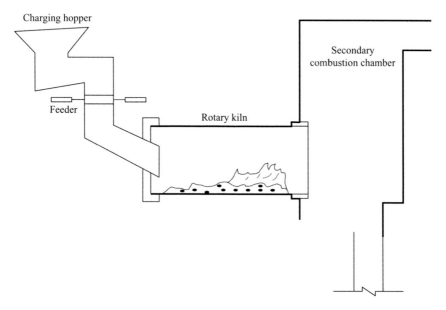

FIGURE 11.10 Structure of a rotary kiln. [From Akiyama, H., in *Sangyou Nenshou Gijutsu,* Energy Conservation Center, Japan, 2000, pp. 141–150. With permission.]

set up on one side, and in the opposed-firing type, burners are set up on both sides. Burners are arranged at the four corners of a furnace in the corner-firing type. At the upper part of the burners, two-stage air ports are mounted to inject a part of the combustion air into the furnace. NO_x concentration decreases in the region from the burners to the two-stage air ports, because a reduction atmosphere is formed in this region.

Figure 11.9[7] represents NO_x emission and unburned carbon concentration in fly-ash when altering the position of the two-stage air port. By increasing the distance from a burner to the two-stage air port, NO_x emission becomes lower as the reduction atmosphere is expanded. However, unburned carbon concentration in fly-ash increases due to the decrease in combustion efficiency. In the actual furnaces, two-stage air ports are mounted at the optimized position.

Kiln

The kiln is used in the cement, ceramic, and refractory material industry and others. The kiln for production of cement is the most common among them. The rotary-type kiln, shown in Figure 11.10,[8] is mainly used as a cement kiln. Materials such as limestone and clay are supplied from the upper ports into the kiln, and fuels such as heavy oil are fed from the lower ports to heat and burn the materials. The temperature of materials reaches around 1500 °C. In some sites, refuse fuels are used instead of fossil fuels. While the kiln type is classified into the wet type and the dry type, the dry type is popular at present.

REFERENCES

1. Miyasaka, T., *Thermal Nucl. Power Jpn.*, 32, 1056–1060, 1981.
2. Makino, H., Tsuji, H., Kimoto, M., Hoshino, T., Kiga, T., and Otake, Y., in *Ninth Australian Coal Science Conference*, distributed on CD-ROM, 2000.
3. Makino, H., Kimoto, M., Kiga, T., and Endo, Y., *Thermal Nucl. Power Jpn.*, 48, 64–72, 1997.
4. Makino, H., Kimoto, M., *Kagaku Kogaku Ronbunshu*, 20, 747–757, 1994.
5. Makino, H., Kimoto, M., and Tsuji, H., *J. Soc. Powder Technol. Jpn.*, 37, 380–388, 2000.
6. Makino, H., Kimoto, M., and Ikeda, M., *J. Soc. Powder Technol. Jpn.*, 37, 457–463, 2000.
7. Makino, H., *Nensho Kenkyu*, 111, 55–68, 1998.
8. Akiyama, H., in *Sangyou Nenshou Gijutsu*, Energy Conservation Center, Japan, 2000, pp. 141–150.

1.12 Dust Collection

Chikao Kanaoka

Ishikawa National College of Technology, Tsubata, Ishikawa, Japan

Various types of dust removal equipment are used to recover valuable materials or as emissions control equipment for air pollution. They adopt various mechanisms to remove particles, depending on particle and gas conditions. Dust removal equipment can be classified into dry and wet collectors. They are also classified into industrial dust collectors and air filters, depending on inlet dust concentration and the pressure drop. Industrial collectors are usually two or three orders of magnitude higher than those of air filters. When they are classified according to gas flow and dust collection mechanism, every dust collector belongs to one of the following categories, shown in Figure 12.1.

(1) Flow-through-type dust collector: This type of collector utilizes the external force perpendicular to the mean gas flow to remove particles outward from the system. Hence the flow does not change with time, and thus removal performance and the pressure drop of the system are stable with time. Gravitational dust collectors, cyclones, and electrostatic precipitators belong to this type.

(2) Obstacle-type dust collector: This type of collector captures dust on obstacles in the equipment. When solid particles are collected by a solid obstacle, captured particles remain on the obstacles, and thus it results in the increase in both collection performance and the flow resistance with operation time. This also makes the release of captured particles from the obstacles difficult, so that this type of dust collector is used as a disposable type of collector. Air filters, granular bed filters, louver-type dust collectors, and the venturi-scrubber belong to this type.

(3) Barrier-type dust collector: This type of collector uses permeable media for gas, not for dust, so that dust is captured and accumulates on the media surface. To maintain the continuous operation of this type of dust collector, accumulated dust has to be removed repeatedly when the pressure drop becomes high. Bag filters and ceramic filters are the typical dust collectors in this type.

Table 12.1 summarizes the general features of dust collectors. Among them, the bag filter shows the highest collection performance, but the collector size has to be large, as its filtration velocity is as low as 1 m/min. Furthermore, its highest operating temperature is limited to 250°C because of the poor thermal durability of the filter material.

On the other hand, the electrostatic precipitator can treat gases moving as fast as 1 m/s at a very low pressure drop. However, its collection performance depends on the electrical property of dust particles. Figure 12.2 and Figure 12.3 show, respectively, the collection efficiency of dust collectors in relation to gas velocity and the initial capital cost of the equipment as a function of the handling gas volume. The total annual cost to operate a dust collector is the lowest at an economical gas velocity, which usually differs from the gas velocity at its best performance. Hence, one has to have a criterion for the selection of dust collectors.

1.12.1 FLOW-THROUGH-TYPE DUST COLLECTORS

The collection efficiency of the collector in this type is expressed by $E = vS/Q$, if the air stream does not have turbulence, by $E = 1 - \exp(-vS/Q)$, if the particle concentration in a vertical cross section

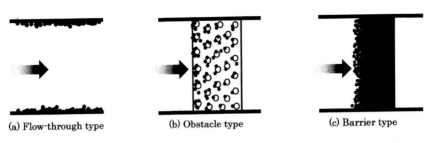

(a) Flow-through type   (b) Obstacle type   (c) Barrier type

FIGURE 12.1   Classification of dust collectors according to gas flow and collection mechanism.

is assumed to be uniform because of turbulence, and by $E = 1/(1 + Q/vS) - 1$ if complete mixing is assumed in the collector. Here S is the total collection surface, Q is the total gas flow rate, and v is the migration velocity to the wall, and it is determined on the external forces acting on the particle such as gravity, centrifugal force, electrostatic force, and so on.

The separation efficiency of this type is shown in Figure 12.4. As one can see from the figure, the existence of turbulence decreases the collection efficiency.

Gravitational Collectors

The gravitational dust collector separates particles by the difference of gravitational settling velocity, which is suited for large particles. In most cases, it is used as a predust collector. Depending on the flow direction of the gas stream, the collectors are classified into single and multiplate types. In both cases, to improve the gravitational effect, the cross-sectional area of the collector is enlarged to operate at a low gas velocity. A Howard-type separator has many horizontal trays; hence, its separation efficiency is quite high because of the short settling distance. However, discharging separated particles and cleaning the equipment are difficult.

Centrifugal Collectors

This type of dust collector separates particles by the difference in centrifugal force created by the change in flow direction and driving particles to the wall. The cyclone is the most widely used centrifugal dust collector because of its simple structure. It has no moving parts and the initial cost is inexpensive. It can be divided into two types: tangential and axial cyclones, according to the type of gas intake. Although the latter type of cyclone is customarily used in parallel so as to deal with a large volume of gas at a low pressure drop, the collection efficiency is not as high. The pressure drop in a cyclone of similar shape depends on the inlet velocity but not on the size. However, the collection efficiency becomes better with decreasing cyclone size. When the same volume of air is dealt with, both collection efficiency and pressure drop decrease in general at elevated temperatures. When dust loading becomes high, the pressure drop decreases, but the collection efficiency improves somewhat. Blowing some air down into the dustbin improves collection performance because it prevents or minimizes reentrainment of dust from the bin. Figure 12.5 shows typical cyclone dimensions, and Figure 12.6 shows the calculated correlation between cut size and cyclone diameter.

Electrostatic Precipitation

An electrostatic precipitator (ESP) separates particles from the gas stream by utilizing Coulombic force acting on charged particles. Coulombic force acts directly on particles so that the structure of the ESP is very simple and different from that of other types of collectors. This feature makes it

TABLE 12.1 General Features of Dust Collectors

	Type	Removable Particle Size (μm)	Pressure drop (Pa)	Maximum temperature (°C)	Inlet concentration (g/m³)	Initial capital cost	Remarks
	Gravitational collector	> 20	50–150	1,000	50 >	Low	Predust collector
Flow through type	Cyclone	> 1	2,000 >	1,000	500 >	Middle	Convenient
	Electrostatic precipitator	> 0.02	300 >	400	20 >	Highest	Not for explosive material
	Scrubber	> 0.2	10,000 >	1,000	100 >	Middle	Require waste water treatment
	Air filter	> 0.01	500 >	100	0.01 >	Low	For building and house
Obstacle type	Air cleaner	> 1	1,000 >	50	1 >	Low	Engine and compressor
	Granular bed	> 1	10,000 >	1,000	50 >	High	Good for hot gas cleaning
Barrier type	Bag filter	> 0.01	2,000 >	250	20 >	High	Not good for condensable material
	Ceramic filter	> 0.01	50,000 >	1,000	50 >	Highest	Good for hot gas cleaning

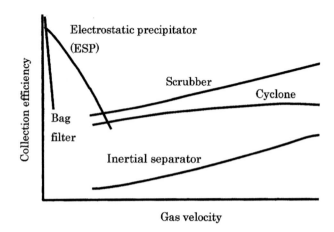

FIGURE 12.2   Dependence of collection efficiency on gas velocity.

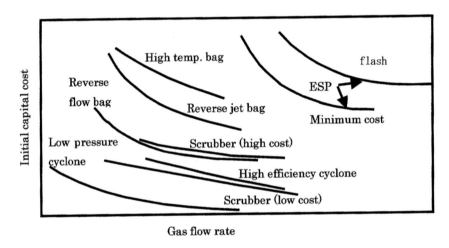

FIGURE 12.3   Relation between initial capital cost and flow rate.

possible to handle a large amount of gas at high temperature and/or at high humidity with a very low pressure drop. An ESP usually consists of particle charging and collection parts. Both charging and collection of particles are accomplished by applying a high voltage between a discharging wire and a plate or cylindrical electrode. Hence, there are two types of ESP: the one-stage type, in which charging and collection of particles take place simultaneously in one set of discharge and collection electrodes, and the two-stage type, in which charging and collection are carried out separately using two sets of electrodes in series. Figure 12.7 shows the structure of the wire and plate-type ESPs.

Approximate collection efficiency of ESPs can be estimated by the following famous Deutsch equation, which is the same equation described before:

$$E = 1 - \exp\left(-\frac{v_e S}{Q}\right) \tag{12.1}$$

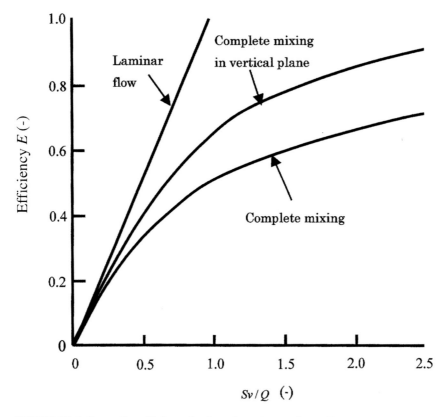

FIGURE 12.4 Separation efficiency for flow-through-type dust collector.

where v_e is the migration velocity of a particle to the collecting electrode, S is the area of the collecting electrode, and Q is the volumetric gas flow rate.

Figure 12.8 shows the measured collection efficiency of an ESP as a function of particle size. Collection efficiency has a minimum around 0.5 μm because the predominant charging mechanism changes from diffusion to field charging around this size range, and the number of charges by diffusion charging is proportional to particle size but is proportional to the second power of size by field charging, whereas drag force is proportional to particle size. The decreasing trend of collection efficiency below 0.05 μm is also considered to be caused by the increase in the fraction of uncharged particles.

The collection performance of ESPs also depends on the electric resistivity of the particle, ρ:

1. $\rho < 5 \times 10^2$ Ωm: Charges on a particle are released immediately after arriving at the collection electrode so that particles are not retained firmly on the electrode, and thus reentrainment of particles happens frequently.
2. $5 \times 10^2 < \rho < 5 \times 10^8$ Ω·m: Charges on a particle are released at a reasonable rate so that captured particles do not reentrain but form a stable dust layer on the electrode.
3. 5×10^8 Ωm $< \rho$: Because of the extremely high resistivity of dust, the accumulation rate of charges from a dust layer exceeds their release rate so that an electric field is formed between a thin dust layer and a collection electrode, and the field strength gets stronger with the accumulation of particles. Finally, back discharge occurs. Furthermore, accumulated dust particles are splashed from the collection electrode when back discharge occurs.

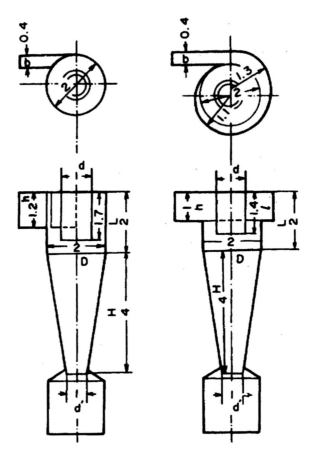

FIGURE 12.5 Size of standard tangential cyclone. [From
Iinoya, K., *Shujin-Sochi,* Nikkan Kogyo Shinbun, Tokyo,
1965. With permission.]

For cases 1 and 3, the collection efficiency of an ESP decreases considerably. Hence, it is very
important to control the electrical resistivity in the range between 5×10^2 and $5 \times 10^8 \, \Omega$m. Among
many factors, humidity and the temperature of the gas are the most influential to the resistivity.
Figure 12.9 shows a measured relationship between resistivity of fly-ash and gas temperature with
humidity as a parameter. The resistivity decreases monotonically against temperature for dry gas,
whereas it has a maximum in humid conditions, and it decreases as the water content increases.
The appearance of maximum resistivity can be explained as follows. Although the resistivity due
to the carriage of charges through a particle, which is referred to as volume resistivity, decreases
as the temperature rises, the resistivity due to the carriage on a particle surface, which is referred
to as surface resistivity and is thought to be related to the thin water layer absorbed on a particle,
decreases as temperature decreases. Accordingly, it is clear that to attain favorable collection performance, an ESP has to be operated either higher or lower than at the temperature giving maximum
resistivity. Several methods have been contrived to reduce the resistivity: spraying water, adding a
small amount of chemicals to the gas, and so on.

On the higher-temperature side, ESPs have not been used because high temperature can cause
problems and can push up the construction cost. However, this is now being reevaluated from the
energy-saving point of view, and pilot plants have been constructed for feasibility studies.

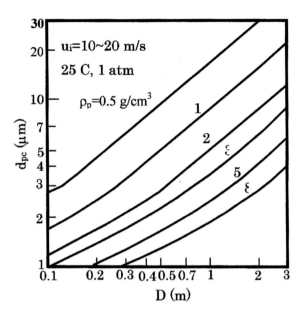

FIGURE 12.6 Cut size of standard cyclone. [From Iinoya, K., *Shujin-Sochi,* Nikkan Kogyo Shinbun, Tokyo, 1965. With permission.]

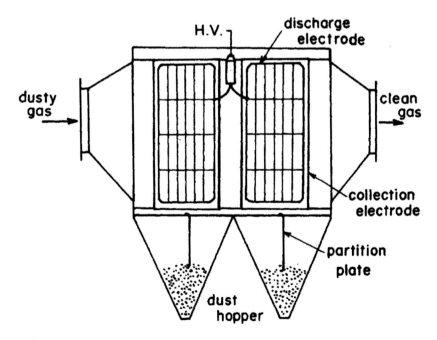

FIGURE 12.7 Parallel-plate electrostatic precipitator.

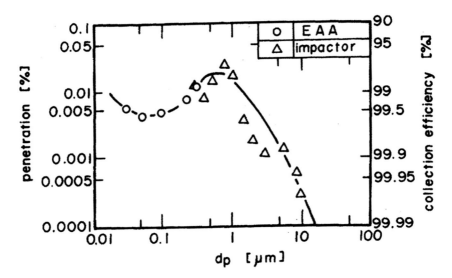

FIGURE 12.8 Measured fractional collection efficiency of ESP; measured by EAA and electrical aerosol analyzer, and impactor.

Charging particles, especially charging high-resistivity particles, is one of the most important problems to be solved in the field of electrostatic precipitation if high collection performance is to be achieved.

1.12.2 OBSTACLE-TYPE DUST COLLECTORS

Since particles are collected by obstacles in the equipment, collection efficiency for this type of collector is expressed by the following equation:

$$E = 1 - \exp\left(-\frac{A}{v_c}\frac{\alpha}{1-\alpha}\eta L\right) \tag{12.2}$$

where v_c is the volume of one collection obstacle, α the packing density of the obstacle, and L the thickness of the obstacle. And A is the constant determined by the shape of obstacle, that is $4/(\pi D_c)$ for cylindrical collector and $3/(2D_c)$. d_c is the diameter of obstacle.

In the equation, constants other η than are determined from the equipment itself, so that E can be estimated if η is given for a given collector and operation conditions, and thus the collection efficiency of obstacles with different shapes is calculated as shown in Figure 12.10. In the figure, Stk is the Stokes number defined by

$$Stk = \frac{C_c \rho_p d_p^2 u}{9\mu D_c} \tag{12.3}$$

where, C_c is the Cunningham's slip correction factor, ρ_p is the particle density, d_p is the particle diameter, D_c is the representative length of a collecting body, u is the gas velocity, and μ is the gas viscosity.

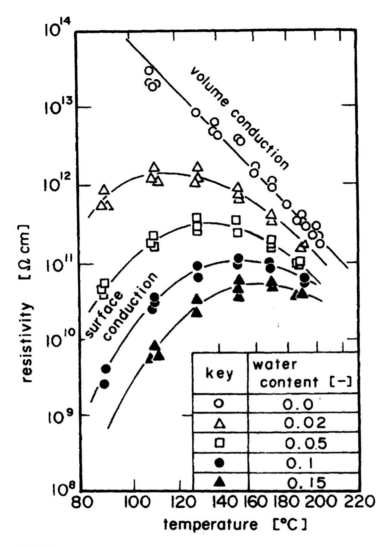

FIGURE 12.9 Effect of gas temperature and humidity on electrical resistivity of dust.

Inertial Dust Collector

The inertial dust collector separates particles by the difference in motion in a curvilinear flow field between the particle and the gas. Although the collection performance is much better than that of gravitational collectors, it is also used as a predust collector. It is classified into two groups, depending on the method of particle collection: separation of particles by many packed obstacles, and separation by flow channels with different curvatures. Figure 12.11 shows some examples of collection bodies and channels.

The collection performance of inertial collectors improves with increasing particle size and gas velocity. Therefore, every inertial collector is designed to intensify the inertial effect by changing the flow direction as often as possible.

Figure 12.12 shows a louver-type dust collector, which intensifies the inertial effect by a large inversion angle and a small gap between blades.

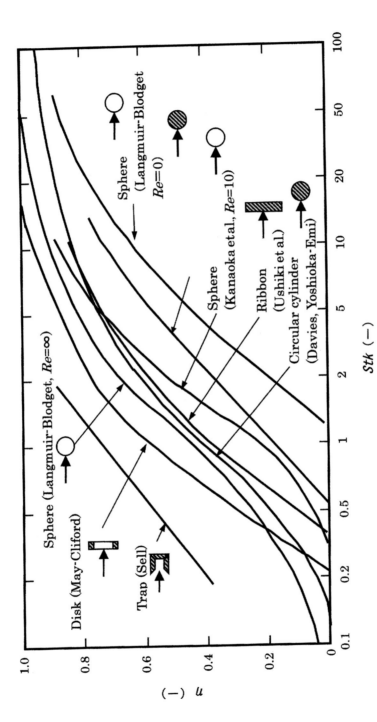

FIGURE 12.10   Calculated collection efficiencies of single collection obstacles.

FIGURE 12.11   Shapes of packing materials for inertial-type dust collector.

In practical applications of inertial collectors, they are used mainly to separate mist rather than solid particles, as a discharge system for collected particles makes their structure complicated and costly.

Air Filters

An air filter is one of the most reliable and efficient methods for dust collection, and particles are mainly used at a low dust concentration so that particles are captured inside the filter, and thus this type of particle collection is called a depth filter.

An air filter is mainly used to purify the air in local environments such as hospitals, manufacturing lines, work spaces, and so on. In those environments, the dust concentration is usually lower than 10 mg/m³ and the particle size is less than several micrometers. Hence, packed filter materials, especially fibrous mats, are used for this purpose. The porosity of a filter mat is generally higher than 85%; hence, particles are captured on individual fibers in the filter. Therefore, the filter efficiency E can be calculated from the single fiber collection efficiency η for a given filtration condition as follows:

$$E = 1 - \exp\left(-\frac{4}{\pi}\frac{\alpha}{1-\alpha}\frac{L}{D_f}\eta\right)$$ (12.4)

where, D_f, α, and L are the fiber diameter, fiber packing density, and filter thickness, respectively.

Although particles are collected on a fiber by a combination of effects, inertia, Brownian diffusion, interception, and gravity are the major factors in mechanical filtration. Electrostatic force becomes important when either fibers or particles or both are electrically charged, or an external electrical field is applied.

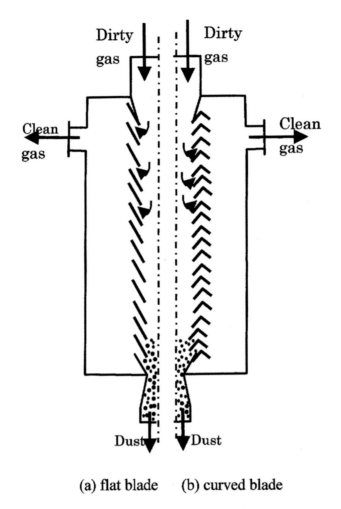

(a) flat blade   (b) curved blade

FIGURE 12.12   Shape of louver-type dust collector.

Many theoretical and empirical expressions have been proposed for each collection mechanism. Figure 12.13 shows a typical single-fiber collection efficiency chart calculated by taking inertia, Brownian diffusion, interception, and gravity into account simultaneously. It is clear from the figure that there exists a particle size or a filtration velocity that gives minimum efficiency, corresponding to the transition region among different mechanisms.

There are two types of air filters: disposable and renewable. For the former type, fiber materials such as glass, metal, and natural and synthetic fibers are packed loosely in a frame or are formed into a matlike structure. Because captured particles accumulate inside the filter, its pressure drop increases with filtration time, so that a zigzag structure is adopted to increase the filtration area to lengthen its service life. It is finally replaced to avoid an increase in running cost when the pressure drop reaches a certain level. The latter type of filter renews part or all of a filtering surface periodically, also to avoid the pressure drop increase. High-efficiency particulate air (HEPA) and ultra-low-penetration air (ULPA) filters are types of paper filters composed of very fine fibers, less than 1 μm on average. They are defined as the filters that collect 0.3 μm particles at the efficiency higher than 99.7% and 0.1 μm particles at higher than 99.9997%, with a pressure drop that does not exceed 12.7 mm H_2O at their specified flow rate, usually 2–5 cm/s. For this reason, HEPA and ULPA filters are used to create highly purified environments, which are necessary in the semiconductor industry, the precision machine industry, and so

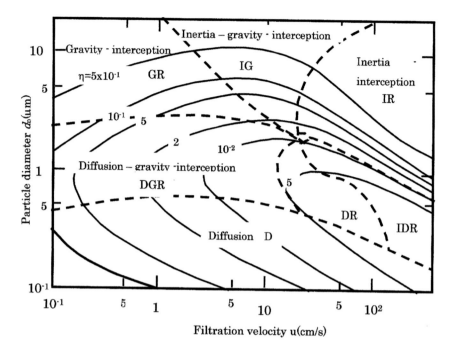

FIGURE 12.13   Total single-fiber collection efficiency ($D_f = 10\ \mu m$, $\alpha = 0.03$).

on. Penetration curves of HEPA and ULPA filters are convex against particle size, and the maximum appears around 0.1 μm (i.e., the collection efficiency of the filter is poorest around this particle size at the conventional filtration condition and decreases as filtration velocity increases). Hence, the filter collection performance can be roughly evaluated by measuring the efficiency in this size range.

An electrostatic fibrous filter is another type of air filter that utilizes electrostatic force to remove particles. Electrostatic attractive force becomes effective when at least either fiber or particle is charged, except that both of them are charged in the same polarity. Therefore, several types are designed, depending on the charging state of fiber and particle. In general, an electrostatic fibrous filter is superior in collection performance to a mechanical fibrous filter but is difficult to operate stably. The electric filter is a stable electrostatic fibrous filter because it is composed of permanently charged fibers and is capable of collecting particles at a high efficiency in the initial filtration stage. However, the efficiency decreases with filtration time because of dust loading. Therefore, this drawback has to be eliminated in practical use.

1.12.3 BARRIER-TYPE DUST COLLECTORS

The most pronounced feature of dust collectors in this type is the almost perfect particle collection regardless of size except just after the removal of accumulated particles on the barrier. This is because most of the particle collection is performed by the already captured accumulated particles by a sieving effect. This kind of collection is usually called surface or cake filtration.

Bag Filters

A bag filter is usually made of fabric and in a cylindrical "bag" shape, hence the name "bag" filter. They are popularly called "fabric" filters, also. Although any type of fiber with a different weave can be used as a filter material, synthetic fibers such as polyester, polypropylene, nylon, and glass

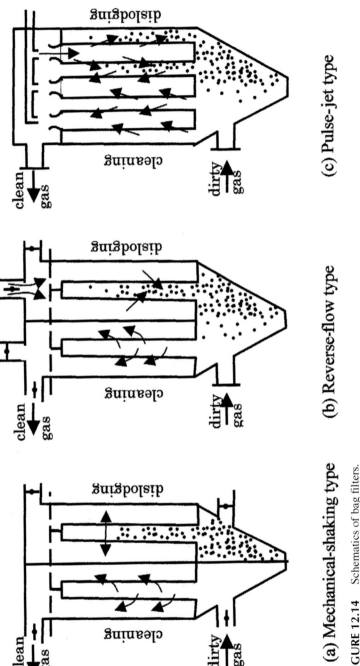

FIGURE 12.14   Schematics of bag filters.

(a) Mechanical-shaking type   (b) Reverse-flow type   (c) Pulse-jet type

are often employed because of their strong resistivity to gas and dust. Their lifetime is usually more than 2 years. However, they cannot be used at a temperature higher than 250°C despite the efforts to develop a high-resistivity fiber. Even the most resistible of glass fibers cannot stand such a high temperature. Because of the small filtration area of each bag (10–300 mm in diameter and length less than 10 m for the cylindrical type and about 1 m × 2 m for the envelope shape), and low filtration velocity (0.5 to 5 m/min), many bags are assembled in a unit, called a bag house, when a large amount of gas must be handled. As mentioned before, collected dust particles accumulate on a fabric surface and thus raise the pressure drop. Hence, the accumulated dust has to be dislodged to maintain continuous operation. Bag filters are categorized into mechanical-shaking, reverse-flow, and pulse-jet types, depending on the method of dislodging dust (i.e., mechanical-shaking and reverse-flow filters have to stop the gas flow during dislodging, thus the bag house of these types is divided into several compartments so that the entire house does not have to be stopped. A schematic drawing is shown in Figure 12.14. In the pulse-jet type, a filter can dislodge dust without stopping the house because it injects strong cleaning air for a short time.

A new application of the fabric filter is the simultaneous separation of gas and dust. This technique is mainly applied to remove HCl and SO_x gases from municipal incineration flue gas by injecting lime slurry or $Ca(OH)_2$. Particles upstream in the bag filter form solid particles. The resulting solid particles are separated by a filter.

1.12.4 MISCELLANEOUS

Scrubbers are widely used as dust collectors because of their relatively high collection efficiency and low initial cost. The venturi scrubber, especially, operating with a very high pressure drop (about 1000 mm H_2O), shows a collection efficiency as high as that of a bag filter. However, because a large amount of water is consumed and has to be treated afterward, it is not popular nowadays. Acoustic, magnetic, and thermal forces also have the potential to separate particles from a gas stream. However, they have not been put into practical use because of economic drawbacks. Other dust collection methods under development are a ceramic filter, and a granular bed filter for hot-gas cleaning. Although they show about the same performance level as that of a bag filter, there remain problems to be solved, especially durability to thermal shock and service life.

REFERENCES

1. Iinoya, K., *Shujin-Sochi,* Nikkan Kogyo Shinbun, Tokyo, 1965.
2. McCain, J. D., Gooch, J. P., and Smith, W. B., *J. Air Pollut. Control Assoc.,* 25, 117–121, 1975.
3. Sullivan, K. M., *EPA Report,* EPA-600/9/82–005b, 1982.
4. Langmuir, I. and Blodgett, K. B., *Army Air Forces Technical Report,* 5418, 1946.
5. Davies, C. N., *Air Filtration,* Academic Press, London, 1973.
6. Yoshioka, N., Emi, H., and Fukushima, M., *Kagaku Kogaku,* 31, 157–163, 1967.
7. Sell, W., *VDI Forsch.,* 347, 1931.
8. Kanaoka, C., Yoshioka, N., Iinoya, K., and Emi, H., *Kagaku Kogaku,* 36, 104–108, 1971.
9. May, K. R. and Clifford, A., *Ann. Occup. Hyg.,* 10, 83–95, 1967.
10. Lewis, W. and Brun, R. J. *NACA,* TN3658, 1954.
11. Masliyah, J. H. and Duff, A., *J. Aerosol Sci.,* 6, 31–43, 1975.
12. Brun, R. J. and Dorsch, R. G., *NACA,* TN3147, 1954.
13. Dorsch, E. G., Brun, R. J., and Gregg, J. K., *NACA,* TN3587, 1955.
14. Ushiki, K., Kubo, K., and Iinoya, K., *Kagaku Kogaku Ronbunshu,* 3, 172–178, 1977.
15. Emi, H., Okuyama, K., and Yoshioka, N., *J. Chem. Eng. Jpn.,* 6, 349–354, 1973.

1.13 Electrostatic Separation

Ken-ichiro Tanoue

Yamaguchi University, Ube, Yamaguchi, Japan

Hiroaki Masuda

Kyoto University, Katsura, Kyoto, Japan

Electrostatic separation based on the difference of electric conductivity of particles has been widely utilized for the 200 μm to 3 mm size range. Figure 13.1 shows electric conductivity[1] for many kinds of materials.

Electrostatic force, which works on a charged particle, is given by

$$F = qE \qquad (13.1)$$

where q is the charge of particle and E is the strength of electric field. Electrostatic separation utilizes this force. Although there are many kinds of methods for particle charging, induced charging, corona charging, and tribocharging are mainly applied in particle separation. Table 13.1 shows the relationship between separating materials and charging method.

1.13.1 SEPARATION MECHANISM

Induction Charging

As electric conductivity differs from the materials, the motion of a particle under an electrostatic field changes dramatically, as shown schematically in Figure 13.2. If a particle is conductive, the particle contacting the surface under an electrostatic field is charged up as high voltage is applied, and it becomes equipotential with the electrode surface. Then the particle jumps up from the surface because of the electrostatic repulsive force. On the other hand, a particle of insulating material needs a much longer time in order to get an induction charge. So, electrostatic force hardly works on it. The induction charge Q is given by[2]

$$Q = C\Delta V \left[1 - \exp\left(-\frac{\sigma t}{C} \right) \right] \qquad (13.2)$$

where C is the capacitance of the particle, ΔV is the voltage difference between electrodes, and σ is the electric conductivity of the particle.

Corona Charging and Induction Charging

When negative high voltage is applied to a metal of large curvature, ionization of gas occurs around it, and then negative ions are accelerated to the grounded surface. Figure 13.3 shows the motion of particles under an electrostatic field. The particles, which are made of high or low electric conductivity, electrify negatively on the grounded surface. As the charge of the particle having high electric conductivity leaks easily to the grounded surface, it is electrified positively by induced electrification. It jumps from the grounded surface by electrostatic repulsive force. On the other

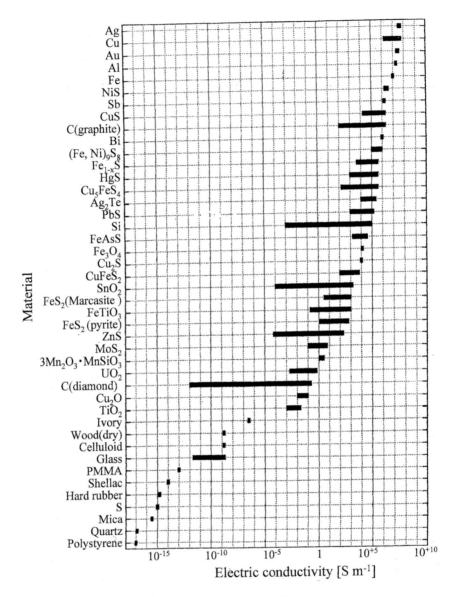

FIGURE 13.1 Electric conductivity for various materials.

TABLE 13.1   Relationship between Separating Materials and Methods

Separating material	Method	Apparatus for electrification
1) Powdered coal	tribo-charge	pipe, fluidized bed, cyclone, pulverizer
2) Farm produce	corona and induced charge	sliding belt, rotating drum
3) Wasted electric wires	induced charge	rotating drum
4) Plastics	tribo-charge	fluidized bed, vibrating transport

hand, the particle with low electric conductivity hardly moves, because the charge of the particle does not leak; therefore, the particle remains on the plate by image force.

Tribocharging

It is hardly practical to separate special plastics from mixed plastic materials or to separate ash and minerals from coal powder by using a corona charge or an induction charge, because the difference in electric conductivity is usually very small. In these cases, tribocharging is used on the basis of contact potential difference (CPD). The CPD that shows the work function difference between two materials is the most direct physical property in relation to particle electrification. Table 13.2 shows the typical work functions[3,4] for many kinds of materials.

In order to separate particles effectively, it is very important to select suitable materials for tribocharging. Charged particles were introduced into the electrostatic field and then separated according to the polarity of the particles.

Various Factors in Relation to Electrostatic Separation

There are three processes in separation of particles by use of electrostatic force: selective charging of particles, charge transfer between particles and electrode, and dynamic motion of particles near

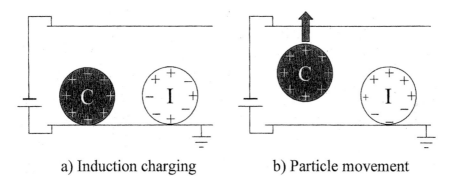

a) Induction charging   b) Particle movement

FIGURE 13.2   Motion of particles under an electrostatic field. C, particle made of conductive material; I, particle made of insulating material.

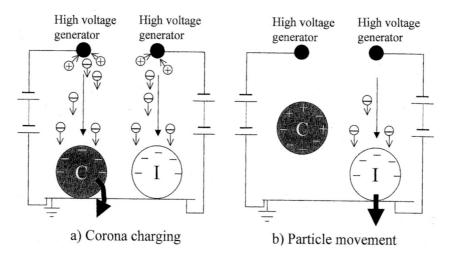

a) Corona charging   b) Particle movement

FIGURE 13.3   Motion of particles under corona charging. C, particle made of conductive material; I, particle made of insulating material.

TABLE 13.2   Work Functions for Various Materials[3,4]

Material	Work function (eV)	Material	Work function (eV)	Material	Work function (eV)
Zn	3.63	BaO	1.1	Polyethylene	5.24±0.24
C	4	CaO	1.60±0.2	Polyethylene	6.04±0.47
Al	4.06–4.26	Y_2O_3	2	Polypropylene	5.43±0.16
Cu	4.25	No_2O_3	2.3	Polypropylene	5.49±0.34
Ti	4.33	ThO_2	2.54	Polystyrene	4.77±0.20
Cr	4.5	Sm_2O_3	2.8	Polyvinyl chloride	4.86±0.73
Ag	4.52–4.74	UO_2	3.15	Polycarbonate	3.85±0.82
Si	4.60–4.91	FeO	3.85	PMMA	4.30±0.29
Fe	4.67–4.81	SiO_2	5	Polytetrafluoroethylene	6.71±0.26
Co	5	Al_2O_3	4.7	Polyimide	4.36±0.06
Ni	5.04–5.35	MgO	4.7	Polyethylene Terephthalate	4.25±0.10
Pt	5.12–5.93	ZrO_2	5.8	Niron66	4.08±0.06
Au	5.31–5.47	TiO_2	6.21	Pylex7740	4.84±0.21

the electrode surface. These processes depend strongly on relative humidity. If moisture adsorbs on the surface of a particle of insulating material, the apparent electric conductivity of the particle may increase. Especially, if ionic concentration in the moisture is high, the electric conductivity may dramatically increase. Generally, the relationship between the conductivity and the humidity is given by[5]

$$\sigma_r = \exp(-K_1 + K_2\psi) \tag{13.3}$$

where σ_r is the relative value of electric conductivity, ψ is the relative humidity, and K_1 and K_2 are constants. In Japan, electrostatic separation has not been widely introduced because it strongly depends on relative humidity.

Electric conductivity of a material changes also with temperature. For metal materials, the conductivity is given by[6]

$$\sigma_{Met.} = \frac{1}{a + bT} \tag{13.4}$$

where T is the ambient temperature, and a and b are constants.

On the other hand, for semiconductors, the conductivity is given by

$$\sigma_{Semi} = \sigma_0 \exp\left(-\frac{E}{KT}\right) \tag{13.5}$$

where σ_0, E, and K are constants. Therefore, the temperature dependence of electric conductivity differs considerably from material to material.

1.13.2 SEPARATION MACHINES

Electrostatic separation has been utilized in industries for refining mineral resources, treatment of pulverized coal, treatment of waste plastics, recycling of electrical appliances, and so on. In this section, some practical separation machines, which have been reported in recent years, are presented.

Refining of Mineral Resources[7-13]

As fewer and fewer high-grade mineral ore deposits are discovered and developed and more become depleted, it becomes increasingly important to develop processes whereby the available low-grade ores may be mined profitably.

The benefaction of mineral ores[11] has been conducted by use of tribo- and induction charging, as shown in Figure 13.4. The apparatus consists of a horizontal aluminum plate with a vibrator, an inverted metallic roof sloped at a certain degree to the aluminum plate, an alternating voltage power supplier, and collecting bins. Figure 13.4a shows that ore particles were supplied on the vibrating aluminum plate, where the particles were charged by a combination of tribocharging and induction charging. They moved in the alternating current field. In Figure 13.4b, centrifugal force acts on the particles due to their complex circular motion induced by the force field, resulting in an outward movement of the particles. The highest-charged particles move the farthest and tend to collect in the lateral bins.

Treatment of Pulverized Coal[3,14-16]

In order to use low-quality coals in a thermal power plant, it is important to remove the ash and mineral contents from pulverized coal. As charcoal contents and ash contents have high conductivity, electrostatic separation has been conducted by use of tribocharging.

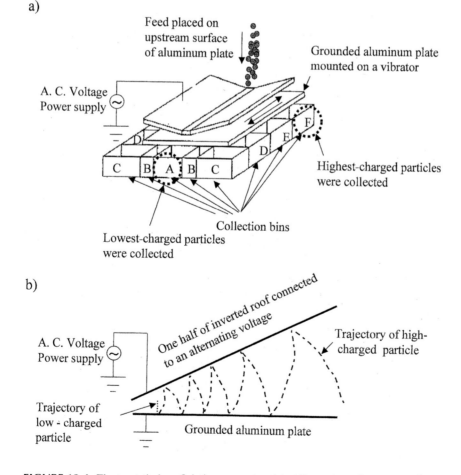

FIGURE 13.4 Electrostatic beneficiation apparatus: (a) oblique view of apparatus showing inverted roof and collection bin locations; (b) end view showing trajectories of charged particles.

Figure 13.5 shows the separation machine used to eliminate ash from pulverized coal. The coal was conveyed by high-speed gas flow to the tribocharging section. Charcoal contents were electrified positively while the remaining ash was charged negatively. In order to electrify them effectively, a spiral copper pipe was introduced in the charging section.

Separation Technologies, Inc.[16] has developed a new electrostatic separation machine, as shown in Figure 13.6. Two parallel plates are set horizontally. The upper plate has a slit to introduce powdered coal. Positive and negative voltages are applied on the upper and lower plate, respectively. A meshed polymer belt was set between the plates and moved by drive rollers. The particles fed into the plates were tribocharged by the belt. Negatively charged particles with ash approached the upper electrode, while positively charged particles with carbon approached the lower electrode; each moved in different directions and then separated.

Treatment of Waste Plastics[17–19]

The development of a process to separate different types of plastics from each other can significantly improve the possibilities to recycle municipal waste. The waste plastics were crushed to pieces and separated to PMMA, ABS, PS, PE, PP, PET, and PVC materials by use of tribocharging.

Hitachi Zosen Corp.[19] has developed a new electrostatic separation machine for plastics, as shown in Figure 13.7. Plastics are fed into a mixer having many pins inside the vessel, where tribocharging of the plastics occurred. The charged plastics were supplied to a grounded rotating drum and were separated under an electrostatic field between the drum and the positive electrode where high voltage was applied.

Treatment of Waste Electrical Wires[20]

It is important to collect the copper and aluminum contents of waste electrical wires for recycling. In this case, waste wires must be stripped of their covering materials and then crushed to the desired size. They have been separated by use of induction charging or induced and corona combined charging.

Recycling of Electrical Appliances[21]

Waste electrical appliances were broken down, pulverized, and then sorted out into sizes of a few millimeters. The materials were composed of many kinds of compounds. Furthermore, these

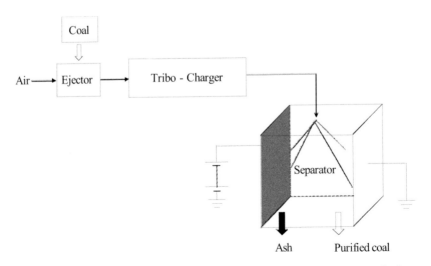

FIGURE 13.5 Electrostatic separation machine for purified coal by use of tribo-charging.

compounds were separated by wind force, magnetic force, and eddy-current force and then collected. However, plastics having PVC and metal wires were not separated. Recently, a separation method for metal wires and PVC has been developed by Higashihama recycle center.[21] The area enclosed by the solid line in Figure 13.8 shows the developed treatment method. The plastics with metal wires were pulverized again. Most metal wires were collected by use of a dry gravitational separator, while the powdered metal was collected by use of an electrostatic separator due to the induction charge. Finally, an electrostatic separator based on tribocharging was utilized for eliminating the PVC materials. The metal compounds were recycled as an ingredient of copper, while the plastics without PVC were utilized as a reducing agent in a blast furnace. Therefore, the amount of reclamation was reduced by one fifth to one tenth of the previous amount.

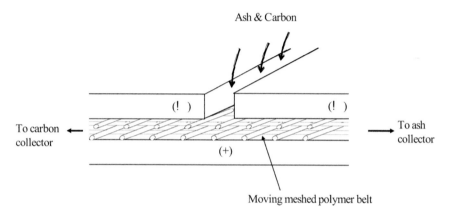

FIGURE 13.6 Electrostatic separation system for pulverized coal in Separation Technologies, Inc.[5]

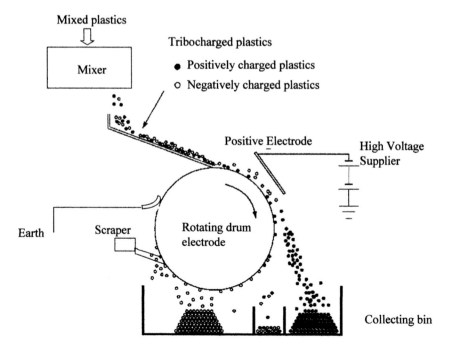

FIGURE 13.7 Electrostatic separation machine for waste plastics.

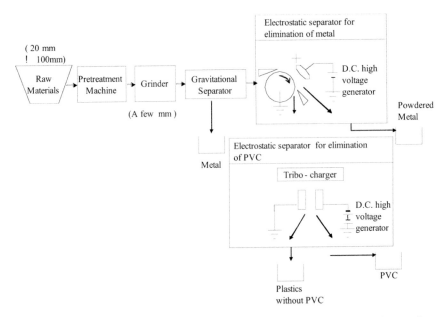

FIGURE 13.8 Application of electrostatic separation for treatment of plastics in recycling plant for electrical appliances.

REFERENCES

1. Kelly, E. G. and Spottiswood, D. J., *Miner. Eng.*, 2, 33–46, 1989.
2. Ito, M. and Owada, S., *Powder Sci. Eng.*, 35, 40–52, 2003 (in Japanese).
3. Gupa, R., Gidaspow, D., and Wasan, D. T., *Powder Technol.*, 75, 79–87, 1993.
4. Chemical Society of Japan, in *Kagaku Binran,* Maruzen, 1984, pp. 493–494 (in Japanese).
5. Chemical Society of Japan, in *Kagaku Binran,* Maruzen, 1984, pp. 494–495 (in Japanese).
6. Kakovsky, I. A. and Revnivtzev, V. I., in *Fifth International Mineral Proceedings Congress,* London, 1960, pp. 775–786.
7. Dance, A. D. and Morrison, R. D., *Miner. Eng.*, 5, 751–765, 1992.
8. Yongzhi, L., in *Proceedings of the First International Conference on Modern Process Mineralogy and Mineral Processing,* Beijing, 1992, pp. 385–390.
9. Okur, E. and Onal, G., in *Proceedings of the First International Conference n Modern Process Mineralogy and Mineral Processing,* Beijing, 1992, pp. 391–396.
10. Zhou, G. and Moon, K., *Can. J. Chem. Eng.*, 72, 78–84, 1994.
11. Celik, M. S. and Yasar, E., *Miner. Eng.*, 8, 829–833, 1995.
12. Inculet, D. R., Quigley, R. M., and Inculet, I. I., *J. Electrostat.*, 34, 17–25, 1995.
13. Li, T. X., Ban, H., Hower, J. C., Stencel, J. M., and Saito, K., *J. Electrostat.*, 47, 133–142, 1999.
14. Thomas, A. L., Rochard, P. K., Robert, H. E., and Nicholas, H. H., *DOE-PETC-TR*, 90–11, 23, 1990.
15. Bouchillon, C. W. and Steele, W. G., *Part. Sci. Technol.*, 10, 73–89, 1992.
16. Tondu, E., Thompson, W. G., Whitlock, D. R., Bittner, J. D., and Vasiliauskas, A., *Miner. Eng.*, 48, 47–50, 1996.
17. Inculet, I. I. and Castle, G. S. S., Tribo-electrification of commercial plastics, in *Air. Inst. Phys. Conf. Ser.,* No. 118, Section 4, Electrostatics '91, Oxford, 1991.
18. Inculet, I. I., Castle, G. S. S., and Brown, J. D., *IEEE Trans.*, IAS, 1397–1389, 1994.
19. Maehata, H., Inoue, T., Tsukahara, M., Arai, H., Tamakoshi, D., Tojyo, C., Nagai, K., and Sekiguchi, Y., *Hitachi Zosen Technical Information*, 59, 222–226, 1998 (in Japanese).
20. Dascalescu, L., Iuga, A., and Morar, R., *Magn. Electr. Separ.*, 4, 241–255, 1993.
21. Matsumura, T., *Hyomengijyutsu,* 52, 244–249, 2001 (in Japanese).

1.14 Magnetic Separation

Toyohisa Fujita
University of Tokyo, Japan

Many kinds of forces are used to separate or filter materials of different quality and size. Mineral processing especially includes the separation of solids, recovery of valuable solids from waste materials, and water or air purification. Materials have electric and magnetic properties; therefore, electromagnetic fields are employed to separate solids with different electric and magnetic properties. Consequently, for efficient separation it is important to increase the electromagnetic forces and decrease the interaction forces between different solid particles. Here various magnetic separation methods and unit operations to separate particles by virtue of differences in their magnetic properties are described.

1.14.1 CLASSIFICATION OF MAGNETIC SEPARATORS[1]

The magnetic separation technique is classified as shown in Table 14.1. In a narrow sense, magnetic separation is a separation technique whose goal is to concentrate a magnetic material, to remove magnetic impurity, or to extract valuable magnetic materials. This is accomplished by discharging the particles captured by a magnet at a position depending on their magnetic properties. Magnetic filtration is the method to separate magnetic particles by capturing them with a filter. When suspended particles are magnetic, the direct method, in which particles are captured directly on the filter, is employed. When the separation target is nonmagnetic or an ion, the magnetic reagent method,[2] in which the separation target is sprinkled with a magnetic reagent, is employed. This magnetite seed method can also be used in the magnetic separation[3] of various diamagnetic particles using the surface charge difference in water. Separation by electromagnetic induction is known as eddy-current separation. Magnetohydrostatic separation is a method used to separate nonmagnetic particles immersed in a magnetic fluid by adjusting the magnetic buoyant force on particles with a magnetic field gradient.

Magnetic separators are classified in terms of the difference of the separation mechanism (Table 14.2). A look at Table 14.2 shows that they are classified into static magnetic field type and AC magnetic field type. The first type is further classified into two main types on the basis of intensity, that is, a low-intensity and a high-power type. Moreover, the second type is classified into three further types, that is, a traveling magnetic field type, a vibrating magnetic field type, and an electromagnetic induction type. Drum-type, belt-type, and high-gradient-type separators have been extensively employed. The low-intensity type uses less than 0.2 T, the medium-intensity type uses from 0.3 to 1.0 T, while the high-intensity one employs more than 1 T.

The magnetic capture condition is related to the magnetic field strength multiply by the gradient of magnetic field strength, and it is given by the following formula:

$$H\nabla H \text{ or } B\nabla B \tag{14.1}$$

The magnetic field strength, field gradient, and conditions are listed in Table 14.3. A look at Table 14.3 shows that the capture condition of a medium-intensity-type separator is about 10 times larger than that of a low-intensity type, while the magnetic capture condition of the high-intensity type is very large (i.e., is more than 10 to 10^4 times higher than that of a medium-intensity-type separator).

TABLE 14.1   Magnetic Separation[1]

Magnetic separation (in a broad sense)

1. Magnetic separation (in a narrow sense)

       Concentration, purification, extraction

2. Magnetic filtration

    2.1 Direct method

    2.2 Magnetic reagent method

    Magnetic seed method,[2] ferrite coprecipitation method, iron flocculants method, adsorption method, organic reagent containing iron addition method

3. Separation by electromagnetic induction

    Eddy-current separation

4. Magnetohydrostatic separation

    Separation using magnetic fluid

TABLE 14.2   Classification of Magnetic Separation

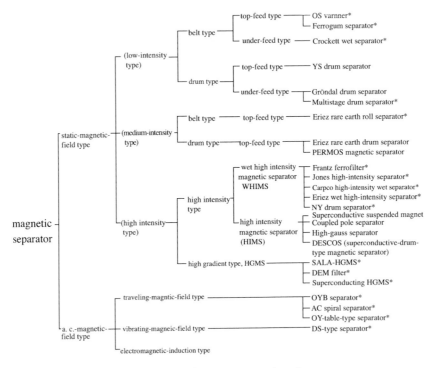

*Matrixes like wire or balls are loaded between magnetic poles.

TABLE 14.3   Magnetic Capture Conditions of Various Magnetic Separators

Classification		Magnetic separator	H (kA/m)	$\partial H/\partial x$ (MA/m²)	Magnetic condition $H(\partial H/\partial x)$ (10^{12}A²/m²)
Conventional low intensity		Drum separator	40	4	0.16
		Davis tube tester	320	32	10
Medium intensity		Rare earth drum or roll separator	360~720	10~100	3.6~72
High intensity	High intensity	Frantz ferrofilter	800	800	640
		NY type drum separator	1600	440	700
		Jones separator	1600	1600	2560
		DESCOS	2550	10~50	25~130
	High gradient	Solenoid separator (Kolm-Marston)	1600	16000~160000	25600~256000
		Superconducting HGMS	4000	16000~200000	64000~800000

1.14.2 STATIC MAGNETIC FIELD SEPARATORS

Open Gradient Magnetic-Field-Type Magnetic Separation

The main application of low-intensity magnetic separators is either removal of strongly magnetic impurities of tramp iron or concentration of a strongly magnetic valuable component. The Groendal drum wet separator has been widely used for the concentration of different materials. A permanent magnet is used instead of electromagnets, and raw material is fed from the lower part, as shown schematically in Figure 14.1.[4] The drum with fixed magnets inside rotates in contact with the floating or suspended particles in compartment a_1. Then magnetic particles are removed as they are attracted and captured on the drum surface by the magnetic force. Nonmagnetic particles are carried over to the next compartment, a_2. It was reported that the highest capturing efficiency of particles on the drum surface is obtained when the peripheral velocity of the drum is equal to the feeding velocity.

Figure 14.2 illustrates a drum separator of the dry type, in which raw material is fed on the top of the rotating drum with a fixed permanent magnet or electromagnet inside. Particles are classified by the balance in magnitude between magnetic and centrifugal forces. When rare-earth permanent magnet blocks are used in the drum, they are fixed inside the drum in such a way that different poles are placed alternately. In this case the magnetic flux density on the surface of the drum is 0.7 T at the maximum. This separator is used for purification of zircon, quartz, feldspar, glass cullet, coal, and so forth.[5] It is effective in separating relatively large particles (from 75 μm to 25 mm).

The drum-type magnetic separator with a superconducting magnetic system[6] is shown in Figure 14.3. The magnetic flux density on the drum surface is approximately 3 T. The drum, which as an exterior diameter of 1.2 m and an external length of 1.5 m, is made of carbon-reinforced plastics. The drum rotates at a speed of 2 to 30 rpm. The evaporating helium is led back to the liquefying plant, which is connected to the drum. For example, the magnetizable gangue serpentine could be completely separated from magnesite in the 1 to 100 mm range size. Also the results of separation of hematite and goethite from refractory crude bauxite (1–10 mm) and the separation of clay from potassium raw salts (1–4 mm) were reported, indicating a very effective process. On the other hand, in China, a suspended superconductive magnet is employed to remove iron impurities from coals. The magnetic flux density is 1.7 T on the surface of the bottom magnet and 1 T at 300 mm from bottom.[7]

The coupled-pole dry separator, high-gauss separator, Dings-induced roll separator, and Wetherilll–Rowand separator belong to the High-Intensity Magnetic Separation (HIMS) type.

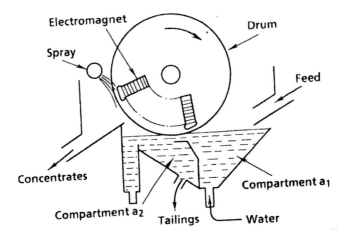

FIGURE 14.1 Groendal drum wet magnetic separator. [From Gaudin, A. M., *Principle of Mineral Dressing*, McGraw-Hill, New York, 1939, p. 450. With permission.]

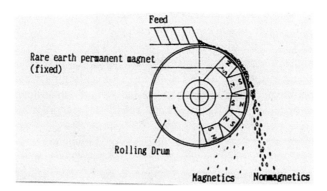

FIGURE 14.2 Drum magnetic separator of the dry type.

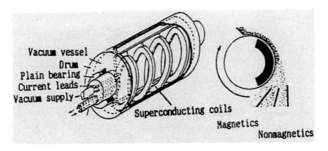

FIGURE 14.3 High-intensity drum-type magnetic separator with superconductive magnetic system (DESCOS).

A coupled-pole dry magnetic separator is shown in Figure 14.4. It is designed to generate a magnetic flux at the gap between two rotors, creating a high-intensity magnetic field of more than 2 T. This type of separator is used for the concentrated recovery and the purification of material.

Matrix-Loaded-Type Magnetic Separation

When matrixes such as ferromagnetic wire or balls are set between magnetic poles, a high-gradient magnetic field can be produced. The Frantz Ferrofilter separator (Figure 14.5), Jones high-intensity wet magnetic separator (Figure 14.6), Carpco high-intensity wet magnetic separator (Figure 14.7), and New York (NY) drum magne separator (Figure 14.8) belong to the Wet High Intensity Magnetic Separator (WHIMS) type. The Frantz Ferrofilter separator, which is the oldest model, is equipped with metal screens or grids as the matrix. Magnetic particles are separated as they are captured by

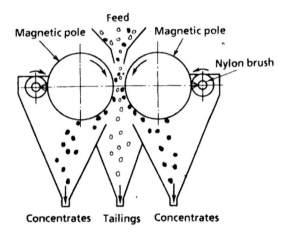

FIGURE 14.4 A coupled-pole dry magnetic separator (Nippon Magnetic Dressing Co., Ltd.).

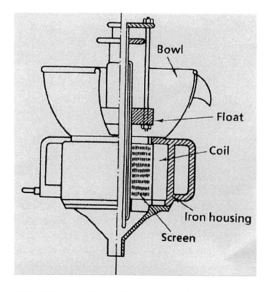

FIGURE 14.5 Frantz Ferrofilter (U.S. Patent 20,704,085).

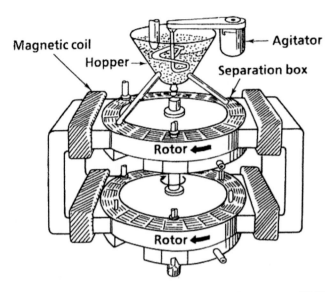

FIGURE 14.6 Jones high-intensity wet magnetic separator (U.K. Patents 768,451 and 767,124).

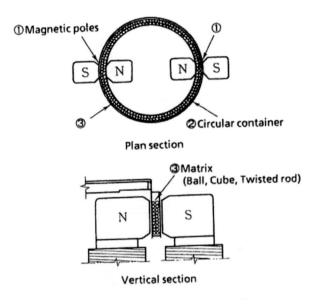

FIGURE 14.7 Carpco high-intensity wet magnetic separator.

the matrix and placed in the magnetic field during the processing of pulp through the matrix. This type of separator has been employed to remove iron particles from clay. The Jones-type separator, which has a matrix of corrugated iron sheets, has been used to dress hematite ore. Balls, cubes, and twisted rods are employed as the matrix in the Carpco-type separator, while expanded metal or steel wool is used in the Eriz-type separator. The NY drum separator is a Yashima (YS) magnetic separator with many steel balls on the drum surface. Magnetic particles are captured by the high-intensity magnetic field created around the contact points of the balls.

The high-gradient magnetic separator has been known since 1968 when the device was first installed at a Georgia kaolin plant to remove weakly magnetic particles less than 2 μm in size from

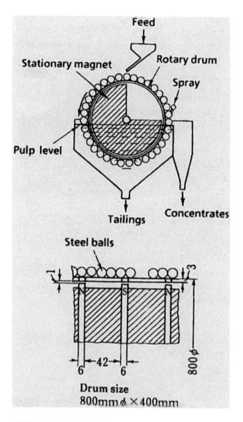

FIGURE 14.8 NY drum separator.

clay. When the matrices are arranged between the solenoid coil and the slurry, flow is parallel to the magnetic field direction, and the captured amount is much larger when compared to the device with the matrices between the magnetic poles.[8] The continuous high-gradient magnetic separator (SALA-HGMS)[9] is shown in Figure 14.9. Expanded metal or steel wool compose ferromagnetic matrices that are filled in the area under the influence of the magnetic field generated by a solenoid coil with a yoke. Moreover, it is desirable that matrices be corrosion and abrasion resistant. Matrices, filled in the magnetic field, increase sharply the magnetic force, that is, $H\nabla H$ in Equation 14.1. The separation box with matrices rotates continuously, with the magnetic flux density in the separator less than 2 T (Figure 14.9). This high-gradient magnetic separator has been applied to the separation of iron ores, purification of industrial minerals, separation of rare-earth minerals, water treatment, and so forth. Also, the continuous high-gradient magnetic separation, which uses high-quality permanent magnets (Fe-Nd-B), has been produced to capture magnetic particles while the magnetic flux density between magnetic poles is about 0.4 T. On the other hand, a high-gradient magnetic separator using superconducting magnets has been developed. Compared to the ordinal HGMS, the magnetic force condition is very much larger. Therefore, finer and lower magnetic susceptibility paramagnetic particles can be captured. In addition, the high magnetic flux density enables an increase of the separation capacity as the flow rate also increases. One example is the superconducting high-gradient magnetic separator that is shown in Figure 14.10. 5 T of magnetic flux density is generated in the space of 500 mm height and 100 mm diameter, which is filled with steel-wool-type matrices. The superconducting coil is placed in liquid helium and is connected to a liquefying plant of the vaporized helium gas. The slurry is fed from the bottom of the canister and discharged from the top. This separator has been employed to remove ferric impurities of micron

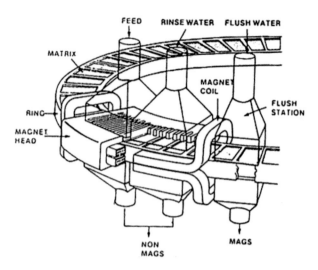

FIGURE 14.9 Continuous high-gradient magnetic separator (SALA-HGMS).

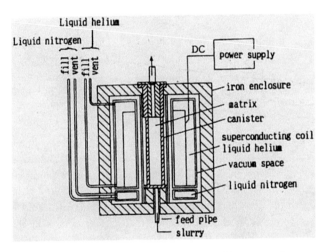

FIGURE 14.10   Superconducting high-gradient magnetic separator (Eriez).

size from kaolin clay. The 2 T separator using a superconducting magnet can operate on 80 to 90% of the power required for conventional HGMSs.

1.14.3 MAGNETOHYDROSTATIC SEPARATION

A colloidal solution of surfactant-coated ferromagnetic particles of 10 nm size, which remains uniform and stable even under a magnetic or centrifugal field, is called a magnetic fluid or ferrofluid. When nonmagnetic materials are immersed in the magnetic fluid under a magnetic field gradient, a magnetic force acts on the nonmagnetic materials toward the direction of low magnetic intensity.

If the magnetic field gradient is directed in the gravity direction, the net force on a nonmagnetic material F is given by

$$F = V[(\rho_p - \rho)g - I\nabla H] \qquad (14.2)$$

where V is the volume of nonmagnetic material immersed in the fluid, ρ_s and ρ represent the densities of the nonmagnetic material and magnetic fluid, respectively, I is the average magnetization of magnetic fluid, ∇H is the magnetic field gradient positioned nonmagnetic material, and g is the gravitational acceleration. The nonmagnetic material sinks if $F_G < 0$ and floats if $F_G > 0$ when F_G is the direction of g. This separator enables even a nonmagnetic material of density 20 g/cm³ to float; thus materials can be separated by adjusting the magnitude of F_G. For example, $I = 0.025$ T and $\nabla H = -8 \times 10^6$ A/m² give $F_G/V = 2 \times 10^5$ N/m³, which is enough to float a nonmagnetic particle of specific gravity of 20. For the magnetohydrostatic separator, a full equipped pilot plant for sink–float separation for recovering nonmagnetic metal scraps has been developed. The separator that recovers aluminum as a float product is shown in Figure 14.11. The use of rare-earth permanent magnets serves the purpose of reducing capital investment and energy consumption, while the use of water-based magnetic fluid is nontoxic, easy to wash out from the products, and easily reused.[10] Shredded automobile nonferrous metal scraps can be separated effectively; for example, 99.9 % recovery of aluminum at a grade of 98% at 500 kg/h capacity can be achieved. The distance between magnetic poles can be regulated to float zinc from a mixture of zinc, copper, and lead. This separator is convenient for recovering particles of 0.5 to 100 mm size. For industrial operation, De Beers Consolidated Mines, Ltd. uses a sink–float separator, which employs a kerosene-based magnetic fluid to separate diamond minerals.[11]

On the other hand, the Mag-sep Corporation (MC) process[12] with a magnet-hydrodyamic separator has been developed to separate finer particles ranging from about 5 µm to 1 mm in size by using a more diluted magnetic fluid of low cost. The mineral particles are fed through the duct as a slurry in a magnetic fluid. As the slurry and duct are rotated, low- and high-density particles are discharged from the inner and outer cylinders respectively. Large oppositely directed force can be applied to particles as light as 0.3 g/cm³, leading to rapid and precise separation.

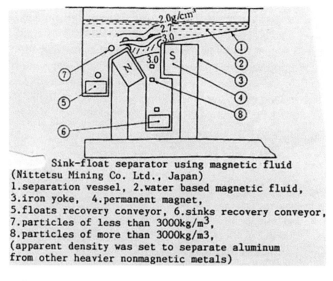

```
      Sink-float separator using magnetic fluid
(Nittetsu Mining Co. Ltd., Japan)
1.separation vessel, 2.water based magnetic fluid,
3.iron yoke,  4.permanent magnet,
5.floats recovery conveyor, 6.sinks recovery conveyor,
7.particles of less than 3000kg/m3,
8.particles of more than 3000kg/m3,
(apparent density was set to separate aluminum
from other heavier nonmagnetic metals)
```

FIGURE 14.11 Sink–float separator using magnetic fluid (Nittetsu Mining Co., Ltd., Japan).

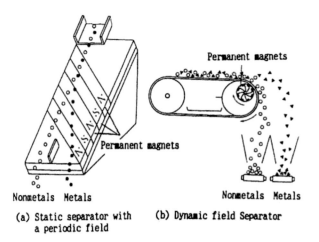

Nonmetals   Metals   Nonmetals   Metals

(a) Static separator with   (b) Dynamic field Separator
    a periodic field

FIGURE 14.12   Eddy-current separator.

1.14.4 ELECTROMAGNETIC-INDUCTION-TYPE SEPARATION

Eddy-current separation is used for the recovery of nonferrous metals from material mixtures whose magnetic fractions are removed by a magnetic separator. A time-dependent change of a magnetic field exerts repulsive forces on electrically conductive particles, which can be used to separate metallic particles from nonmetallic particles. Many eddy-current separators were developed for the recovery of nonferrous metals from shredded car scrap, granulated power cables, municipal solid waste, and so on. The alternating magnetic fields can be generated by the utilization of a linear motor or permanent magnet configurations. Two types of eddy-current separators with permanent magnets are shown in Figure 14.12. By using permanent magnets, the generation costs of eddy-current separation can be reduced considerably and the construction of equipment can be less complex. Figure 14.12a shows a static separator where particles are moving through a field distribution generated by the configuration of permanent magnets mounted on stable ramps or walls. Figure 14.12b shows a dynamic separator where the magnetic field is generated by machinery with moving magnets and the repulsive forces can be enhanced by the motion of the magnetic fields. Two belt drums support the conveyer belt. One drum is driven, and inside the other the magnet system rotates at an appreciably greater speed, for example, 2500 rpm. The repulsive force acting on the particle can become larger as the value of conductivity divided by density increases. Also, the separation condition depends on particle dimensions and shape and the field distribution in the separation area.[13] In particular, recent use of high-quality rare-earth permanent magnets in eddy-current separation has improved the separation efficiency. For example, 3 to 100 mm sizes of aluminum from a mixture can be separated effectively at 5 m^3/hr.[14]

REFERENCES

1. Yashima, S. and Fujita, T., *J. Soc. Powder Technol. Jpn.*, 28, 318–328, 1991.
2. Fujita, T. and Mamiya, M., *J. Mining Metall. Inst. Jpn.*, 103, 35–40, 1987.
3. Fujita, T, Wei, Y., and Mamiya, M., *J. Mining Metall. Inst. Jpn.*, 103, 513–518, 1987.
4. Gaudin, A. M., in *Principle of Mineral Dressing*, McGraw-Hill, New York, 1939, p. 450.
5. Marinescu, M., *IEEE Trans. Magn.*, MAG-25, 2732–2738, 1989.
6. Wasmuth, H. D., *Aufbereitungs-Technik*, 30, 753–760, 1989.
7. Homma, T., *Proc. Fall Annual Meeting MMIJ Ube*, C2-18, 2003.
8. Kim, Y. S., Fujita, T., Hashimoto, S., and Shimoiizaka, J., *Proc. 15th Int. Mineral Processing Congress (IMPC)*, Cannes, 1, 381–390, 1985.

9. Oberteuffer, J. A. and Wechsler, I., *Proc. Fine Part. Process.*, Las Vegas, 1178–1216, 1980.
10. Fujita, T., in *Magnetic Fluids and Applications Handbook,* Berkovski, B., Ed., Begell House, New York, 1996, pp. 755–789.
11. Sbovoda, J., *Proc. MINPREX,* Melborne, 297–301, 2000.
12. Walker, M. S., *Proc. 15th Int. Miner. Process. Congr. (IMPC),* Cannes, 1, 307–316, 1985.
13. Fujita, T., Sotojima, Y., and Kuzuno, K., *J. Mining Mater. Process. Inst. Jpn.,* 111, 177–180, 1995.
14. Warlitz, G., *Aluminium.* 65, 1125–1131, 1981.

1.15 Gravity Thickening

Eiji Iritani

Nagoya University, Chikusa-ku, Nagoya, Japan

Sedimentation is a separation of solid particles from a suspension due to the effect of a body force, which may be either gravity or centrifugal, on the buoyant mass of the particle. The sedimentation operation is often referred to as either thickening or clarification. If the main purpose of the operation is to concentrate solids into a denser slurry, the operation is generally called thickening, whereas, if the major concern is to produce a relatively clear liquid phase, the operation is usually called clarification. Because both thickening and clarification occur in any sedimentation basin, both functions have to be considered in the thickener design.

1.15.1 PRETREATMENT

The sedimentation rate of fine particles can be artificially increased by the addition of coagulants or flocculants, which causes precipitation of colloidal particles and the formation of flocs. For these to be effective, it is often necessary to adjust the pH of the slurry. Coagulants are usually inorganics that neutralize the surface charges on the particles, thus allowing them to collide and adhere. Common examples include aluminum sulfate, ferric chloride, and ferric sulfate. Flocculants are organic polyelectrolytes or long-chain polymers, which cause a physical linkage by bridging and enmeshment, and sometimes by particle–charge interaction. Examples include polyacrylamide and polyacrylate. Performance of coagulants or flocculants proposed for a given case can be determined beforehand in jar tests. A very useful test for filtration applications is the capillary suction time test, which measures the permeability of a flocculated suspension by using the uniform capillary suction of a filter paper.[1] The chemicals are combined with the slurry by rapid mixing, then allowed to settle at a lower shear rate consistent with the plant unit.

1.15.2 IDEAL SETTLING BASIN

A slurry of discrete particles settling in an ideal basin will be considered. The paths followed by the particles are straight lines determined by the vector sums of the two velocity components, as illustrated in Figure 15.1. The horizontal velocity component of the particle, due to the fluid flow across the basin, is equal to v. A particle with settling velocity u_0, starting at the top of the inlet, will reach the bottom at the outlet. Consequently, particles with a settling velocity less than u_0 will be only partially removed. A particle with settling velocity u_s will reach the bottom at the outlet if it enters the settling basin at height h from the bottom. Only particles initially at heights less than h will be removed. Therefore, the fraction η_p of particles with settling velocity u_s that are removed from an ideal basin is given by

$$\eta_p = \frac{h}{H} = \frac{u_s}{u_0} = \frac{u_s}{Q/A} \qquad (15.1)$$

where H is the liquid depth, Q is the volumetric flow rate of slurry, and A is the surface area of the basin.

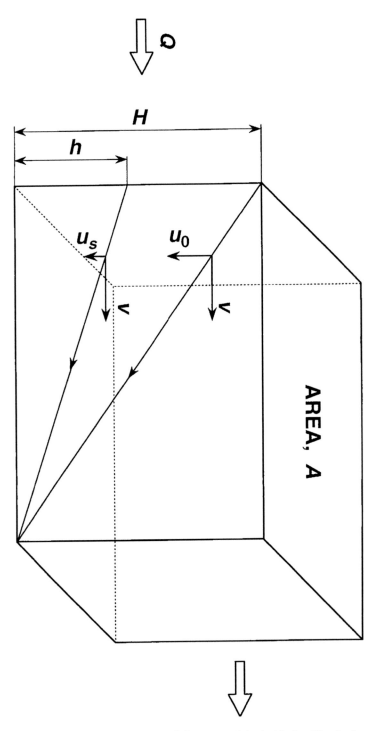

FIGURE 15.1 Settling behavior of discrete particles in ideal settling basin.

The analysis of sedimentation in an ideal basin clearly demonstrates that the slurry depth should be as shallow as practical. Gravity settlers that use inclined surfaces in the form of flat plates (lamellae) or tubes have the effect of increasing the settling rate of the particulates.[2] When flat plates or tubes are inclined at a steep angle, the settled sludge slides downward. This induces an upward flow of the clarified effluent, and the solids can be effectively collected at the bottom.

1.15.3 SETTLING CURVE

A typical batch settling curve is illustrated in Figure 15.2. The height of the interface between the settling particles and the clear supernatant liquid is plotted as a function of time. In zone settling, the slurry settles as a mass with a clear interface between the slurry and supernatant because of the relatively high solids concentration. In this stage, the slurry moves downward at a uniform velocity, which is dependent on the initial solids concentration. As the settling solids approach the bottom of the basin, they develop a layer with a higher concentration of suspended solids. After the subsiding interface reaches the rising layer of higher concentration, the sedimentation rate of the interface decreases gradually. During this transition zone, the subsidence rate of the interface is closely related to the instantaneous concentration at the interface. As the concentration continues to increase at the bottom, a zone of compacting sediment forms where solids are supported partially by the solids beneath them. The interface continues to subside in compression until an equilibrium condition is attained.

The sedimentation rate in the consolidation region is given approximately by

$$-\frac{dH}{d\theta} = k(H - H_\infty) \tag{15.2}$$

where H is the height of the sediment at time θ, H_∞ is the final height of the sediment, and k is the empirical constant. Integrating Equation 15.2, the time taken for the slurry line to fall from a height

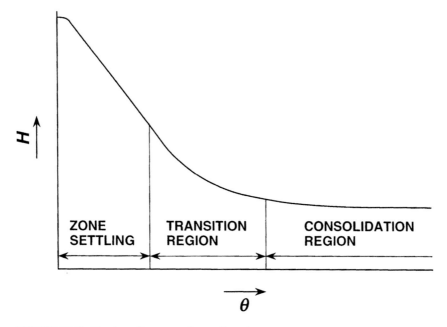

FIGURE 15.2 Batch settling curve for settling slurry.

H_c, corresponding to the critical settling point where the compression zone begins, to a height H can be obtained by

$$\ln(H - H_\infty) = -k\theta + \ln(H_c - H_\infty) \qquad (15.3)$$

Consequently, if $\ln(H–H_\infty)$ is plotted against θ, a straight line with slope $-k$ is obtained.

1.15.4   KYNCH THEORY

Kynch[3] has analyzed the behavior of concentrated suspensions during batch sedimentation using a continuity approach. The theory is mainly based on the postulate that the settling velocity of solids is a function only of the local solids concentration. When the settling velocity is u at some horizontal level where the volumetric concentration of particles is C, the volumetric sedimentation rate G per unit area (i.e., the solids flux) is given by

$$G = Cu \qquad (15.4)$$

Consider a material balance of an elemental layer between a height H above the bottom and a height $H + dH$. Since in a time interval $d\theta$ the accumulation of particles in the layer is given by the difference in flux, between $G(H + dH)$ into the upper layer and $G(H)$ out through the lower layer, one obtains

$$\frac{\partial C}{\partial \theta} = \frac{\partial G}{\partial H} = \frac{dG}{dC} \cdot \frac{dC}{dH} \qquad (15.5)$$

Because the concentration of particles is generally a function of both position and time, one obtains

$$dC = \frac{\partial C}{\partial H} dH + \frac{\partial C}{\partial \theta} d\theta \qquad (15.6)$$

Constant concentration conditions ($dC = 0$) are therefore defined by

$$\frac{\partial C}{\partial H} = -\frac{\partial C}{\partial \theta} \left(\frac{dH}{d\theta} \right)^{-1} \qquad (15.7)$$

Substituting Equation 15.7 into Equation 15.5, one obtains

$$-\frac{dG}{dC} = \frac{dH}{d\theta} = V \qquad (15.8)$$

Because Equation 15.8 refers to a constant concentration, dG/dC must also be constant. Consequently, V ($= dH/d\theta$) is constant too for any given concentration and is the propagation velocity of a zone of constant concentration C. While during sedimentation the interface between the clear liquid and the settling solids is moving downward, layers of constant concentration appear to move upward at the base of the vessel.

   This result is widely employed to obtain the relation between solids flux and concentration needed for thickener design from a single batch sedimentation test. It is possible to obtain the plot of G versus C over the concentration range C_0 to C_{max} (concentration at the compression point) from only a single experiment of a slurry with an initial solids concentration C_0. The sedimentation curve has a decreasing negative slope reflecting the increasing concentration of solids at the interface after point B, as illustrated in Figure 15.3. Line OP represents the locus of points of some concentration C ($C_0 < C < C_{max}$). It corresponds to the propagation of a wave at the upward velocity V from the bottom of the sediment. When the wave propagates the interface of point P, all the particles in the slurry must have passed through the plane of the wave. Therefore, one obtains

$$C(u+V)\theta_1 = C_0 H_0 \tag{15.9}$$

By drawing a tangent to the settling curve ABPD at P, the intercept T of the tangent on the H axis is located. Then, $u\theta_1$ is equal to QT because $-u$ is the slope of a tangent to the settling curve at P, and $V\theta_1$ is equal to OQ because V is the slope of line OP. Consequently, the concentration C corresponding to the line OP is given by

$$C = C_0 \frac{OA}{OT} \tag{15.10}$$

and the corresponding solids flux is obtained from

$$G = Cu = C_0 \frac{OA}{OT} u \tag{15.11}$$

By drawing the tangent at a series of points on the curve BPD and measuring the corresponding slope $-u$ and intercept OT, it is possible to calculate the solid flux G for any concentration C ($C_0 < C < C_{max}$).

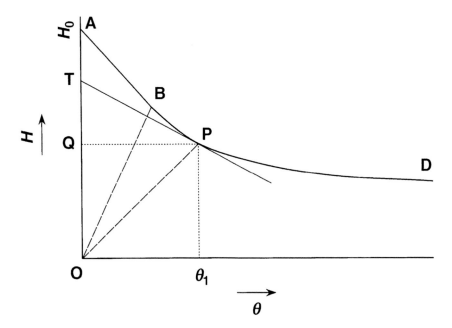

FIGURE 15.3 Construction of Kynch theory.

1.15.5 DESIGN OF A CONTINUOUS THICKENER

A thickener is a sedimentation basin that is employed to concentrate a slurry prior to filtration or centrifugation. Normally a larger fraction of the total liquid is removed in thickening than in subsequent operations. In most cases, the particles in a thickener settle collectively in the zone settling regime. As shown in Figure 15.4, the bottom part of a thickener is filled with a layer of settled solids, which increases in concentration with greater depths. A clarified liquid separates from the thickened liquid at the interface and is taken off at the top. A thickener therefore fulfills the dual function of providing a concentrated underflow and a clear liquid overflow. In designing a thickener, the areas needed for clarification and thickening are therefore examined separately. The larger of the two areas determines the size required to achieve the specified performance.

Clarification Area

The cross-sectional area needed for clarification can be calculated from the initial sedimentation velocity obtained from a batch sedimentation test. The area must be large enough so that the rising velocity of overflow liquid is less than the batch sedimentation velocity. The minimum area required in a thickener for clarification is, therefore, determined by

$$A_c = \frac{Q_e}{u_s} \tag{15.12}$$

where A_c is the surface area needed for clarification, Q_e is the overflow rate of the clear liquid, and u_s is the initial sedimentation rate of slurry at the feed concentration.

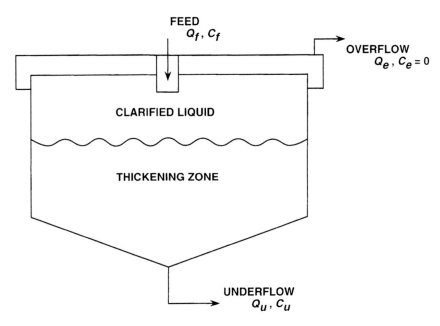

FIGURE 15.4 Schematic view of continuous thickener.

Coe and Clevenger Method

In the thickening region, solids move toward the underflow both by gravity sedimentation and by bulk movement resulting from underflow withdrawal. The total flux G of solids at concentration C_i can, therefore, be given by

$$G = C_i u_i + C_i u_b \qquad (15.13)$$

where u_i is the sedimentation velocity at concentration C_i, and u_b is the downward velocity of slurry arising from the removal of the underflow. The form of solids flux terms is shown in Figure 15.5. The flux of solids by bulk transport is a linear function of solids concentration with slope u_b. The gravity flux of solids goes through a maximum as the concentration increases, as obtained from batch sedimentation tests for various concentrations. For most zone settling slurries, the combination of gravity and bulk flow flux terms produces a total flux curve with a maximum and minimum. The minimum in the total flux curve exists at some limiting concentration between the feed and underflow concentrations and represents the limiting solids-handling capacity of the slurry. The total solids contained in the feed must be less than the solids-transmitting capability of this limiting concentration layer. Because the gravity flux becomes zero at the bottom of the thickener, all solids are taken off by bulk flow only. Consequently, if a horizontal line is drawn tangent to the minimum in the total flux curve, it intersects the bulk flux line at the underflow concentration C_u. Assuming that all solids in the feed leave in the underflow, the cross-sectional area required for thickening can be obtained from[4]

$$A = \frac{C_f Q_f}{G_l} = \frac{C_u Q_u}{G_l} \qquad (15.14)$$

where C_f is the solids concentration of the influent, Q_f is the flow rate of the influent, G_l is the limiting solids flux at concentration C_l, and Q_u is the flow rate of the underflow.

An alternative approach employs the batch flux curve directly.[5] The settling velocity u_i by gravity is the slope of line from the origin to any point on the batch flux curve, as illustrated in Figure 15.6

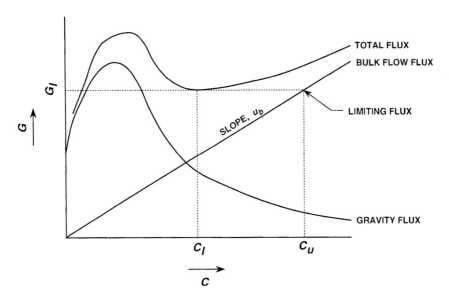

FIGURE 15.5 Total solids flux curve in continuous thickener.

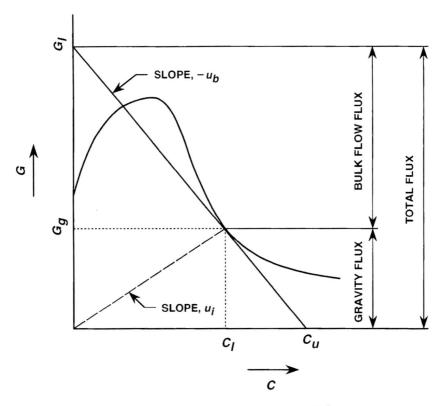

FIGURE 15.6 Solids flux in continuous thickener based on batch flux curve.

(often known as a Kynch flux plot). If this line intersects a tangent to the batch flux curve at the point of tangency, the intersection corresponds to the limiting solids concentration C_l, and gravity flux G_g. By using Equation 15.14, the bulk velocity is given by

$$u_b = \frac{Q_u}{A} = \frac{G_l}{C_u} \tag{15.15}$$

Thus, the bulk downward velocity is the slope of a tangent operating line connecting solids flux G_l on the flux axis to the corresponding underflow solids concentration C_u on the concentration axis. The distance from G_g to G_l is the solids flux by bulk downward transport when solids are removed at concentration C_u. Consequently, the limiting solids-handling capacity G_l in Equation 15.15 can be determined for a given underflow concentration C_u.

Talmage and Fitch Method

The procedure of Talmage and Fitch[6] requires data from only a single batch settling curve, on the basis of the method of Kynch. On the basis of the material balance, the values of H_2 and H_u shown in Figure 15.7 are given by

$$C_0 H_0 = C_2 H_2 = C_u H_u \tag{15.16}$$

where C_2 is the concentration at the compression point θ_2. An "underflow" line is drawn parallel to the time axis at $H = H_u$ on a plot of height H versus sedimentation time θ. By drawing a tangent to

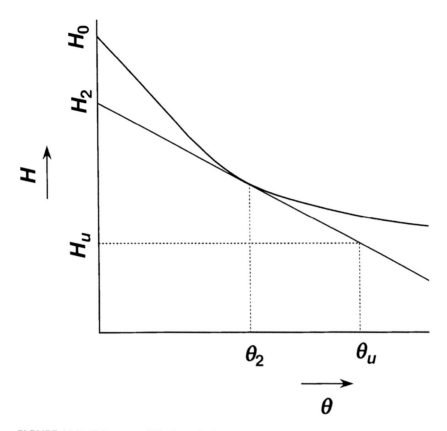

FIGURE 15.7 Talmage and Fitch method.

the settling curve through point H_2 on the H axis, the value of θ_u can be determined from the intersection of the tangent and the underflow line. Consequently, the required thickener area A can be found from

$$A = C_f Q_f \left(\frac{\theta_u}{C_0 H_0} \right) \tag{15.17}$$

REFERENCES

1. Gale, R. S., *Filtr. Separ.,* 8, 531–538, 1971.
2. Nakamura, H. and Kuroda, K., *Keijo J. Med.,* 8, 256–296, 1937.
3. Kynch, G. J., *Trans. Faraday Soc.,* 48, 166–176, 1952.
4. Coe, H. S. and Clevenger, G. H., *Trans. Am. Inst. Mining Eng.,* 55, 356–384, 1916.
5. Yoshioka, N., Hotta, Y., Tanaka, S., Naito, S., and Tsugami, S., *Kagaku Kogaku,* 21, 66–74, 1957.
6. Talmage, W. P. and Fitch, E. B., *Ind. Eng. Chem.,* 47, 38–41, 1955.

1.16 Filtration

Eiji Iritani

Nagoya University, Nagoya, Chikusa-ku, Japan

Filtration is the operation of separating a dispersed phase of solid particles from a fluid by means of a porous filter medium, which permits the passage of the fluid but retains the particles. Filtration is probably one of the oldest unit operations. The old forms of filtration by straining through porous materials were described by the earliest Chinese writers. A gravity filter used in a chemical process industry was described in an Egyptian papyrus which has its origin in about the third century A.D. In recent years, many developments have increased the application of filtration. Filtration steps are required in many important processes and in widely divergent industries. The importance of filtration technique has been emphasized by the increased need for protection of the environment. Recently, membrane filtration of colloids has become increasingly important in widely diversified fields.

The operations are divided into two broad categories: "cake" and "depth" filtration. From the viewpoint of a driving force, cake filtration is further divided into pressure, vacuum, gravity, and centrifugal operations. In cake filtration, particles in a slurry form a deposit as a filter cake on the surface of the supporting porous medium while the fluid passes through it. After an initial period of deposition, the filter cake itself starts to act as the filter medium while further particles are deposited. In depth filtration (sometimes called filter medium filtration or clarifying filtration), particles are captured within the complex pore structures of the filter medium, and the cake is not formed on the surface of the medium. In many processes, a stage of depth filtration precedes the formation of a cake. The first particles can enter the medium, and with very dilute slurries, there can be a time lag before a cake begins to form. Smaller particles enter the medium, whereas larger particles bridge the openings and start the buildup of a surface layer. Depth filtration is generally used to remove small quantities of contaminants. Cake filtration is primarily employed for more concentrated slurries. In practice, cake filtration is employed in industry more often than depth filtration. The following discussion will be concerned with cake filtration.

For purposes of mathematical treatment, cake filtration processes are classified according to the variations of both pressure and flow rate with time. The pumping mechanism generally determines the flow characteristics and serves as a basis for division into the following categories: (a) constant-pressure filtration (the actuating mechanism is compressed gas maintained at a constant pressure, or a vacuum pump), (b) constant-rate filtration (positive displacement pumps of various types are employed), and (c) variable-pressure, variable-rate filtration (the use of a centrifugal pump results in the rate varying with the back pressure on the pump). The relation between filtration pressure p and time θ for the three types of filtration is illustrated in Figure 16.1. The constant-pressure curve is represented by a horizontal line. The pressure increases with time linearly for the constant-rate filtration of an incompressible cake. For a compressible cake formed in constant-rate filtration, the p versus θ curve is concave upward. Variable-pressure, variable-rate filtration is conducted by using a filter actuated by a centrifugal pump. Depending on the characteristics of the centrifugal pump, widely differing curves might be encountered.

The structure of the cake formed and, consequently, its resistance to liquid flow depend on the properties of the solid particles and the liquid-phase suspension, as well as on the conditions of filtration. The cake structure is first established by such hydrodynamic factors as cake porosity, mean particle size, size distribution, particle-specific surface area, and sphericity. It is also strongly influenced by some factors that can be denoted conditionally as physicochemical. The influence of

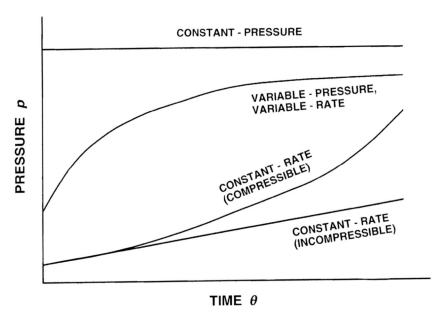

FIGURE 16.1 Relation between filtration pressure and time for different operations.

physicochemical factors is closely related to surface phenomena at the solid–liquid boundary. For fine particle suspensions, colloidal forces control the nature of the filter cake. The repulsive electrostatic forces will vary with the surface charge of the suspended particles, which varies with the solution environment. Therefore, the filtration behaviors of the colloids are affected significantly by the solution properties, including pH and electrolyte strength.[1]

1.16.1 BASIS OF CAKE FILTRATION THEORY

Equations of Flow through Porous Media

Basic laws governing the flow of liquids through uniform, incompressible beds serve as a basis in developing formulas for more complex, nonuniform, compressible filter cakes formed on the filter medium during cake filtration. Darcy's law can be expressed in the form

$$\frac{dp_L}{dx} = \frac{\mu}{K_p} u \tag{16.1}$$

where p_L is the local hydraulic pressure, x is the distance from the filter medium, μ is the viscosity of liquid, K_p is the permeability, and u is the apparent liquid velocity relative to the solids. This velocity is expressed as the volumetric flow rate per unit area, which is defined by

$$u = q - \frac{\varepsilon}{1-\varepsilon} r = q - er \tag{16.2}$$

where q is the apparent flow rate of liquid, ε is the local porosity, r is the apparent migration rate of solid particles, and e is the local void ratio. In filtration, it is customary to use the volume ω of dry solids per unit medium area instead of the distance x from the medium. Thus, the incremental volume $d\omega$ is given by

$$d\omega = (1-\varepsilon)dx \qquad (16.3)$$

Substituting Equation 16.3 into Equation 16.1, one obtains

$$\frac{dp_L}{d\omega} = \frac{\mu}{K_p(1-\varepsilon)}u = \mu\rho_s\alpha u \qquad (16.4)$$

where ρ_s is the true density of solids and α is the local specific filtration resistance.

Drag on Particles

When suspended solids are deposited during cake filtration, liquid flows through the interstices of the compressible cake in the direction of decreasing hydraulic pressure. The solids forming the cake are compact and relatively dry at the filter medium, whereas the surface layer is in a wet and soupy condition. Thus, the porosity is minimum at the point of contact between the cake and the filter medium where $\omega = 0$ (Figure 16.2) and maximum at the cake surface ($\omega = \omega_0$, where ω_0 is the solids volume of the entire cake per unit medium area) where the liquid enters. The drag imposed on each particle is communicated to the adjacent particles. Consequently, the net solid compressive pressure increases as the filter medium is approached, thereby accounting for the decreasing porosity.

On the assumption that inertial forces are negligible, the force balance over the increment $d\omega$ can be described by

$$\frac{\partial p_L}{\partial \omega} + \frac{\partial p_s}{\partial \omega} = 0 \qquad (16.5)$$

or upon integration,

$$p_L + p_s = p \qquad (16.6)$$

where p_s is the local solid compressive pressure. Equation 16.5 and Equation 16.6 clearly state that the solid compressive pressure increases as the hydraulic pressure decreases, as shown in Figure 16.2.

Porosity and Specific Filtration Resistance

It is generally assumed in compressible cake filtration theory that the local porosity and specific filtration resistance are unique functions of the solid compressive pressure. These relations can be determined accurately by use of the compression-permeability cell.[2] Such relations can be also determined directly from the filtration experiments in which a filter is subjected to a sudden reduction in its filtration area.[3] As a fair approximation of compression-permeability cell data for a number of substances, the power functional relations among the porosity, specific filtration resistance, and solid compressive pressure can be employed as follows:

$$\begin{aligned}
\varepsilon &= \varepsilon_i, & p_s &\leq p_i \\
\varepsilon &= \varepsilon_0 p_s^{-1}, & p_s &> p_i
\end{aligned} \qquad (16.7)$$

$$\begin{aligned}
\alpha &= \alpha_i, & p_s &\leq p_i \\
\alpha &= \beta_p + \alpha_{0p}p_s^n \approx \alpha_{0p}p_s^n, & p_s &> p_i
\end{aligned} \qquad (16.8)$$

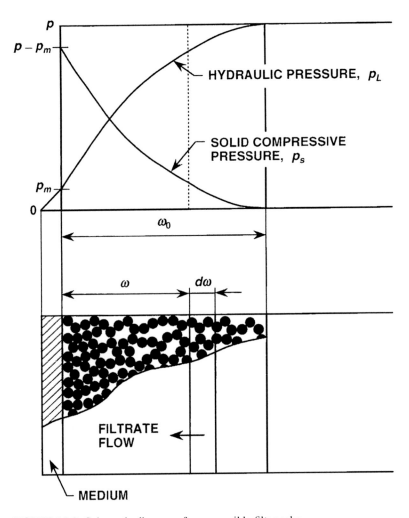

FIGURE 16.2 Schematic diagram of compressible filter cake.

where n is a compressibility coefficient, which is equal to zero for incompressible substances. The local specific filtration resistance α is related to the local porosity ε in the form

$$\alpha = \frac{kS_0^{\,2}\left(1-\varepsilon\right)}{\rho_s\varepsilon^3} \tag{16.9}$$

where S_0 is the effective specific surface area per unit volume of the particles and k is the Kozeny constant which normally takes the value of 5.

Average Specific Filtration Resistance

Combining Equation 16.4 and Equation 16.5, the basic flow equation can be rewritten as

$$\frac{dp_L}{d\omega} = -\frac{dp_s}{d\omega} = \mu\rho_s\alpha u \tag{16.10}$$

Integration of Equation 16.10 is carried out between the limits (see Figure 16.2) of $\omega = 0$ at the filter medium and $\omega = \omega_0$ at the cake surface. In addition, p_s is taken as $p - p_m$ at the filter medium and 0 at the cake surface. The pressure p_m at the exit of the filtrate from the cake is related to the resistance R_m of the filter medium by

$$p_m = \mu u_1 R_m \qquad (16.11)$$

where u_1 is the filtration rate. Integration of Equation 16.10 on the postulate that u is constant $(= u_1)$ throughout the cake leads to

$$\frac{dv}{d\theta} = u_1 = \frac{p}{\mu(\alpha_{av}w + R_m)} \qquad (16.12)$$

where v is the filtrate volume per unit medium area, ω is the filtration time, w ($= \rho_s\omega_0$) is the solids mass of the entire cake per unit medium area, and α_{av} is the average specific filtration resistance, which is defined by

$$\alpha_{av} = \frac{p - p_m}{\int_0^{p-p_m}(1/\alpha)dp_s} \qquad (16.13)$$

The total cake resistance changes as the mass of cake grows with time. Many analyses of filtration start with Equation 16.12. Recently, the cake filtration theory can be also used to analyze membrane separation such as ultrafiltration of protein solutions, and it has the potential for analyzing the membrane fouling during ultrafiltration.[4]

The above-mentioned derivation is not rigorous in that it has been assumed that u and the filtration area are constant throughout the cake. Shirato et al.[5] presented a more complex equation, which took the variation of u with distance into account. Equation 16.13 is only approximately correct; hence, the average specific filtration resistance must be modified for a general expression. This is apparent, especially for thick slurries, as

$$\alpha'_{av} = J_s\alpha_{av} \qquad (16.14)$$

where α'_{av} is the average specific filtration resistance that accounts for the internal flow variations in filter cake, and J_s is the correction factor. The factor J_s depends on both the filtration pressure and the slurry concentration. Although the pressure has relatively little effect, J_s can change remarkably for the concentrated slurry of compressible materials, as illustrated in Figure 16.3.[5] Whenever the filtering area varies, as in radial flow filtration, Equation 16.12 must be modified. In the external two-dimensional filtration onto the cylindrical element, the filtration rate under constant-pressure conditions can be given by

$$\frac{dv}{d\theta} = \frac{p\{(r_o/r_i)^2 - 1\}}{2\mu\alpha_{av}w\ln(r_o/r_i)} \qquad (16.15)$$

where r_i is the radius of the medium surface and r_o is the radius of the cake surface.

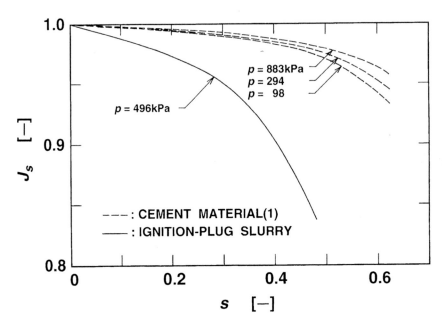

FIGURE 16.3 Effect of slurry concentration s on correction factor J_s.

Material Balance

From an overall viewpoint of filtration, a material balance can be written on a unit medium area basis in the form

$$\frac{w}{s} = mw + \rho v \tag{16.16}$$

where s is the average mass fraction of solids in the slurry, m is the ratio of wet to dry cake mass, and ρ is the density of the filtrate. This material balance assumes that all feed slurry is filtered to form a cake. Solving for w in Equation 16.16, one gets

$$w = \frac{\rho s}{1 - ms} v \tag{16.17}$$

Frequently, the quantity m is related to the average porosity ε_{av} of the cake by

$$m = 1 + \frac{\rho \varepsilon_{av}}{\rho_s (1 - \varepsilon_{av})} \tag{16.18}$$

It is important to relate the cake thickness L to the filtrate volume v per unit medium area. For the entire cake, Equation 16.3 in combination with Equation 16.17 yields

$$w = \frac{\rho s}{1 - ms} v = \rho_s (1 - \varepsilon_{av}) L \tag{16.19}$$

1.16.2 CONSTANT-PRESSURE AND CONSTANT-RATE FILTRATION

Constant-Pressure Filtration

An accurate solution of Equation 16.12 requires numerical techniques. However, on the postulate of constant-pressure filtration that α_{av} is constant during filtration and a function of p alone, a simple relation between v and θ can be obtained. Combining Equation 16.12 with Equation 16.17, one obtains

$$\frac{d\theta}{dv} = \frac{2}{K_v}(v + v_m)$$ (16.20)

where v_m is the fictitious filtrate volume per unit medium area required to form a filter cake of resistance equal to the medium resistance R_m, and K_v is Ruth coefficient of constant-pressure filtration[6] defined by

$$K_v = \frac{2p(1 - ms)}{\mu \alpha_{av} \rho s}$$ (16.21)

Integration of Equation 16.20 yields

$$(v + v_m)^2 = K_v(\theta + \theta_m)$$ (16.22)

where θ_m is the fictitious filtration time corresponding to the medium resistance. Equation 16.22 yields a parabolic relation between v and θ. Constant-pressure filtration has long been the favorite method for obtaining experimental data in the laboratory because of its simplicity. Interpretation of data of constant-pressure filtration test is generally based on Equation 16.20. It might appear easier to use Equation 16.22 in the form

$$\frac{\theta}{v} = \frac{1}{K_v}v + \frac{2}{K_v}v_m$$ (16.23)

Thus, plotting θ/v against v produces a straight line. Equation 16.20 is, in fact, better because it does not require identification of the precise time at which $\theta = 0$. Plotting the reciprocal filtration rate $(d\theta/dv)$ against v may lead to a straight line having the slope of $2/K_v$ in accordance with Equation 16.20. Knowing the value of K_v, it is possible to obtain the value of α_{av} from Equation 16.21 and then to construct a logarithmic graph of α_{av} versus p. The medium resistance can be calculated from the intercept. Empirically, α_{av} can be represented as functions of p by

$$\alpha_{av} = \alpha_0(p - p_m)^{n'} \approx \alpha_0 p^n$$ (16.24)

Constant-Rate Filtration

Substituting Equation 16.11 and Equation 16.17 into Equation 16.12, one obtains

$$u_1 = \frac{1 - ms}{\mu \alpha_{av} \rho s v}(p - p_m)$$ (16.25)

In constant-rate filtration, the filtrate volume is related to the time by the simple relation

$$v = u_1 \theta$$ (16.26)

Substituting Equation 16.26 into Equation 16.25 and solving it with respect to time θ, one obtains the relation between p and θ at constant-rate filtration as

$$\theta = \frac{1-ms}{\mu \alpha_{av} \rho s u_1^2}(p - p_m) \tag{16.27}$$

If α_{av} is represented by Equation 16.24, Equation 16.27 leads to

$$\theta = \frac{1-ms}{\mu \alpha_0 \rho s u_1^2}(p - p_m)^{1-n'} \tag{16.28}$$

In constant-rate filtration, the pressure-time relationship is measured. A logarithmic plot of the pressure drop across the filter cake against time yields a straight line. When the cake is incompressible ($n' = 0$), Equation 16.28 implies a linear relationship between filtration time and pressure drop, as illustrated in Figure 16.1.

1.16.3 INTERNAL STRUCTURE OF FILTER CAKE

In filter cakes, the variation of porosity with distance from the cake surface is important from both theoretical and industrial viewpoints. In the development of filtration theory, the porosity plays a fundamental role in its relation to flow rates, pressure, and other parameters involved in the differential equations of flow through compressible, porous material. Porosity variation determines the average porosity and liquid content of the filter cake in commercial operation.

Equation 16.10 is integrated over a portion of the cake and then the entire cake, assuming that u is constant ($= u_1$) throughout the cake. Also, it is assumed that the medium resistance is negligible. Integrating from ω to ω_0 and 0 to ω_0 yields respectively

$$\frac{1}{\rho_s} \int_0^{p_s}(1/\alpha)dp_s = \mu u_1(\omega_0 - \omega) \tag{16.29}$$

$$\frac{1}{\rho_\sigma} \int_0^{p}(1/\alpha)dp_s = \mu u_1 \omega_0 \tag{16.30}$$

Equation 16.30 provides a relationship among the filtration rate u_1, the solids volume ω_0, and the pressure drop p across the cake. Dividing Equation 16.29 by Equation 16.30, one obtains

$$\frac{\int_0^{p_s}(1/\alpha)dp_s}{\int_0^{p}(1/\alpha)dp_s} = 1 - \frac{\omega}{\omega_0} \tag{16.31}$$

To convert ω/ω_0 in Equation 16.31 into x/L, the following equation can be employed:

$$\frac{x}{L} = \frac{\int_0^{\omega/\omega_0} \frac{d(\omega/\omega_0)}{1-\varepsilon}}{\int_0^{1} \frac{d(\omega/\omega_0)}{1-\varepsilon}} \tag{16.32}$$

Consequently, Equation 16.31 with aid of Equation 16.32 gives the fractional distance through the cake as a function of the upper limit p_s of integration. It indicates that p_s versus x/L curves are independent of flow and total thickness. In turn, p_L (equal to $p - p_s$) and ε can be related to x/L through their functional relationships to p_s. Equation 16.10 with aid of Equation 16.3 can be used in an analogous manner to relate p_s to the distance through the cake directly by the form

$$\frac{\int_0^{p_s} \dfrac{dp_s}{\alpha(1-\varepsilon)}}{\int_0^{p} \dfrac{dp_s}{\alpha(1-\varepsilon)}} = 1 - \frac{x}{L} \tag{16.33}$$

Because u_1 varies markedly throughout the cake for highly concentrated slurries, Equation 16.31 and Equation 16.33 must be modified. The power relation can be used to approximate the porosity versus solid compressive pressure data as follows:

$$1 - \varepsilon = E' p_s^{\lambda'} \tag{16.34}$$

Substituting Equation 16.34 and Equation 16.8 into Equation 16.33, the integral in Equation 16.33 can be replaced by approximate formulas to yield[7]

$$\left(1 - \frac{p_L}{p}\right)^{1-n-\lambda'} = 1 - \frac{x}{L} \tag{16.35}$$

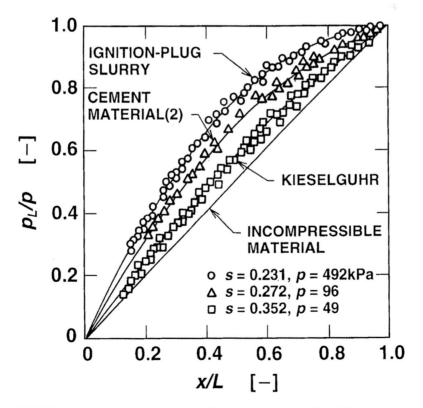

FIGURE 16.4 Relation between hydraulic pressure and normalized distance.

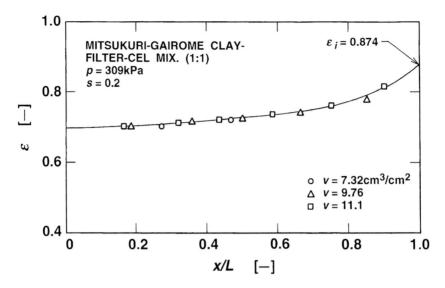

FIGURE 16.5 Relation between local porosity and normalized distance.

In Figure 16.4, the fractional hydraulic pressure drop is shown as a function of the fractional distance through the cake for a number of substances.[8] As the compressibility increases, the hydraulic pressure drop becomes large near the filter medium. In Figure 16.5, the porosity is plotted against the fractional distance through the cake.[9] Under constant-pressure filtration, the local porosity depends on the normalized distance x/L alone.

1.16.4   NON-NEWTONIAN FILTRATION

In filtration of non-Newtonian fluid–solid mixtures, an empirical relation known as a power law is widely used in representing the rheological behavior of the filtrate in the form

$$\tau = K\dot{\gamma}^{N} \tag{16.36}$$

where τ is the shear stress, K is the fluid consistency index, $\dot{\gamma}$ is the shear rate, and N is the flow behavior index, which is a measure of the degree of the non-Newtonian behavior of the fluid; the greater the departure from unity of N, the more remarkable is the non-Newtonian behavior of the fluid. On the basis of Equation 16.36, the equation for describing the filtration rate of power-law non-Newtonian fluids can be obtained in the form[10]

$$\left(\frac{1}{u_1}\right)^{N} = \left(\frac{d\theta}{dv}\right)^{N} = \frac{K\gamma_{av}\rho s (v + v_m)}{p(1 - ms)} \tag{16.37}$$

where γ_{av} is the average specific filtration resistance for power-law fluids.

In Figure 16.6, $d\theta/dv$ and $(d\theta/dv)^{N}$ are plotted against v for constant-pressure filtration. The curve of $d\theta/dv$ versus v is concave upward, whereas the plot of $(d\theta/dv)^{N}$ versus v shows a linear relationship in accordance with the theory indicated by Equation 16.37. From the analysis for the internal structure of filter cake, it is derived that the cake formed by non-Newtonian filtration of pseudoplastic fluids ($N < 1$) is denser in structure than that formed by usual Newtonian filtration.

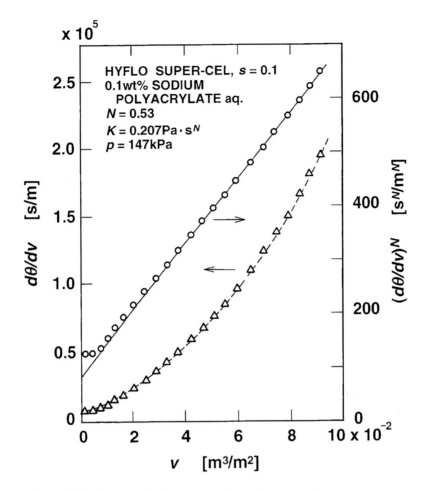

FIGURE 16.6 $d\theta/dv$ and $(d\theta/dv)^N$ versus v in non-Newtonian filtration.

1.16.5   FILTRATION EQUIPMENT

A wide variety of filters is commercially available in many industries. The operating principles and important features of typical filtration equipment are described below.[11,12]

Filter Press

The filter press has been the best known and perhaps the most widely used of all batch pressure filters. The common filter press is the plate-and-frame design, which is held together as a pack in a strongly constructed framework, as illustrated in Figure 16.7. A varying number of filter chambers are assembled, which consist of medium-covered plates alternating with frames that provide space for the cake. The chambers are closed and tightened by a screw or hydraulic ram, which forces the plates and frames together, making a gasket of the filter cloth. The slurry feed is introduced into the chambers under pressure and fills each chamber approximately simultaneously. The liquid passes through the filter medium, which in turn retains the solids. The filtrate is removed at a discharge outlet. The filter cake forms until the frames are full. Completion of the cake formation is judged by

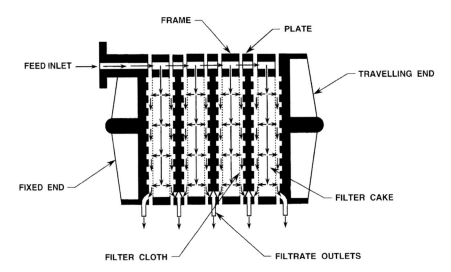

FIGURE 16.7 Schematic view of plate-and-frame filter press.

the filtering time, decrease in the rate of the feed, or rise in back pressure. Once the frames are full of cake, filtration is stopped and a wash liquid is applied if the solids are to be recovered. This can be followed by deliquoring of the filter cake.

Rotary Drum Filter

The rotary drum filter is the most common continuous filter. A standard drum filter consists of three main parts: a drum mounted horizontally with an automatic filter valve; a slurry reservoir with a stirring device; and a scraper for cake discharge, as illustrated in Figure 16.8. The cylindrical drum, having a porous wall and covered on the outside with the filter media, rotates about a horizontal axis with a portion immersed in the slurry. The filter operates continuously through stages of cake formation due to filtration, washing, drying with air or steam, and discharge with an air blower. The normal driving force is vacuum, although enclosed pressurized units are sometimes built. The drum is usually subdivided into a number of separate compartments so that the various stages can be performed. The principal advantage of this filter is the continuity of its operation.

Thin-Cake or Dynamic Filter

The thin-cake or dynamic filter has appeared on the market in recent years. Crossflow filtration operates by recirculating the feed flow parallel to the filter medium. The velocity of the feed in recirculation sweeps away particles deposited on the filter medium, thereby limiting the cake thickness. Crossflow filtration is especially effective for membrane filtration of colloids. The continuous pressure filter shown in Figure 16.9 is a kind of filter thickener, consisting of fixed filtering plates alternating with thin rotating disk. The filtering plates are covered with a filter medium on both sides, and the thin disks are rotated at a high speed sufficient to prevent the cake growth on the filter medium. The feed slurry is dewatered as it flows through the stages, and it emerges continuously from the discharge valve as a concentrate of a pastelike consistency without the danger of filter blocking because the slurry is continuously moving. Little cake formation occurs in the intensive shearing induced by high slurry velocities parallel to the filter medium. As a result, high filtration rates are accomplished. Very thin cake formation allows such a filter to be designed of relatively small size.

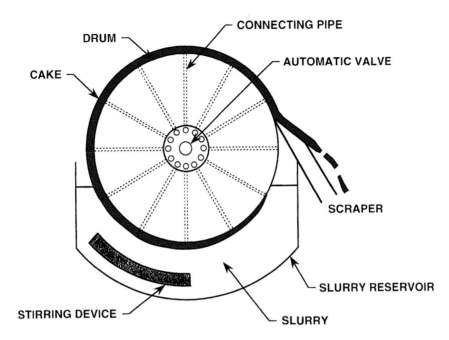

FIGURE 16.8 Schematic view of rotary drum filter.

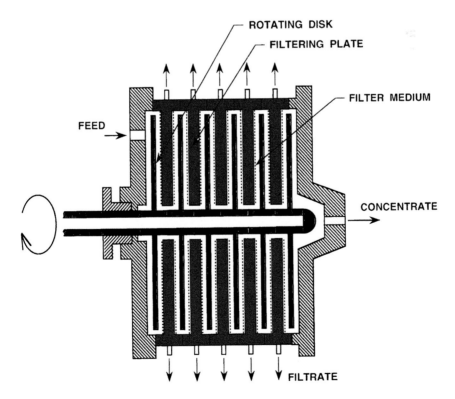

FIGURE 16.9 Schematic view of thin-cake filter.

REFERENCES

1. Iritani, E., Toyoda, Y., and Murase, T., *J. Chem. Eng. Jpn.,* 30, 614–619, 1997.
2. Grace, H. P., *Chem. Eng. Prog.,* 49, 303–318, 1953.
3. Iritani, E., Nakatsuka, S., Aoki, H., and Murase, T., *J. Chem. Eng. Jpn.,* 24, 177–183, 1991.
4. Iritani, E., Hattori, K., and Murase, T., *J. Membrane Sci.,* 81, 1–13, 1993.
5. Shirato, M., Sambuichi, M., Kato, H., and Aragaki, T., *AIChE J.,* 15, 405–409, 1969.
6. Ruth, B. F., *Ind. Eng. Chem.,* 27, 708–723, 1935.
7. Tiller, F. M. and Cooper, H. R., *AIChE J.,* 8, 445–449, 1962.
8. Tiller, F. M. and Shirato, M., *AIChE J.,* 10, 61–67, 1964.
9. Shirato, M., Aragaki, T., Ichimura, K., and Ootsuji, N., *J. Chem. Eng. Jpn.,* 4, 172–177, 1971.
10. Shirato, M., Aragaki, T., Iritani, E., Wakimoto, M., Fujiyoshi, S., and Nanda, S., *J. Chem. Eng. Jpn.,* 10, 54–60, 1977.
11. Cheremisinoff, N. P., in *Liquid Filtration,* Butterworth-Heinemann, Woburn, MA, 1998, pp. 88–117.
12. Shirato, M., Murase, T., Iritani, E., Tiller, F. M., and Alciatore, A. F., in *Filtration: Principles and Practices,* Matteson, M. J. and Orr, C., Eds., Marcel Dekker, New York, 1987, pp. 309–324.

1.17 Expression

Eiji Iritani

Nagoya University, Nagoya, Chikusa-ku, Japan

Expression is the operation of squeezing the liquid from solid–liquid mixtures by compression under conditions that allow the liquid to pass while the solid particles are retained between the supporting media. Expression is distinguished from filtration in that the pressure is applied by movement of the retaining walls such as pistons, membranes, rollers, or belts rather than by pumping a slurry into a fixed-volume chamber. In filtration, the material is sufficiently fluid to be pumpable, whereas in expression, the material can appear to be either entirely semisolid or a slurry. Operational expressions have become increasingly important in such widely divergent fields as food processing and fermentation industries, wastewater treatment, and chemical process industries because the energy required to express liquid from cakes is negligible compared to the heat required for thermal drying.

1.17.1   BASIS OF EXPRESSION

Figure 17.1 shows the schematic view of the compression-permeability cell employed to investigate the deliquoring mechanism of the compressed cake during an expression operation. In the experiments, the solid–liquid mixture is introduced into a cylinder, and a constant mechanical load is applied through the piston. The liquid squeezed from the mixture is allowed to drain from both the top and bottom media. The variation with time θ of the thickness L of the mixture or compressed cake is measured by the dial gauge.

The expression is made up of two stages, filtration and consolidation, according to the flow mechanism within the porous materials. When the original mixture in the cylinder is able to flow as a slurry, applying a load to the piston causes a sudden, uniform increase of hydraulic pressure in the slurry, and the resulting hydraulic pressure is equal to the applied pressure, and consequently deliquoring proceeds on the principle of filtration. Slurry filtration terminates in cake consolidation when the whole slurry forms a filter cake. In the consolidation stage, the bulk volume of the cake is reduced. On the basis of a mass balance, the maximum thickness L_1 of the filter cake can be calculated from

$$L_1 = \left(\frac{m-1}{\rho} + \frac{1}{\rho_\sigma} \right) \rho_s \omega_0 \qquad (17.1)$$

where m is the ratio of the mass of wet cake to the mass of dry cake, ρ is the density of liquid, ρ_s is the true density of solids, and ω_0 is the net solid volume per unit medium area. Deliquoring due to consolidation is distinguished from that due to filtration in that the hydraulic pressure distribution through a compressed cake varies significantly. In contrast, the hydraulic pressure distribution through a filter cake during constant-pressure filtration essentially does not change. Provided the original concentration of the mixture is larger than a limiting value, deliquoring of the mixture proceeds from the beginning on the principle of consolidation.

1.17.2   MODIFIED TERZAGHI MODEL

The equations for describing the consolidation behaviors are derived from the basic equation of flow through porous media. It is convenient to use a solid particle distribution represented by ω divided by the cross-sectional area, where ω is a moving plane that represents the solid volume

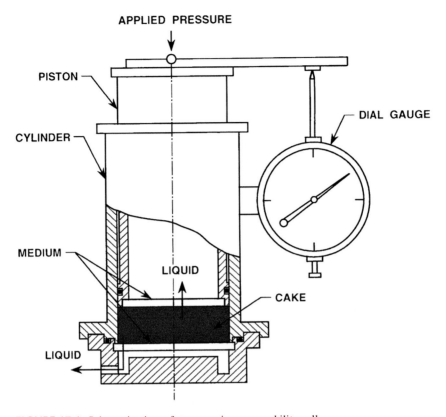

FIGURE 17.1 Schematic view of compression-permeability cell.

between the plane and the medium. The apparent velocity u of the liquid viewed from this moving material can be described by

$$u = \frac{1}{\mu \alpha \rho_s} \cdot \frac{\partial p_L}{\partial \omega} = -\frac{1}{\mu \alpha \rho_s} \cdot \frac{\partial p_s}{\partial \omega} \tag{17.2}$$

where μ is the viscosity of liquid, α is the local specific flow resistance, p_L is the local hydraulic pressure, and p_s is the local solid compressive pressure. The mass balance of liquid in the infinitesimal layer leads to the equation of continuity in the form

$$\frac{\partial e}{\partial \theta_c} = \frac{\partial u}{\partial \omega} \tag{17.3}$$

where e is the local void ratio and θ_c is the consolidation time. Combination of Equation 17.2 with Equation 17.3 leads to

$$\frac{\partial p_s}{\partial \theta_c} = C_e \left\{ \frac{\partial^2 p_s}{\partial \omega^2} - \frac{1}{\alpha} \cdot \frac{d\alpha}{dp_s} \left(\frac{\partial p_s}{\partial \omega} \right)^2 \right\} \tag{17.4}$$

where C_e is the modified average consolidation coefficient defined by

$$C_e = \left\{ \mu \alpha \rho_s \left(-\frac{de}{dp_s} \right) \right\}^{-1} \tag{17.5}$$

Combining Equation 17.2 with Equation 17.3 on the assumption that C_e is constant, one obtains the well-known form of the diffusion equation

$$\frac{\partial p_s}{\partial \theta_c} = C_e \frac{\partial^2 p_s}{\partial \omega^2} \tag{17.6}$$

This equation is similar to the basic consolidation equation presented by Terzaghi[1] in the field of soil mechanics with spatial fixed coordinates.

It is important to know the initial condition, specifically distributions of the hydraulic pressure (or the solid compressive pressure) in the cake, in order to obtain the solution of Equation 17.6. The hydraulic pressure distributions within filter cakes with moderate compressibility can be generally approximated by a sinusoidal curve. Thus, the solution of Equation 17.6 for a constant-pressure expression of filter cake is given by[2]

$$U_c = \frac{L_1 - L}{L_1 - L_\infty} = 1 - \exp\left(-\frac{\pi^2 T_c}{4}\right) \tag{17.7}$$

where U_c is the average consolidation ratio, and L_1, L, and L_∞ are the thickness of the cake at $\theta_c = 0$, θ_c, and ∞, respectively. The dimensionless consolidation time T_c is defined by

$$T_c = \frac{i^2 C_e \theta_c}{\omega_0^2} \tag{17.8}$$

where i is the number of drainage surfaces. This equation clearly indicates that the consolidation time θ_c required for attaining a specified value of the degree of consolidation is directly proportional to ω_0^2 and is inversely proportional to i^2. The solution of Equation 17.6 for a constant-pressure expression of homogeneous semisolid materials is given by

$$U_c = \frac{L_1 - L}{L_1 - L_\infty} = 1 - \sum_{N=1}^{\infty} \frac{8}{(2N-1)^2 \pi^2} \exp\left\{-\frac{(2N-1)^2 \pi^2}{4} T_c\right\} \tag{17.9}$$

Figure 17.2 compares the experimental data for consolidation of a homogeneous semisolid of the Korean kaolin[1] with calculations using Equation 17.9. The agreement is rather poor, whereas the calculated curve is similar in shape to the experimental curve. In addition, although Equation 17.4 can be solved numerically by considering variations of C_e, the agreement is still poor.

The practical and important quantity C_e can be determined by the "fitting method" by consideration of the similarity in shape of the theoretical curve of U_c versus $\sqrt{\theta_c}$, and experimental observations. In constant-pressure expression of the filter cake, the time θ_{90} for attaining 90% of U_c can be obtained from the experimental data, as shown in Figure 17.3, and thus the value of C_e is calculated from

$$C_e = \frac{0.933 \omega_0^2}{i^2 \theta_{90}} \tag{17.10}$$

The quantity C_e can also be estimated using the compression-permeability cell data shown in Figure 17.4. The specific flow resistance α is plotted against the solid compressive pressure p_s, at the left in the figure, and the void ratio e is plotted against p_s, at the right, for a number of materials with widely different compressibility. Compression-permeability cell data can be represented by

$$\alpha = \alpha_0 + \alpha_1 p_s^n \tag{17.11}$$

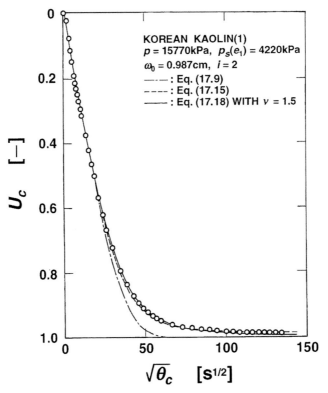

FIGURE 17.2 Constant-pressure expression of semisolid material.

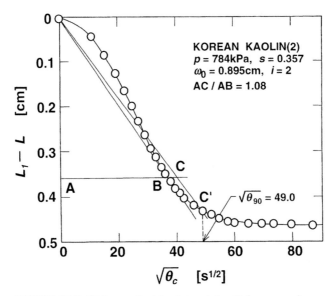

FIGURE 17.3 Fitting method for determining C_e from experimental data.

$$e = e_0 - C_c \ln p_s \qquad (17.12)$$

On the basis of the empirical constitutive equations 17.11 and 17.12, the modified average consolidation coefficient C_e can be calculated from

$$C_e = \frac{p_{s,av}}{\mu \rho_s C_c \left(\alpha_0 + \alpha_1 p_{s,av}{}^n \right)} \qquad (17.13)$$

where $p_{s,av}$ is the average solid compressive pressure, which is often approximated by the arithmetic mean value of the initial average solid compressive pressure $p_s(e_{1,av})$ within the cake and the applied pressure p. The symbol $e_{1,av}$ denotes the initial average void ratio. Since this equation essentially affords an estimation of C_e, the consolidation process can be theoretically predicted from calculations based on compression-permeability cell data.

1.17.3 SECONDARY CONSOLIDATION

Secondary consolidation has to be taken into account in order to obtain more rigorous equations for describing the consolidation behavior. Because the variation of the void ratio e within the cake is caused by both the change in the local solid compressive pressure p_s and the simultaneous creep effect of materials, e is a function of both p_s and the consolidation time θ_c. Assuming that the rheological behavior of secondary consolidation is described by the Voigt element shown in Figure 17.5, the average consolidation ratio U_c is given by

$$U_c = \frac{L_1 - L}{L_1 - L_\infty} = (1 - B)\left\{1 - \exp\left(-\frac{\pi^2 T_c}{4}\right)\right\} + B\{1 - \exp(-\eta \theta_c)\} \qquad (17.14)$$

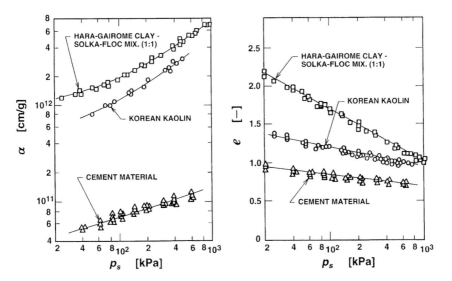

FIGURE 17.4 Compression-permeability cell data.

for constant-pressure expression of filter cakes, and by

$$U_c = \frac{L_1 - L}{L_1 - L_\infty} = (1 - B)\left[1 - \sum_{N=1}^{\infty} \frac{8}{(2N-1)^2 \pi^2} \exp\left\{-\frac{(2N-1)^2 \pi^2}{4} T_c\right\}\right] \\ + B\{1 - \exp(-\eta\theta_c)\}$$

(17.15)

for constant-pressure expression of semisolid materials.[3]

The quantity B is an empirical constant defined by

$$B = v_{sc.max} / v_{c.max}$$

(17.16)

where $v_{sc.max}$ and $v_{c.max}$ are the total liquid volumes squeezed by the secondary consolidation and until final equilibrium, respectively. The quantity η is the empirical constant due to creep of the materials. Because $-\pi^2 T_c/4 >> \eta$ for the large values of θ_c, Equation 17.14 and Equation 17.15 become approximately

$$U_c = \frac{L_1 - L}{L_1 - L_\infty} \approx 1 - B\exp(-\eta\theta_c)$$

(17.17)

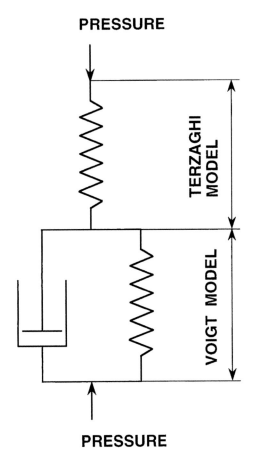

FIGURE 17.5 Schematic view of Terzaghi–Voigt combined model.

Consequently, the values of both B and η can be determined from the later stage of the plot of $\ln(1 - U_c)$ versus θ_c. In Figure 17.2, the prediction based on Equation 17.15 is compared with the experimental data. The data are in relatively good agreement with the predictions. However, with biological and cellular materials, the consolidation mechanism is much more complex because the liquid is stored mainly in cells.[4]

1.17.4 SIMPLIFIED ANALYSIS

The solutions based on the modified Terzaghi model give less satisfactory values, especially in the later stage of consolidation. Sophisticated solutions based on the Terzaghi–Voigt combined model coincide very well with the experimental results. However, the solutions might not be useful in normal industrial practice because they include the three empirical constants C_e, B, and η, which have to be determined from two graphical plots.

It is clear from Figure 17.2 that the relation between U_c and $\sqrt{\theta_c}$, calculated by Equation 17.9, is linear and nearly coincides with experimental data in the early stage of consolidation of semi-solid material under constant-pressure conditions. However, agreement is poor in the later stage, and the theoretical curve approaches the final equilibrium more rapidly than the experimental results. Because U_c is directly proportional to $\sqrt{\theta_c}$ for small values of θ_c and asymptotically approaches unity for large values of θ_c, one of the simplest equations, instead of Equation 17.9, can tentatively be derived as[5]

$$U_c = \frac{L_1 - L}{L_1 - L_\infty} = \left[\frac{4}{\pi} T_c \left\{ 1 + \left(\frac{4}{\pi} T_c \right)^v \right\}^{-1/v} \right]^{1/2} \qquad (17.18)$$

where v is the consolidation behavior index which takes secondary consolidation effects into account. When U_c is plotted against $\sqrt{T_c}$, the maximum percentage of error calculated by Equation 17.18 with $v = 2.85$, compared to Equation 17.9, is only 0.60. It was previously indicated that Equation 17.9 gives less satisfactory values. However, it can be seen from Figure 17.2 that Equation 17.18 with $v < 2.85$ gives quite satisfactory results, and the effect of secondary consolidation is considered in this range.

1.17.5 EXPRESSION EQUIPMENT

Various types of equipment for expression are available in industry.[6] The automated filter press with a compression mechanism was developed in the 1960s in order to reduce the cake moisture content. A lateral, recessed plate and an expression plate with the impermeable membrane comprises a filter chamber, as shown in Figure 17.6. After feed slurry is forced into the chamber under pressure, filtration occurs and the membrane is pressed against the expression plate. Once the chamber is filled with filter cake, the membrane is expanded by pumping compressed air or water into the cavity between the membrane and the plate. As the membrane moves toward the opposite plate, deliquoring of the filter cake due to consolidation takes place. On completion of expression, the chamber is automatically opened and the cake is discharged.

As shown in Figure 17.7, the belt press as an example of continuous expression equipment has a horizontal gravity dewatering zone of belt preceding the pressure dewatering zone. The slurry is sandwiched between the carrying belt and the cover belt at the end of the gravity dewatering zone. From this point to the cake discharge point, the sludge is compressed by compressive and shear forces that are applied through the belt.

In recent years, a screw press, used in the oil milling industry as an expeller, has been widely employed in the chemical industry and sewage sludge treatment. In the screw press, materials are fed to a perforated barrel containing a rotating worm. Materials trapped between the worm and the inside of the cylindrical barrel pass through a gradually reducing flow area, experiencing an increasing pressure. The compressed cake leaves the unit through an adjustable discharge port.

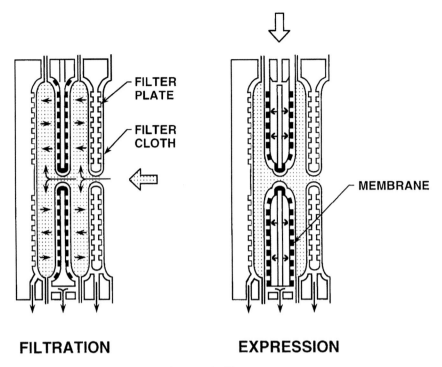

FILTRATION   **EXPRESSION**

FIGURE 17.6 Schematic view of automatic filter press.

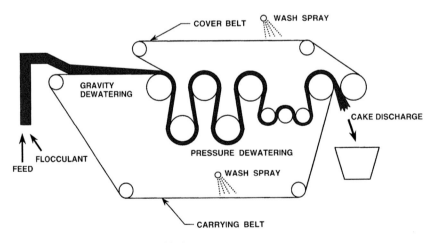

FIGURE 17.7 Schematic view of belt press.

REFERENCES

1. Terzaghi, K., in *Theoretical Soil Mechanics*, John Wiley, New York, 1948, pp. 265–296.
2. Shirato, M., Murase, T., Kato, H., and Fukaya, S., *Kagaku Kogaku*, 31, 1125–1131, 1967.
3. Shirato, M., Murase, T., Tokunaga, A., and Yamada, O., *J. Chem. Eng. Jpn.*, 7, 229–231, 1974.
4. Lanoisellé, J. L., Vorobyov, E. I., Bouvier, J. M., and Piar, G., *AIChE J.*, 42, 2057–2068, 1996.
5. Shirato, M., Murase, T., Atsumi, K., Aragaki, T., and Noguchi, T., *J. Chem. Eng. Jpn.*, 12, 51–55, 1979.
6. Shirato, M., Murase, T., Iritani, E., Tiller, F. M., and Alciatore, A. F., in *Filtration: Principles and Practices*, Matteson, M. J. and Orr, C., Eds., Marcel Dekker, New York, 1987, pp. 371–377.

1.18 Flotation

Hiroki Yotsumoto

National Institute of Advanced Industrial Science and Technology, Tsukuba, Ibaraki, Japan

Flotation is one of the most important methods for mineral separation. Because of its capability of fine particle processing, flotation has been applied to low-grade ores that other separation methods cannot handle economically. In flotation, ore is finely ground and is mixed with water to make a suspension called pulp. After the addition of flotation reagents such as a collector, frother, and so on, the pulp is subject to violent agitation with aeration in a flotation cell for sufficient collision between minerals and bubbles. The targeted minerals coated with hydrophobic collector film are captured at the air–water interface of bubbles and are lifted to the surface of the pulp to form a froth layer. The froth is skimmed off from the flotation cell as a concentrate, whereas unwanted minerals remaining in the pulp are discharged as tailing.

Plain flotation was first practiced to obtain one type of concentrate from simple sulfide ores. Differential flotation was later developed to recover progressively more than two kinds of concentrates from complex sulfide ores bearing many valuable minerals. With the development of flotation reagents as well as flotation machines, flotation has been applied successfully to recovering many kinds of minerals including soluble-salts minerals. Recently it has also been applied to recovering various substances such as chemically precipitated colloids, ions, microorganisms, and so on in chemical and food industries and in wastewater and seawater treatment.

1.18.1 PRINCIPLES OF FLOTATION

Flotation is a complex physicochemical process and its theory is not completely understood. One of the theoretical approaches is to correlate the contact angle of mineral with its floatability. The static state in which a mineral particle is attached to a large bubble in water is similar to that of a particle at a free surface as illustrated in Figure 18.1.[1] If the particle is assumed to be a cylinder at a free water surface as shown in Figure 18.1b, the lifting force, f, acting on the particle is expressed in the form

$$f = 2\pi r T \sin \theta + \pi r h \rho g \qquad (18.1)$$

where r is the radius of the cylinder, T is the surface tension of water, ρ is the density of water, g is the acceleration of gravity, h is the depth of the dimple, and θ is the contact angle as shown in Figure 18.1b. On the right-hand side of Equation 18.1, the first term is associated with the force due to the surface tension of water, and the second term is the force equal to the weight of the cylindrical mass of liquid displaced by air. The contribution of both terms to the lifting force, f, varies with the mineral size, r. For large particles, the lifting force can be given only by the second term. For small particles, on the other hand, the second term becomes negligible, and the lifting force is regarded as

$$f = 2\pi r T \sin \theta \qquad (18.2)$$

When T is constant, Equation 18.2 shows that the lifting force increases with the increase of θ.

The contact angle at the solid–water–air interface is a measure of the adhesion force between bubble and particle and also is a measure of the hydrophobicity of particulate solid. θ in Equation

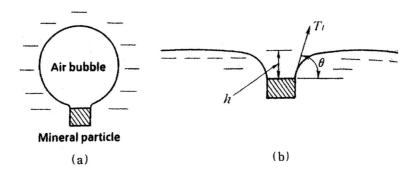

FIGURE 18.1 Suspension of a mineral particle at an air–water interface.

TABLE 18.1   Relationship between Contact Angle and Floatability for Various Kinds of Minerals

Mineral	Contact Angle (deg)	Floatability (%)
Galena	70–75	90
Sphalerite	71–72	95–98
Pyrite	58–73	89–92
Quartz	55–58	78–79
Calcite	45	11–56
Slate	13	6
Sandstone	0	1

18.2 is closely related to the contact angle of the suspended particulate solid. Such minerals as fresh galena, native sulfur, graphite, talc, and so on are generally hydrophobic, with very large contact angles. On the other hand, quartz, calcite, dolomite, corundum, and so on are considered to be hydrophilic, which means that their contact angles are relatively small. Peterson[2] presented the relationship between contact angle and floatability for several kinds of minerals under certain experimental conditions, as shown in Table 18.1.

Another approach is to explain the bubble–particle attachment with DLVO theory,[3] which predicts the coagulation of two solid particles considering the balance of opposing surface forces, namely, electrostatic repulsion and molecular (van der Waals) attraction between the particles. In a slow coagulation condition, an energy barrier is formed at a certain interparticle distance by the interaction of repulsive and attractive potential energies (see Figure 18.2a). If the relative kinetic energy of the two particles is larger than the magnitude of the energy barrier, the particles are supposed to collide with each other. Assuming that a bubble is one of the particles, one can calculate the interaction energy between the bubble and particle using DLVO theory. In the case of bubble–particle attachment, molecular force is always repulsive, unlike the case of particle–particle attachment. When the bubble and particle are unequally charged, for instance, +100 mV (bubble) and +50 mV (particle), electrostatic repulsion turns into attraction at a very short bubble–particle distance.[4] As a result, a very large, sometimes infinitely large, energy barrier is formed between bubble and particle. The potential curve $V_t = V_e + V_d$ in Figure 18.2b shows this large energy barrier. Such a barrier is supposed to prevent babble–particle attachment, whereas flotation occurs even if an infinitely large barrier is predicted.

To solve this discrepancy, some researchers assume the existence of "hydrophobic attraction" between a bubble and a particle. The force was found by Israelachvili and Pashley in 1982 between

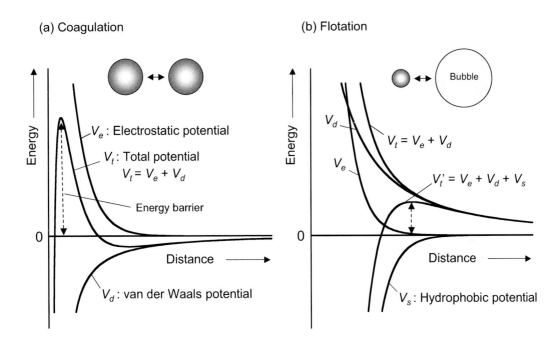

FIGURE 18.2 Concept of DLVO theory for coagulation and flotation.

mica surfaces coated with a surfactant monolayer.[5] The hydrophobic attraction was of a long-range nature and was much larger than the van der Waals attraction. If this force exits between a bubble and a particle, it overcomes the van der Waals attraction and electrostatic repulsions and reduces the energy barrier to a certain magnitude, which corresponds with the floatability of mineral. The potential curve $V_t' = V_e + V_d + V_s$ in Figure 18.2b shows this small barrier. The hydrophobic attraction may be the major driving force of flotation. However, there are arguments about its existence in a bubble–particle system.

1.18.2 CLASSIFICATION OF MINERALS ACCORDING TO THEIR FLOTATION BEHAVIOR

In general, minerals can be classified into several groups depending on their surface property such as hydrophobicity or hydrophilicity. There are some minerals which have high natural floatability. Minerals of this type are graphite, sulfur, molybdenite, diamond, coal, and talc, as shown in group A in Table 18.2. They are sometimes called nonpolar minerals. The surfaces of nonpolar minerals are characterized by strong hydrophobicity depending on their chemical composition and bonding type. The nonpolar minerals are composed of atoms bonded with nonpolar covalent bonds or covalent molecules held together by van der Waals force.

Minerals with polar covalent or ionic bonding are known as polar types. These minerals are naturally hydrophilic. Minerals belonging to the polar group are roughly divided into various subclasses B to F, as listed in Table 18.2, depending on the magnitude of hydrophilicity. Minerals in group B are sulfides and native metals. As can be seen in Table 18.2, the degree of hydrophilicity increases from sulfide minerals, through sulfates in group C, to carbonates, phosphates, and so on in group D, oxides in group E, and, finally, silicates and quartz in group F.

TABLE 18.2 Classification of Minerals

Group	Surface Property	Minerals
A	Nonpolar	Graphite, sulphur, molybdenite, diamond, coal, talc
B	Polar	Sulphides: bornite, chalcopyrite, galena, pyrite, sphalerite stibnite, argentite, arsenopyrite, etc.
		Native metals: Au, Pt, Ag, Cu
C	Polar	Sulfates: barite, anhydrite, gypsum, anglesite
D	Polar	Carbonates: cerrusite, malachite, azurite, smithsonite, siderite, calcite, witherite, agnesite, dolomite
		Other minerals: fluorite, apatite, monazite, scheelite
E	Polar	Oxides: hematite, magnetite, goethite, chromite, ilmenite, rutile, corundum, pyrolusite, limonite, wolframite, cassiterite
F	Polar	Silicates: zircon, willemite, hemimorphite, beryl, feldspar, garnet, sillimanite, quartz

1.18.3 FLOTATION REAGENTS

Although a great number of chemicals have been known to be flotation reagents so far, no more than 50 chemicals seem to be employed in practical flotation plants at present. Flotation reagents can be classified into four categories based on their action in flotation: collectors, frothers, activators, and depressants, as indicated by Wark.[6] A reagent can belong to more than one of these categories. For example, soluble organic compounds, besides being frothers, are collectors for some minerals as well. Lime and soda ash used to regulate the pH of the pulp are often called pH regulators. Further, a sufficient flotation of a desired mineral sometimes requires a good dispersion of gangue slime in the flotation pulp. The reagents used for this purpose are called dispersants.

Collectors

A collector is generally an organic reagent that adsorbs on a mineral surface and makes it hydrophobic. Thus, it serves as a bridge between the mineral and air bubble. Common collectors are listed in Table 18.3.

Frothers

Frothers are used to help the dispersion of air bubbles throughout the pulp in the flotation cell, to promote the stability of attachment between mineral and air bubble, and also to stabilize the froth formation at the surface of pulp. Table 18.4 shows the types of chemical reagent used as frothers in flotation. The type and molecular structure of frothers affect their frothing power, the stability of froth, and the physical characteristics of froth layer. Pine oil, whose main constituent is terpineol, is one of the most popular frothers in the flotation of various minerals.

Depressants

When ore contains more than two valuable minerals to separate as individual concentrates, it is necessary to regulate the floatability of minerals. The reagents used to depress the flotation of desired minerals are called depressants. Reagents such as lime, caustic soda, and sodium carbonate are often used to regulate the pH of flotation pulp in the alkaline range. They are also depressants for sulfide minerals, because most sulfide minerals are depressed in high-pH solutions. Sodium cyanide, sodium sulfide,

TABLE 18.3 Classification of Common Collectors

Class	Typical Collectors	Aimed Minerals
Oils	Coal tar oil, fuel oil, creosote	Natural sulfur, graphite, molybdenite, coal, etc.
Acids containing a hydrocarbon group and the salts of their acids	Alkyl dithiocarbonates: $S=C\begin{smallmatrix}\diagup OR \\ \diagdown SM\end{smallmatrix}$ Dialkyl dithiocarbamates: $S=C\begin{smallmatrix}\diagup NR_2 \\ \diagdown SM\end{smallmatrix}$ Dialkyl dithiophosphates: $S=P\begin{smallmatrix}\diagup OR \\ -OR \\ \diagdown SM\end{smallmatrix}$ (R: hydrocarbon chain, C = 2–5) Fatty acids: RCOOM	Sulfide minerals: chalcopyrite, chalcocite, convellite, sphalerite, galena, pyrite, etc. Phosphate ores: barite, chromite, fluorite, scheelite, hematite, calcite, etc.
	Alkyl sulfonates: $R-SO_3M$ (R: hydrocarbon chain, $C \geq 8$)	Iron oxides: kyanite, talc, garnet, chromite, barite, rhodochrosite, etc.
Bases containing a hydrocarbon group and the salts of their bases	Amines derivatives: RNH_3X (primary) R_2NH_2X (secondary) R_3NHX (tertiary) R_4N-X (quaternary) (R: hydrocarbon group, $C \geq 8$)	Silica, mica, vermiculite, smithsonite, sylvite, feldspar, etc.

sodium sulfite, and sulfur dioxide gas are well-known depressants in sulfide flotation. Sodium silicate, fluoric acid, starch, tannin, lignite, and so on are depressants in the flotation of nonsulfide minerals.

Activators

Activators promote the adsorption of a collector onto a mineral surface by altering the chemical nature of the mineral surface. Copper sulfate is one of the most important activators for spharlerite when a sulfydril collector is used. It is also an activator for arsenopyrite, cobaltite, and stibnite. Sodium sulfide is, in general, an activator for several nonsulfide minerals, such as oxides, carbonates, and sulfates of heavy metals. For such minerals, sodium sulfide reacts with metal constituents at the solid–water interface to form a kind of metal sulfide film at the mineral surface. Then the minerals thus treated are amenable to flotation by using conventional collectors for sulfide minerals. Instead of sodium sulfide, sodium hydrosulfide and ammonium sulfide can be used. Lead nitrate

TABLE 18.4   Classification of Common Frothers

Chemical Type	Type Formula	Representative Frothers
Alcohols	R–OH (R: hydrocarbon group)	Aliphatic alcohols: R is a straight or branched hydrocarbon chain (C = 5–8)
		Cresylic acids: R is a benzene ring with short-chain alkyl substituents
		Pine oils: R is a terpene ring structure
Ethers	R–O–R' (R, R': hydrocarbon groups)	Alkoxy substituted paraffins (triethoxybutane)
		Polypropylene glycols
		Methoxytripropylene glycol
Ketones	R–C–R' 　　$\mid$ 　　O (R, R': hydrocarbon groups)	Camphor

is used as an activator for stibnite in flotation with sulfydril collector, and also as an activator for halite in the fatty acid flotation of this salt in the presence of sylvite. There are some other activation instances: the activation of mica by lead salts for flotation by an oleic acid collector and the activation of feldspar by hydrofluoric acid for flotation by an amine collector.

1.18.4   FLOTATION MACHINES

Although many different machines have been developed and discarded in the past and many machines are currently being manufactured, it can be said that there are two groups: pneumatic machines and mechanical agitation or subaeration machines.

Pneumatic Machines

Pneumatic machines use air blown in by means of pipes, nozzles, or perforated plates, in which case the air must be dispersed by baffles to create a great deal of bubbles in the pulp and to give sufficient aeration or agitation. Figure 18.3 shows one of the old types of pneumatic machines, called the SW cell. It consists of a long tank with a V-shaped base into which many vertical pipes deliver compressed air. Internal baffles provide sufficient agitation and aeration. The Davcra cell is shown in Figure 18.4. The cell consists of a tank segmented by a vertical baffle. Air and feed slurry are injected into the tank through a cyclone-type dispersion nozzle. The flotation column[7] has been developed for the flotation of fine particles, a possible configuration of which is shown in Figure 18.5. Air is introduced from the lower part of the column while feed slurry enters from the upper part of the column. Therefore, mineral particles countercurrently contact a rising swarm of air bubbles in the section below the feed point. Floatable particles collide with and adhere to the bubbles and are transported to the washing section above the feed point. Wash water cleans the froth and releases particles entrained in the water lifted by rising bubbles. Nonfloatable particles are removed from the bottom of the column as tailing. High-grade concentrates are obtained by the use of wash water. Industrial use of flotation columns has increased in recent years.

Mechanical Agitation or Subaeration Machines

A mechanical agitation machine is equipped with a specially designed agitating device, called an impeller or rotor, which agitates the pulp violently and at the same time introduces natural air into the pulp by centrifugal pressure. This type of machine can either be self-aerating or have air blown in. The

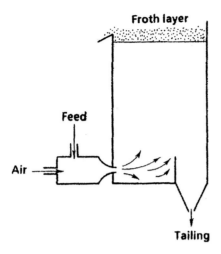

FIGURE 18.3 SW flotation cell.

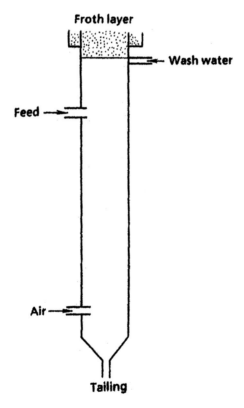

FIGURE 18.4 Davera cell.

flotation is a continuous process. Flotation cells are arranged in series, forming a bank. There are two basic modes of flotation machine operation: cell-to-cell flotation and free-flow flotation. In the second of these modes, as there is no weir between cells; pulp is free to flow through the cells without interference. Figure 18.6 shows schematically the Denver Sub-A cell, which is of the cell-to-cell operation type. The flotation takes place in an individual square cell separated from the adjoining cell by an adjustable

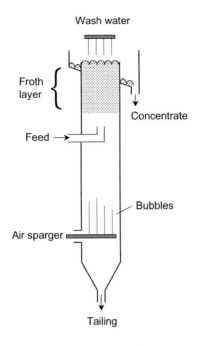

FIGURE 18.5 Flotation column.

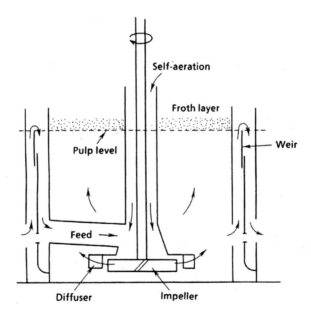

FIGURE 18.6 Denver Sub-A cell.

weir. The feed from the weir of the preceding cell enters the inlet pipe and is carried directly to the next cell, the flow being aided by the suction action of the impeller. The pulp then falls on top of the rotating impeller while air is also drawn down the standpipe. This design assures a favorable air–pulp mixture. The mineral-laden bubbles go upward to form froth. Removal of the froth is accomplished by froth paddles. Tailings that do not float pass on from cell to cell, being subjected over and over to the flotation

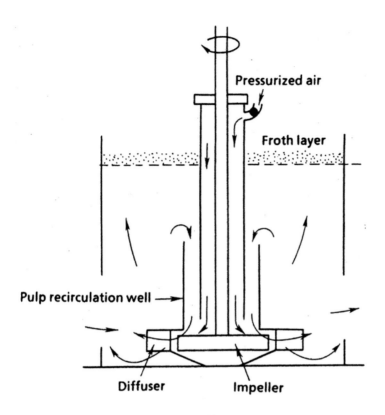

FIGURE 18.7 Denver DR flotation machine.

TABLE 18.5 Critical pH Values for Various Sulfide Minerals by Using 25 mg/l of Potassium Ethyl Dithiocarbonate

Mineral	Critical pH Value (room temperature)
Sphalerite	—
Pyrrhotite	6.0
Arsenopyrite	8.4
Galena	10.4
Pyrite	10.5
Marcasite	11.0
Chalcopyrite	11.8
Covellite	13.2
Cu-activated sphalerite	13.3
Bornite	13.8
Tetrahedrite	13.8
Chalcocite	>14.0

process. The Denver DR free-flow machine shown in Figure 18.7 is designed to handle larger tonnages in the flotation circuit. In individual cells, feed pipes are eliminated. The pulp level is controlled by a single tailing weir at the end on the trough. This machine requires pressurized air. The cell that is most commonly used in industry is of a mechanical agitation type: the Agitair, Fargergren, Warman, and Denver cells. They all have their own special designs in agitation mechanism, air introduction, flow pattern of pulp in the cell, and collision efficiency between particles and bubbles.

1.18.5   DIFFERENTIAL FLOTATION

Separating a complex ore into two or more valuable minerals and gangue by flotation is called differential flotation or selective flotation. The pH of flotation pulp is one of the most important parameters in view of selective flotation. For example, at a fixed addition of a collector, there is a pH value below which any given mineral floats and above which it does not float. This critical flotation pH value depends on the nature of the mineral, the type of collector, the collector concentration, temperature, and the type of existing chemical species, if any, and its concentration. Wark and Cox[8] determined critical pH values for various types of sulfide minerals in terms of collector type, collector concentration, and temperature. One of the results is presented in Table 18.5. This table clearly shows the importance of controlling the pH of the pulp in the selective flotation of sulfide minerals. Besides controlling pulp pH, a successful differential flotation is generally achieved by the use of suitable depressants and activators.

1.18.6   PLANT PRACTICE OF DIFFERENTIAL FLOTATION

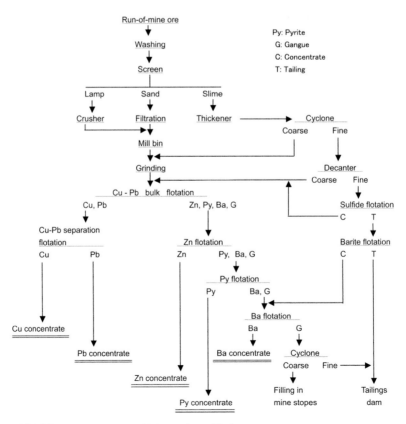

FIGURE 18.8 Flowsheet of Matsumine mill plant.

As a typical example of the differential flotation, the flow sheet of the Matsumine mill plant in the Hanaoka mine in Japan is shown in Figure 18.8. The mill was closed in 1995 because of economic problems, but a number of technological developments were achieved with respect to complex sulfide ore processing. The treated ore was complex sulfide ore, locally called "Kuroko" (black ore), which was composed of fine-grained minerals. The main sulfide minerals are chalcopyrite, pyrite, galena, and sphalerite, whereas some nonsulfide minerals are barite, gypsum, sericite, chlorite, kaolinite, montmorillonite, and quartz. After ore was crushed down to less than 100 mm in size at the underground crushing site, the run-of-mine ore was conveyed up to the mill plant. The ore was washed to remove slime and clay adhering to the ore, using a washing screen. As in Figure 18.8, the washed lamp ore was transported to the secondary crushing section where the size of the ore was reduced to less than 25 mm. Material put through the washing screen was classified into sand (less than 25 mm) and slime fractions. The sand fraction and the product of the secondary crushing section were combined as the feed of the grinding stage. At the grinding section, material was ground down to around 0.05 mm in size by two-stage grinding, namely the rod–ball mill grinding circuit, to produce flotation feed.

In the flotation section, the flotation circuit consisted of five subcircuits to recover the following:

1. Bulk Cu-Pb concentrate in the bulk flotation of copper sulfide minerals and galena using diethyldithiophosphate as a collector after conditioning the flotation feed with sulfur dioxide gas and lime at pH 4–5.5
2. Cu concentrate in the copper mineral flotation by heating the bulk Cu-Pb slurry to about 70°C to depress galena, the tailings being the Pb concentrate
3. Zn concentrate in the activated sphalerite flotation with copper sulfate as an activator and xanthates as collectors when the bulk flotation tailings were used as the feed
4. Py concentrate in the flotation of pyrite with xanthate after adjusting the pH of pulp to 3.5
5. Ba concentrate in the flotation barite with alkyl-sulfonate as a collector at pH 6.0

Generally, each subcircuit consisted of rougher, scavenger, and cleaner steps. The slime obtained at the washing screen stage was thickened in a thickener and then treated through a specially designed slime-process circuit including cyclones, decanters, and flotators. The products of the slime-processing circuit were fed to appropriate sections of the principal concentration circuit, the tailing being discarded at the dam.

As can be seen in Figure 18.8, five kinds of concentrates were recovered by differential flotation at the Matsumine mill plant.

REFERENCES

1. Gaudin, A. M., in *Flotation,* 2nd Ed., McGraw-Hill, New York, 1957, pp. 157–159.
2. Peterson, W., in *Schwimmaufbereitung,* Vertag von Theodor Steinkopff, Dresden, 1936, pp. 48–49.
3. Verwey, E. J. W. and Overbeek, J. Th. G., in *Theory of the Stability of Lyophobic Colloids,* Elsevier, New York, 1948, pp. 106–115.
4. Derjaguin, B. V., Churaev, N. V., and Muller, V. M., in *Surface Forces,* Plenum Publishing, New York, 1987, pp. 198–202.
5. Israelachvili, J., in *Intermolecular and Surface Forces,* 2nd Ed., Academic Press, San Diego, 1991, pp. 282–287.
6. Wark, I. W., in *Principles of Flotation,* Australian Institute of Mining and Metallurgy, Melbourne, 1938, p. 9.
7. Finch, J. A. and Dobby, G. S., in *Column Flotation,* Pergamon Press, New York, 1990, pp. 1–4.
8. Wark, I. W. and Cox, A. B., *Trans. Am. Inst. Mining Metall. Eng.,* 112, 189–302, 1935.

1.19 Electrostatic Powder Coating

Ken-ichiro Tanoue
Yamaguchi University, Ube, Yamaguchi, Japan

Hiroaki Masuda
Kyoto University, Katsura, Kyoto, Japan

Electrostatic powder coating has been widely utilized for surface finishing in many industrial applications.[1,2] Recently, the coating has been making inroads in the automotive industry. Since the coating process does not use solvents, it is becoming increasingly attractive in light of possible new clean air standards for volatile organic compounds. The coating system consists of powder coating guns, powder feeding machine, powder coating booth, and powder collecting machine. It is important to combine effectively the equipment for various objectives.

1.19.1 COATING MACHINES

Different configurations for coating guns have been proposed, and they may use two distinct techniques for powder charging. One is corona-charging and the other is tribocharging. Furthermore, other coating machines have been developed.

Corona-Charging Guns

Figure 19.1 shows a corona-charging gun. Powder particles are assembled with air and fed from a nozzle of the gun as a gas–solid two-phase flow. A needle electrode is set in the tip of the nozzle. Negative high voltage is applied on the needle electrode in a range of about –60 kV to –100 kV. At this time, a high electric field is formed between the needle and grounded objects. The powder is exposed to ionized air flow, and corona charging occurs. With continued increase in charge collection by particles, electric field strength E will increase until a certain value is reached corresponds to a situation where no further transport of ions on the particle is possible. This limiting value of surface charge, the Pauthenier limit,[3] is expressed mathematically by the following relationship;

$$q = 4\pi\varepsilon_0 \left(1 + 2\, \frac{\varepsilon_r - 1}{\varepsilon_r + 1} \right) r_P^2\, E$$

(19.1)

where, ε_0 is the permittivity of free space, 8.854×10^{-12} Fm^{-1}, ε_r is the relative permittivity of the powder particle, r_p is the radius of the particle, and E is the electric field strength. Particles are transported toward the objects by coulomb force and drag force and then deposited. The coating gun has been utilized for almost all powder particles. However, as there are free ions in the booth, the "orange peel phenomenon" will occur sometimes on the surface of a thick powder layer. New corona guns with an earth ring to collect free ions have been developed to prevent the effect.

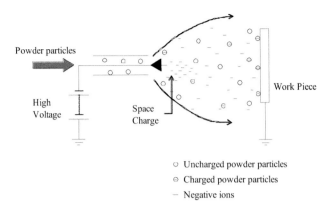

FIGURE 19.1 Corona-charging gun.

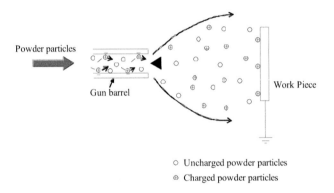

FIGURE 19.2 Tribocharging gun.

Tribocharging Guns

Figure 19.2 shows a tribocharging gun. Powder particles are assembled with air flow and they collide with the inner wall of the gun. At this time, tribocharging of particles occurs. As the inner wall of the gun is commonly made of polytetrafluoroethylene, the powder particles are mainly charged positively on a basis of contact potential difference. Then the charged powder particles are ejected from a nozzle of the gun. As a gun of this kind needs no high-voltage generator, the coating system is very simple and has the following merits:

1. Even if we touch the tip of the nozzle, we don't receive an electric shock.
2. As an electric field hardly occurs between guns and objects, the system can uniformly coat objects with some concave parts.
3. There are no free ions in the coating booth. Therefore, the orange peel phenomenon seldom occurs.

However, it is difficult to control the charge freely because the system is influenced strongly by relative humidity and room temperature,[4] contact potential difference[5,6] between particle and inner wall of the gun, particle size,[7] and air velocity[8] in the gun. Furthermore, the tribocharge depends strongly on the tangential impact velocity of the powder particles.[9–10]

Disc-Type Coating Machine (a Kind of Corona-Tribocharging Gun)

Figure 19.3 shows the disc-type coating machine. In this coating machine, the disc plate, which is made of resin, is installed at the outlet of the powder. The powder particles transported by the powder feeding machine with air flow are spouted out uniformly from the edge of the disc. Work piece objects are carried by a loop-type conveyor around the disc, as shown in Figure 19.3. The coating machine is expected to have six times the ability of the above-mentioned guns.

1.19.2 POWDER FEEDING MACHINE[11]

In order to feed powder particles from a vessel to the gun, a pneumatic conveying system with an air injector is mainly utilized as an easy and low-priced method. However, it is difficult to keep a stable feed of the powder particles using only an air injector. Recently, a few kinds of powder feeders, which are able to supply particles with a constant feed rate, have come onto the market.

Screw Feeder

This type has been mostly applied to the combination with powder coating, as shown in Figure 19.4. The apparatus consists of a powder tank, powder-feeding section with screw feeder, and an air injector. The powder particles are fed to the air injector with high accuracy (about $\pm 2\%$) by controlling the rotational speed of a motor. In this case, the injector has a role that only conveys the powder particles to the coating machine. Therefore, the powder feed rate does not vary even if air flow rate to the air injector is changed. To prevent changing the concentration of the powder in a hopper, the powder tank and the powder feeding section are separated through a control valve. The amount of powder particles is monitored by a level meter and controlled by the gate valve.

Table Feeder

The construction of this apparatus is almost the same as for the screw feeder. The feeder section consists of a turntable, motor, hopper, and air injector. The powder particles on the table are conveyed by air force or a mechanical scraper to the injector. The flow is smoother than in the screw feeder.

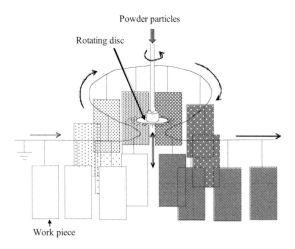

FIGURE 19.3 Disc-type coating machine.

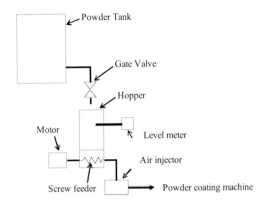

FIGURE 19.4   Powder feeding machine with screw feeder.

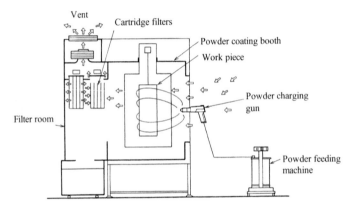

FIGURE 19.5 Powder coating booth using cartridge filters.

1.19.3 POWDER COATING BOOTH

As there are many combinations of powder coating booths and powder collectors, it is necessary to select the booth carefully according to configurations and dimensions of the objects, ensuring space, change of colors, and protection of the working area. In this section, characteristics of representative coating booths are introduced.

Cartridge Filter Type

In Figure 19.5, cartridge filters are set on a wall of the coating booth, and suspended powder particles in the booth are collected directly by suction fan. In order to change colors of powder, it is necessary to prepare the same number of cartridge filters and suction fans as there are colors. Although the coating booth needs no large space, it is difficult to conduct the multicolor powder coating.

Cyclone after Filter Type

In Figure 19.6, precipitation of powder particles are conducted in two stages: cyclones and after-filters. First, most of oversprayed powder particles are collected by the cyclones. At the second stage, small particles passed through the cyclones are collected by after-filters. This coating booth is available for multicolor coating.

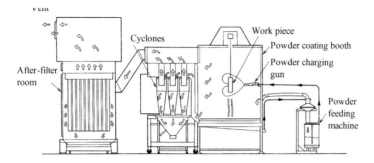

FIGURE 19.6   Powder coating booth using multicyclone and after-filters.

1.19.4   NUMERICAL SIMULATION FOR ELECTROSTATIC POWDER COATING[12-17]

Recently, numerical calculation for the electrostatic powder coating has been investigated. The equation of motion of a particle is given by

$$m_p \frac{dv}{dt} = - \frac{\pi D_P^2 \rho C_D v - u(v - u)}{8 C_c} + m_p g + S F_E \qquad (19.2)$$

The right-hand terms in the equation show drag force, gravity force, and electrostatic force, respectively. Charged particles are affected by coulomb force ($F_{E,C} = qE$) and other electrostatic forces, which are described in Section 2.5.1.

Air Flow Field

To calculate the aerodynamic forces acting on a particle is complex. The Navier–Stokes equations are used in evaluating fluid flow conditions and normally include three scalar equations for momentum, one for mass conservation, and one for energy conservation. These equations are given by,[18]

$$\rho \frac{Du}{Dt} = - \nabla p - [\nabla \cdot \tau] + \rho g \qquad (19.3)$$

$$\frac{D\rho u}{Dt} = 0 \qquad (19.4)$$

$$\rho \frac{De_f}{Dt} = - (\nabla f_e) - (\nabla \cdot pu) - (\nabla \cdot [\tau \cdot u]) \qquad (19.5)$$

where ρ is the fluid density, p is the fluid pressure, τ is the viscous stress, e_f is total fluid energy, and f_e is the energy flux. Furthermore, an equation of state is required in order to couple the mass conservation with momentum conservation via fluid density.

$$\rho = \frac{PM}{RT} \qquad (19.6)$$

where M is the molecular weight of fluid gas, R is the gas constant, and T is the fluid temperature. For common electrostatic powder coating, the air flow field can be treated as incompressible.

Furthermore, physical properties in the equations are assumed to be constant. Therefore, the air flow field can be solved using Equation 19.3 and Equation 19.4.

As the momentum in Equation 19.3 includes both the pressure field and velocity field, the fields must be solved severally. SIMPLE[19] and SMAC[20] schemes can be utilized as the representative algorithm in order to calculate the Navier–Stokes equations.

Electric Field

The electric field that is generated due to the imposed voltage between the emitting electrode and the grounded plate can be described by a following Laplace equation,

$$\Delta \Phi = 0 \qquad (19.7)$$

where the potential Φ is related to the electric filed E according to $E = -\nabla \Phi$. Since a high voltage is applied between the two electrodes, a corona discharge may take place. The presence of a space charge generated by the corona complicates the electric field solution in the computational domain. Such an electric field with a space charge is governed by Poisson's equation:

$$\Delta \Phi = -\frac{\rho_E}{\varepsilon_0} \qquad (19.8)$$

where ρ_E is the space charge density and ε_0 is the electrical permittivity of the gas phase. The Laplace and Poisson equations can be solved by the successive overrelaxation method. A numerical technique proposed by Elmoursi[21] is applicable to obtain a self-consistent solution for an ionic space charge field. The Lagrangian particle-tracking model is coupled with the electric field during the iteration. The procedure is described in the following steps:

1. Solve the electric field under Laplace condition ($\rho_E = \phi$). This provides the initial estimation of the electric field.
2. Calculate particle trajectories with discrete phase model using Equation 19.2.
3. Determine the charge density distribution due to the charged powder in the ion drift region.
4. Solve the electric field with space charge.
5. Repeat steps 2–4 until the solution of the space charge field is convergent.

This procedure can be utilized for the numerical simulation of the electrostatic powder coating with corona charging. On the other hand, for tribocharging, the image force works mainly on the charged particles. If many charged particles are deposited on the work piece object, apparent coulomb force decreases with time and then the particles hardly deposit further. Tanoue et al. have investigated the fact experimentally and numerically.[17,22]

REFERENCES

1. Reddy, V. and Dawson, S., *Powder Coating Applications,* Society of Manufacturing Engineers Powder Coating Institute, Alexandria, VA, 1990.
2. Boochi, G. J., *Powder Coating,* 4, 14–18, 1993.
3. Pauthenier, M. M. and Moreau-Hanot, M., *J. d'Physique Radium,* 7, 590–613, 1932.
4. Nomura, T., Taniguchi, N., and Masuda, H., *J. Soc. Powder Technol. Jpn.,* 36, 168–173, 1999 (in Japanese).

5. Itakura, T., Masuda, H., Ohtsuka, C., and Matsusaka, S., *J. Electrostatics,* 38, 213–226, 1996.

6. Tanoue, K., Morita, K., Maruyama, H., and Masuda, H., *AIChE J.,* 47, 2419–2424, 2001.

7. Masuda, H. and Iinoya, K., *AIChE J.,* 24, 950–956, 1978.

8. Masuda, H., Komatsu, T., Mitsui, N., and Iinoya, K., *J. Electrostatics,* 2, 341–350, 1976/1977.

9. Ema, A., Sugiyama, S., Tanoue, K., and Masuda, H., *J. Inst. Electrostatics Jpn.,* 26l, 130–136, 2002.

10. Ema, A., Yasuda, D., Tanoue, K., and Masuda, H., *Powder Technol.,* 135–136, 2–3, 2003.

11. Yanagida, K., Morita, T., and Takeuchi, M., *J. Electrostatics,* 49, 1–13, 2000.

12. Woolard, D. E. and Ramani, K., *J. Electrostatics,* 35, 373–387, 1995.

13. Chen, H., Sims, R. A., Mountain, J. R., Burnside, G., Reddy, R. N., Mazumder, M. K., and Gatlin, B., *Part. Sci. Technol.,* 14, 239–254, 1996.

14. Adamiak, K., *J. Electrostatics,* 40, 395–400, 1997.

15. Ye, Q., Steigleder, T., Scheibe, A., and Domnick, J., *J. Electrostatics,* 54, 189–205, 2002.

16. Ang, M. L. and Lloyd, P. J., *Int. J. Multiphase Flow,* 13, 823–836, 1987.

17. Tanoue, K., Inoue, Y., and Masuda, H., *Aerosol Sci. Technol.,* 37, 1–14, 2003.

18. Bird, R. B., Stewart, W. E., and Lightfoot, E. N., in *Transport Phenomena,* 3rd Ed., John Wiley, New York, 1960, p. 322.

19. Patanker, S. V., *Numerical Heat Transfer and Fluid Flow,* McGraw-Hill, New York, 1980.

20. Amsden, A. A. and Harlow, F. H., *The SMAC Method: A Numerical Technique for Calculating Incompressible Fluid Flows,* Los Alamos Scientific Laboratory Report LA-4370, 1970.

21. Elmoursi, A. A., *IEEE Trans. Ind. Appl.,* 28, 1174–1181, 1992.

22. Tanoue, K., Yamamoto, M., Ema, A., and Masuda, H., *Kagaku Kogaku Ronbunsyu,* 28, 196–201, 2002 (in Japanese).

1.20 Multipurpose Equipment

Jun Oshitani and Kuniaki Gotoh
Okayama University, Okayama, Japan

Shigeki Toyama[*]
Nagoya University, Aichi, Japan

1.20.1 FLUIDIZED BEDS

If a gas passes upward through a bed of particles, the bed is fixed at low gas velocities, but when the velocity is increased further, the particles separate from each other and are supported by the gas, then the bed starts fluidizing. In the fluidized bed, the particles contact with the gas at high efficiency, and temperature is kept uniform because of good mixing of the particles. The features have been utilized for industrial applications[1]; the first use of the fluidized bed as a large-scale reactor was Winkler's coal gasifier in 1926. Since the 1940s the fluidized bed had played an important role for fluid catalytic cracking using a combination of a reactor and a catalyst regenerator. Recently a circulating fluidized bed, as shown in Figure 20.1, has been widely used as a boiler for solid fuel and waste.

Figure 20.2 illustrates a typical diagram of a pressure drop ΔP versus gas velocity u_0. If the gas velocity is increased, the pressure drop gradually increases with the gas velocity and reaches a maximum in the fixed bed. When the gas velocity is increased further, the pressure drop falls down slightly because of the bed expansion, and then becomes constant independent of the gas velocity in the fluidized bed. The gas velocity at which fluidization occurs is called the minimum fluidization velocity u_{mf}. During the fluidization, the apparent particle weight is equal to the pressure drop as expressed by

$$\Delta P = L(1 - \epsilon)(\rho_p - \rho_f)g \tag{20.1}$$

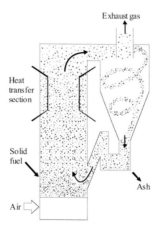

FIGURE 20.1 Schematic drawing of circulating fluidized bed.

[*] Retired

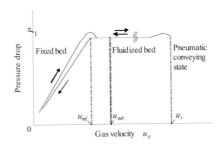

FIGURE 20.2 Typical diagram of a pressure drop versus gas velocity.

where L is the bed height, ϵ is the void fraction, and ρ_p and ρ_f are the densities of the particles and gas, respectively. The minimum fluidization velocity can be estimated by combining Equation 20.1 and the Kozeny–Carman equation as follows:

$$u_{mf} = \frac{\phi_c^2}{180} \cdot \frac{g(\rho_p - \rho_f)D_p^2}{\mu} \cdot \frac{\epsilon^3}{(1-\epsilon)} \tag{20.2}$$

where ϕ_c is the shape factor, D_p is the particle diameter, and μ is the gas viscosity, respectively.

If the gas velocity is increased from u_{mf}, the bed starts bubbling, and the gas velocity at which bubbling first occurs is called the minimum bubbling velocity u_{mb}. Figure 20.3 shows a bubble, which is consistent with regions of low solid density in the fluidized bed. The bubble rises, accompanying the surrounding particles in the wake, and grows by coalescing with other bubbles. If the bubble diameter is almost equal to the bed diameter, the state is called slugging. When the gas velocity is much larger than the terminal velocity u_t, the particles are conveyed by the gas.

Gerdart classified the particles into four groups through careful observations of fluidization using various kinds of particles.[2] The groups A–D are shown in Figure 20.4 by using the particle diameter D_p and the density difference between the particle and gas ($\rho_p - \rho_f$). Geldart C particles are difficult to fluidize because they are fine and cohesive powders. Vibration is sometimes used to help the fluidization of such particles. Geldart A particles having low density ($\rho_p < 1400$ kg/m³) and/or small mean diameter easily fluidize. When the particles are fluidized, the bed expands before bubbling occurs; in other words, $u_{mb}/u_{mf} > 1$. Geldart B particles (1400 kg/m³ $< \rho_p < 4000$ kg/m³ and 40 μm $< D_p < 500$ μm) are sandlike, and for these particles bubbling starts at incipient fluidization ($u_{mb}/u_{mf} > 1$). Geldart D particles are dense and/or large, and bubbles rise more slowly and grow to large size with rapid coalescing.

The fluidized beds of Geldart A, B, and D particles behave like a liquid, having their own density and viscosity.[3,4] Thus, the gas–solid fluidized bed can be used as a separator of different density objects. In the separation technique, the fluidized bed acts like a dry separation medium in a similar manner to the wet medium of dense medium separation.[5,6] When objects are immersed in the fluidized bed, the objects with a density smaller than the apparent density of the fluidized bed float, whereas those of larger density settle in the fluidized bed. Fraser and Yancey first applied fluidized-bed medium separation (FBMS) to coal cleaning.[7] Joy et al. developed a unique separator combining the effects of a fluidized bed and a vibrating table for continuous separation,[8] Coal cleaning with the FBMS has also been investigated in China.[9,10] The FBMS was applied to the separation of agricultural products in the 1980s.[11–13] Oshitani et al. developed a continuous separator and investigated the separation performance of the FBMS for silicastone and pyrophyllite,[14,15] automobile shredder residues,[16,17] and coal cleaning.[18]

The apparent density of the fluidized bed can be controlled by mixing the proper particles into the bed. However, segregation takes place in some combinations of binary particle mixture.

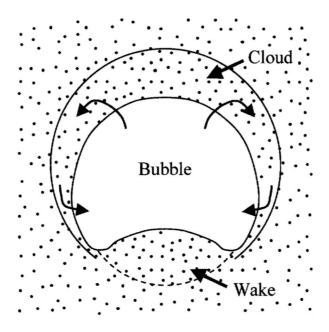

FIGURE 20.3 Bubble observed in fluidized bed.

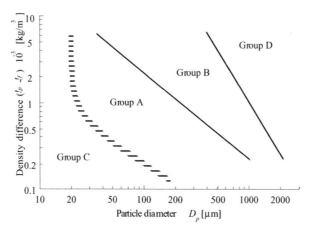

FIGURE 20.4 Particle classification by Geldart (fluidization by air).

Lighter particles (flotsam) float while heavier particles (jetsam) settle in the fluidized bed.[19] Mixing/segregation of a binary particle mixture and the mechanisms of segregation in the gas–solid fluidized bed have been investigated extensively.[20–25]

1.20.2 MOVING BEDS

Type and Application

Iron and Steel

A blast furnace to produce pig iron is shown in Figure 20.5a, and the moving grade shown in Figure 20.5b is applied to the ore sintering as a pretreatment process and also to the dry quenching of coke to recover heat. The simple moving bed shown in Figure 20.5c is also applied to the dry quenching of coke. The type in Figure 20.5d has been developed as a dust collector at a high temperature and is capable of recovering heat simultaneously by inserted pipes.

Cement and Lime

The type in Figure 20.5e is an old standard in which limestone and cokes are alternatively charged. The type in Figure 20.5f has been developed to burn liquid fuels. The cross-current type in Figure 20.5g permits the design of a large scale.

Environments

The cross-current type in Figure 20.5h is used to remove contaminants from the effluent gas by a moving bed of lime or activated carbon. The Herschoff type, in Figure 20.5i has been developed as a calcinator of iron sulfate and applied to an incinerator of swage mud.

Coal

The types in Figure 20.5j,k,l are developed as coal gasifiers.

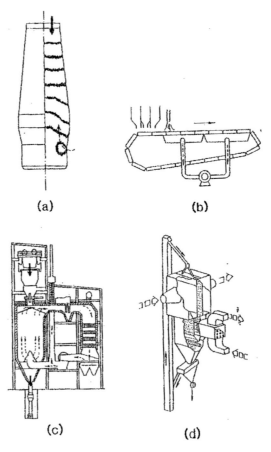

(a)   (b)

(c)   (d)

FIGURE 20.5 Various types of moving beds. [From Association of Powder Process Industry and Engineering, Ed., *Kilns for Processes*, Nikkan-Kougyo-Shinbun, Tokyo, 1985, pp. 6–7. With permission.]

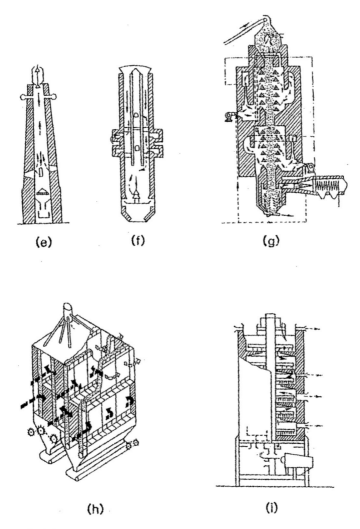

(e)   (f)   (g)

(h)   (i)

FIGURE 20.5 (Continued)

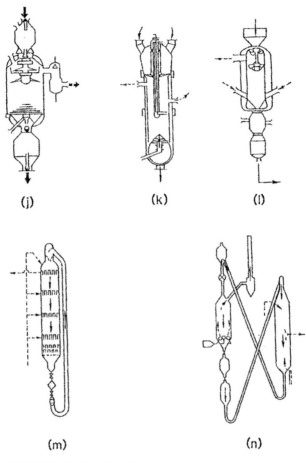

(j)   (k)   (l)

(m)   (n)

FIGURE 20.5 (Continued)

Others

The types in Figure 20.5m,n are used in the petroleum industry.

Treatment of Particulate Materials

Feeding

The formation of a uniform bed is of primary importance because the segregation of particles is apt to take place caused by the difference in size and density. The following methods are recommended to avoid this: (1) multiple legs (Figure 20.5m), a distributing cone (Figure 20.5j,l,m), (3) a rotating bed (Figure 20.5f) or trough feeders (Figure 20.5h), and (4) a vertical pipe having multiple outlets. When isolation from the atmosphere is requested, double or triple dumpers (Figure 20.5j,l,m) or screw types are used.

Discharge of Particles

A uniform moving down is also desired to attain a uniform reaction and to expand the effective volume. The uniformity of the moving bed is broken at any contraction part. To minimize the broken region, the following methods are recommended: (1) installation of a diamond cone or guide vanes (Figure 20.6a,b), (2) insertion of a grizzly sieve or a perforated plate (Figure 20.6c,d), and (3) multiple legs (Figure 20.6e) or multiple plates (Figure 20.6f).

1.20.3 ROTARY KILN

Type and Functions

The standard type is shown in Figure 20.7, and a variation is shown in Figure 20.8. These are quoted from patents for preheaters of cement raw materials having bypaths or dividers.[28] The type

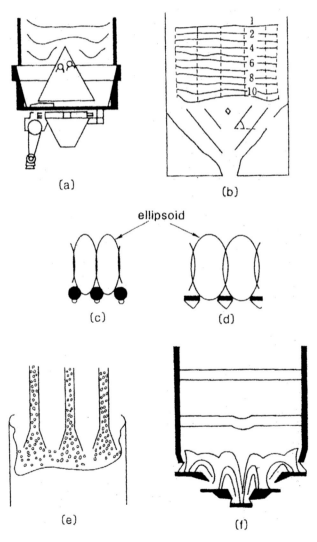

FIGURE 20.6 Techniques to attain uniform discharge. [From Association of Powder Process Industry and Engineering, Ed., *Kilns for Processes,* Nikkan-Kougyo-Shinbun, Tokyo, 1985, pp. 44–48. With permission.]

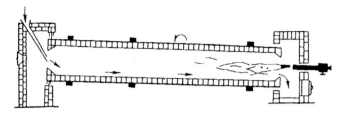

FIGURE 20.7 A standard rotary kiln.

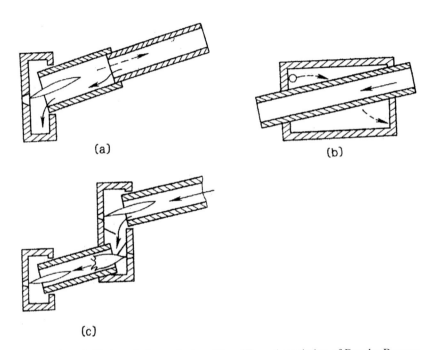

(a)

(b)

(c)

FIGURE 20.8 Various designs of rotary kilns. [From Association of Powder Process Industry and Engineering, Ed., *Kilns for Processes,* Nikkan-Kougyo-Shinbun, Tokyo, 1985, p. 9. With permission.]

in Figure 20.8a, is a coaxial dual-cylinders system with different rotating speeds, and the type in Figure 20.8b is a design of outside heating to isolate the calcining material from the heating gas. Multiple burners in Figure 20.8c are for attaining a uniform temperature distribution.

Retention Time and Bed Depth

Several equations to predict the retention time of particle flow through a rotary kiln have been correlated as shown in Table 20.1.[29] The depth distribution of a particle bed is predicted from the graph shown in Figure 20.9.[30] A uniform depth is attained when the parameter $h_L/RN\phi = 0.193$.

Calcining Process

The advantage of rotary kilns is that they are capable of intensive heating by direct radiation from the flame and uniform calcinations by mixing through kiln action. On the other hand, the

TABLE 20.1 Equations to Predict the Retention Time of a Particle Flowing through Rotary Drums

Reported by	Equation	Remark
Sullivan et al.	$$\tau_1 = 0.03102 \left(\frac{L8^{1/2}}{DSN} \right)$$	Simple equation
Bayard	$$\tau_2 = \frac{0.00308\,(6+24)}{NS^2} \left[\frac{LS-6H}{D} + \frac{12H}{2D-2H-h} \frac{(Hh^{1/2})+H}{h} \right]$$	With rings
Kramers et al.	$$\frac{h_0}{RH_\phi} - \frac{h}{RN_\phi} + 0.193 \ln \frac{h_0/RN_\phi - 0.193}{h/RN_\phi - 0.193} = \frac{L-X}{L} \frac{1}{N_\phi N_K}$$ $$N_\phi = \frac{F \sin\beta}{\rho_S HR^3 \tan\alpha^1} \quad N_K = \frac{R\cos\beta}{\tan\alpha^1}$$	Thickness distribution of particle bed
Vahl et al.	$$\tau_4 = 0.91 \left(\frac{L}{D}\right)^2 \times \left(\frac{D}{h_f}\right)^{0.57} \times \frac{\tan\beta}{N}$$ where D/h_f is given by $$F = 0.74 \left(\frac{h_L}{D}\right)^{2.05} \times \frac{D^4 N \cot\beta - \phi_s}{L}$$	No inclination ($S = 0$)
Baranovskñ	$$\tau_5 = 0.431 \times \frac{L\sin\beta}{DN\alpha^{\bullet}} \times \left(\frac{2\phi - \sin\phi}{\sin^3\phi}\right)$$	Including ϕ
Zablotony	$$\tau_6 = 0.433 \times \frac{L}{DN} \left(\frac{\beta}{\alpha^{\bullet}}\right) 0.85$$	Simple equation

D, inner diameter of rotary drum (m); F, feed rate (kg/min); g, gravitational acceleration (m/min²); h, thickness of particle bed (m); $h_0 = h$ at outlet (m); H, height of ring (m); L, length of rotary drum (m); N, rotational speed (rpm); R, rotation radius (m); S, inclination (m/m); X, volume fraction of particle hold-up; α, inclination of rotary drum (deg); β, angle of repose of particles (deg); x, axial distance (m); ρ_s, bulk density of particles (kg/m3); ϕ, circumferential angle of particle bed (deg); τ, retention time (min).

Source: Toyama S., *Chem. Eng.,* 11, 533, 1966 (in Japanese).

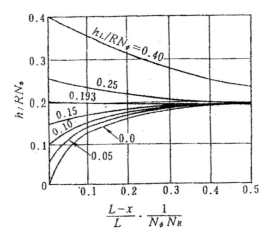

FIGURE 20.9 Diagram to predict a thickness distribution of particle bed. [From Association of Powder Process Industry and Engineering, Ed., *Kilns for Processes*, Nikkan-Kougyo-Shinbun, Tokyo, 1985, pp. 65–72. With permission.]

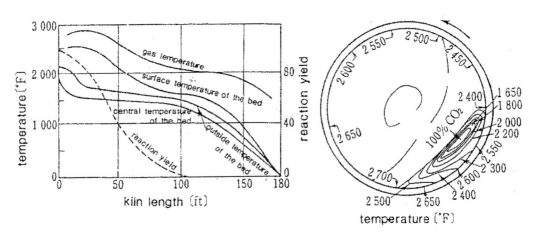

FIGURE 20.10 Axial and cross-sectional distributions of temperature and reaction yield in a rotary kiln for calcining lime. [From Kramers, H. and Crookewit, *Chem. Eng. Sci.*, 1, 259–267, 1952. With permission.]

composition of gas in the kiln cannot be changed to a large extent. Typical temperature distributions in a lime calciner are shown in Figure 20.10.[31]

Mathematical models to simulate their reactions and heat have been abundantly developed.[32,33]

REFERENCES

1. Kunii, D. and Levenspiel, O., in *Fluidization Engineering*, 2nd Ed., Butterworth-Heinmann, Boston, 1991, pp. 15–60.
2. Geldart, D., *Powder Technol.*, 7, 285–292, 1973.
3. Schmilovitch, Z., Zaltzman, A., Wolf, D., and Verma, B. P., *Trans. ASAE*, 35, 11–16, 1992.
4. Bakhtiyarov, S. I. and Overfelt, R. A., *Powder Technol.*, 99, 53–59, 1998.

5. Wills, B. A., in *Mineral Processing Technology,* 6th Ed., Butterworth-Heinmann, Boston, 1997, pp. 237–257.
6. Burt, R., *Miner. Eng.,* 12, 1291–1300, 1999.
7. Fraser, T. and Yancey, H. F., *Coal Age,* 29, 325–327, 1926.
8. Joy, A. S., Douglas, E., Walsh, T., and Whitehead, A., *Filtration Separation,* 9, 532–538 and 544, 1972.
9. Chen, Q. R., Yang, Y., Yu, Z. M., and Wang, T. G., in *Proceedings of Eighth Annual International Pittsburgh Coal Conference, Pittsburgh,* 1991, pp. 266–271.
10. Luo, Z. F. and Chen, Q. R., *Int. J. Miner. Process.,* 63, 167–175, 2001.
11. Zaltzman, A., Feller, R., Mizrach, A., and Schmilovitch, Z., *Trans. ASAE,* 26, 987–990 and 995, 1983.
12. Zaltzman, A. and Schmilovitch, Z., *Trans. ASAE,* 29, 1462–1469, 1986.
13. Zaltzman, A., Verma, B. P., and Schmilovitch, Z., *Trans. ASAE,* 30, 823–831, 1987.
14. Oshitani, J., Kondo, M., Nishi, H., and Tanaka, Z., *J. Soc. Powder Technol. Jpn.,* 38, 4–10, 2001.
15. Oshitani, J., Tani, K., and Tanaka, Z., *J. Chem. Eng. Jpn.,* 36, 1376–1383, 2003.
16. Oshitani, J., Kajiwara, T., Kiyoshima, K., and Tanaka, Z., *J. Soc. Powder Technol. Jpn.,* 38, 702–709, 2001.
17. Oshitani, J., Kiyoshima, K., and Tanaka, Z., *Kagaku Kogaku Ronbunshu,* 29, 8–14, 2003.
18. Oshitani, J., Tani, K., Takase, K., and Tanaka, Z., *J. Soc. Powder Technol. Jpn.,* 41, 334–341, 2004.
19. Rowe, P. N., Nienow, A. W., and Agbim, A. J., *Trans. Inst. Chem. Eng.,* 50, 310–323, 1972.
20. Rowe, P. N. and Nienow, A. W., *Powder Technol.,* 15, 141–147, 1976.
21. Rice, R. W. and Brainovich, J. F., Jr., *AIChE J.,* 32, 7–16, 1986.
22. Hoffmann, A. C., Janssen, L. P. B. M., and Prins, J., *Chem. Eng. Sci.,* 48, 1583–1592, 1993.
23. Wu, S. Y. and Baeyens, J., *Powder Technol.,* 98, 139–150, 1998.
24. Marzocchella, A., Salatino, P., Pastena, V. D., and Lirer, L., *AIChE J.,* 46, 2175–2182, 2000.
25. Formisani, B., De Cristofaro, G., and Girimonte, R., *Chem. Eng. Sci.,* 56, 109–119, 2001.
26. Association of Powder Process Industry and Engineering, Ed., *Kilns for Processes,* Nikkan-Kougyo-Shinbun, Tokyo, 1985, pp. 6–7.
27. Association of Powder Process Industry and Engineering, Ed., *Kilns for Processes,* Nikkan-Kougyo-Shinbun, Tokyo, 1985, pp. 44–48.
28. Association of Powder Process Industry and Engineering, Ed., *Kilns for Processes,* Nikkan-Kougyo-Shinbun, Tokyo, 1985, p. 9.
29. Association of Powder Process Industry and Engineering, Ed., *Kilns for Processes,* Nikkan-Kougyo-Shinbun, Tokyo, 1985, pp. 65–72.
30. Kramers, H. and Crookewit, P., *Chem. Eng. Sci.,* 1, 259–267, 1952.
31. Takatsu, M., *Yogyokyokai-si,* 73, C359–363, 1965.
32. Elgeti, K., *Chem. Eng. Technol.,* 25, 651–655, 2002.
33. Martins, M. A., Oliveira, L. S., and Franca, A. S., *Zement-Kalk-Gips,* 55(4), 76–87, 2002.

1.21 Simulation

Charles S. Campbell
University of Southern California, Los Angeles, California, USA

Ko Higashitani
Kyoto University, Katsura, Kyoto, Japan

Yutaka Tsuji
Yokosuka Research Laboratory, Central Research Institute of Electric
Power Industry, Yokosuka, Kanagawa, Japan

Toshitsugu Tanaka
Osaka University Suita, Osaka, Japan

Shinichi Yuu
Ootake R. & D. Consultant Office Fukuoka, Japan

Kengo Ichiki
The Johns Hopkins University, Baltimore, Maryland, USA

Yoshiyuki Shirakawa
Doshisha University, Kyoto, Japan

1.21.1 COMPUTER SIMULATION OF POWDER FLOWS

This review will describe computational techniques for simulating the flows of powders or granular materials. These techniques simulate systems consisting of discrete particles within which each individual particle is followed as it interacts with other particles and with the system boundaries. In effect this involves simultaneously integrating all the equations of motion for all the particles in the system. This may appear to be a massive task but has been accomplished for several million particles.

This technique is especially useful for studying granular flows due to the difficulty of obtaining detailed results from direct experimentation. That difficulty lies in the relative lack of instrumentation that can operate in a dense flowing particle environment. For example, as a result of the large concentration of particles, granular flows are opaque to optical probing. Ultrasonic probing has been suggested as a possible alternative but has not yet been developed into workable instruments. But even the intrusive probes that have been developed give only limited information. The computer simulation techniques described in this chapter are used to set up simulated flowing powder systems upon which "experiments" are performed. These experiments take the form of averages over the properties of the system. The advantage of a computer simulation is that literally everything about the simulated system is known and is accessible to the computer experimenter. The techniques are fairly basic, and the art of using these computer simulations lies not in setting up the programs but in knowing how to interrogate the systems to provide whatever information is of interest.

These simulations are sometimes called the discrete or distinct element method (but always DEM), a term coined by the originator of the technique, Peter Cundall. That term has come to be used for many different numerical techniques (and was not very descriptive in the first place), so they will simply be referred to here as "computer simulations." The origins actually lie in the field of molecular dynamics, which was the first field to mechanistically follow the motions of molecules under the influence of external forces (e.g., work by Alder and Wainwright[1]). Cundall[2] was the first to use this type of model to study granular flows. Since then the use of these models has become widespread—too widespread to include an exhaustive bibliography of all applications in the space of this review.

The simulations described here will assume that the flow occurs in the absence of an interstitial fluid. The interstitial fluid is ignored because it is not yet possible to model the motion of the fluid with the same degree of accuracy as it is to model the motions of the particles that only respond to contact and body forces. As such, these techniques are applicable to cases in which the interstitial fluid does not play a significant role in determining the overall mechanics of the system. Such is the case when the majority of the forces experienced by the particles are due to the solid–solid contacts with their neighbors and with the system boundaries, and are not strongly influenced by any fluid that might fill the interparticle gaps. Some simulations have added interstitial fluid effects in an approximate manner (e.g., work by Tsuji and Kawaguichi,[3] Haff et al.,[4] and Potapov et al.[5]), but these will not be discussed in detail here.

These computer simulations can roughly be divided into two types, rigid- and soft-particle models, which because of the differences in the ways that the particle interactions are modeled, follow different computational algorithms. Other reviews of computer simulation techniques can be found in work by Campbell[6] and Herrmann and Luding.[7]

Rigid-Particle Models

The stress levels in most granular flows are usually small enough so that the particle surfaces will not elastically deform to any significant degree. It is then a reasonable approximation to assume that the particles are perfectly rigid and cannot deform at all. But perfectly rigid particles have infinite elastic moduli, so that any collision between such particles must occur instantaneously. At their essence, rigid-particle models are synonymous with assuming that all particle interactions are instantaneous collisions. (Similar assumptions are made in all theoretical calculations of granular flow in the so-called rapid-flow regime; see the review by Campbell.[8]) A convenient byproduct of this assumption is that simultaneous collisions between three or more particles will occur with zero probability so that only two-particle or binary collisions need be accounted for. Examples of rigid-particle simulations can be found elsewhere.[9–17]

The assumption of instantaneous collisions dictates the way that the simulation evolves through time and consequently determines the way that the simulation algorithm is structured. Between collisions the particles follow simple kinematic trajectories, which in gravity-free conditions are just straight lines but become parabolic when gravity is present. These trajectories only change as the result of collisions, and a particle's trajectory between collisions is explicitly defined by the result of the particle's last collision. Thus time in the simulation is most efficiently updated from collision to collision, as the state of the system between collisions can be easily determined if needed. (Thus, these are sometimes referred to as "event driven" simulations.) The algorithm is shown in Figure 21.1 and proceeds as follows: When the simulation is started, the time at which the first collision occurs is computed from the particle trajectories. The positions and velocities of all the particles are updated to that time. The collision result is computed, the time of the next collision to occur is found, and the process is repeated. The collision result is determined from a standard center-of-mass collision solution of the type learned in elementary physics classes. (An excellent implementation of a rigid-particle collision solution is described in detail by Walton.[18]) As the particle is assumed to be infinitely rigid, elastic properties do not appear in the equations, and the collision solution is determined solely from

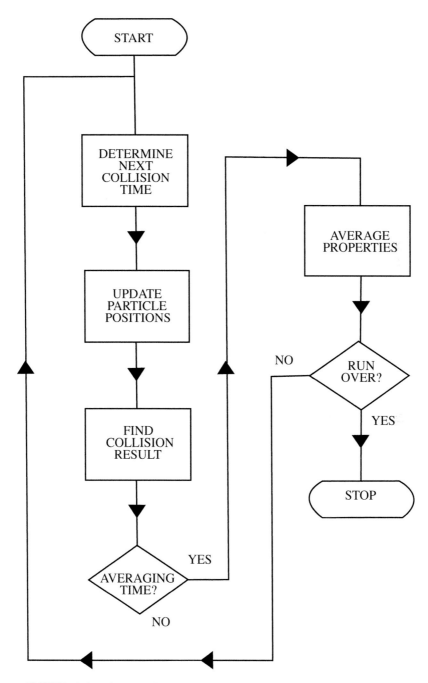

FIGURE 21.1 A flowchart illustrating the rigid-particle algorithm.

momentum and energy considerations. Inelasticity is typically introduced through a coefficient of restitution, which may be constant or a function of the collisional velocity. The particle velocities in the tangential direction are typically coupled frictionally.

The rate at which time progresses in a rigid-particle simulation is inversely proportional to the collision frequency, making this procedure extremely efficient especially at low solid concentrations where collisions are infrequent. As the concentration becomes large, collisions occur so frequently that rigid-particle models become very inefficient. As a further consequence, rigid-particle models cannot be applied to any situation involving stagnant zones or other situation where particles are in contact for long periods of time. Such regions are common in real granular systems, and when they occur, rigid-particle simulations will try to model the long-duration contacts as a number of rapidly occurring instantaneous collisions; as a result, the collision frequency goes to infinity and the time between collisions goes to zero. At that point, the simulation time, which progresses from collision to collision, cannot change, and the simulation effectively stops. Furthermore due to the particle inelasticity, particles form clusters,[19] which eventually may lead to situations where particles in the clusters come into continuous contact. This process has recently been called "inelastic collapse,"[20] and there are approximate methods for avoiding its occurrence. For example, Campbell[9] noted these events occurring for particles that were trying to roll along bounding walls. He solved the problem by designating a particle as "rolling" when collisions with the wall became too frequent. Luding and McNamara[21] used a similar technique, which removed dissipation from that collisions that became to frequent; while this eliminated the inelastic collapse, it in no way approximates the dynamics of a long-duration contact.

Soft-Particle Models

The most realistic solution to this problem is to simply allow long-duration contact between particles. To do this in a consistent manner means that the particles cannot be perfectly rigid, and therefore any contact will be of finite duration. This means that an allowance must also be made for a particle to be in contact with several particles simultaneously. These "soft-particle" simulations were originally developed by Peter Cundall[2,22] and applied primarily to quasi-static situations. (Although to demonstrate its utility, the model was used[2] to simulate flow problems such as the emptying of a hopper.) For reasons that will soon become clear, this is the most common simulation technique and has been used by many other investigators.[23–36]

As before, the structure of the simulation is controlled by how the particle's material is modeled. Most flowing systems are not subject to tremendous stresses (certainly not when compared to geological systems). Consequently, the interparticle forces are generally too small to cause significant changes in the shapes of the particles, and for computational ease, the particle shape in a soft-particle simulation is not allowed to change. Instead when particles come into contact, their surfaces are allowed to overlap somewhat, and the degree of overlap determines the magnitude of the elastic force at the contact. Generally, the force generated normal to the contact point is modeled as a simple spring. (For cohesionless particles, the spring is not allowed to support any force in tension and is eliminated as soon as the particles lose contact.) For most models, a linear Hookean spring model is used, although it presents few problems to use a nonlinear spring. In granular flows, each collision must dissipate energy. In soft-particle models, this is usually accomplished by connecting a viscos dashpot in parallel with the spring. A constant dashpot coefficient coupled to a linear spring can be shown to be the equivalent to a constant coefficient of restitution. To represent the forces generated in the direction tangential to the point of contact, a spring (occasionally coupled with a dashpot) is also employed. In addition, a frictional coupling is used to link the surfaces of the two contacting particles together in the tangential direction that allows the surface to slip tangertially when the frictional strength is overcome. The ubiquity of this particular configuration is probably due to its use in the original model of Cundall.[2]

Each contact exerts both a force and moment on each of the particles involved. The total force and moment on a particle are the sum of those applied by all its contacts possibly combined with a body force such as gravity. The subsequent motion of the particle is governed through Newton's second law by an ordinary differential equation, and the motion of the entire bulk material is determined by the simultaneous solution of all the differential equations for all of the constituent particles. The solution is readily accomplished using any numerical method that is capable of solving systems of ordinary differential equations; however, the integration must be carefully performed,

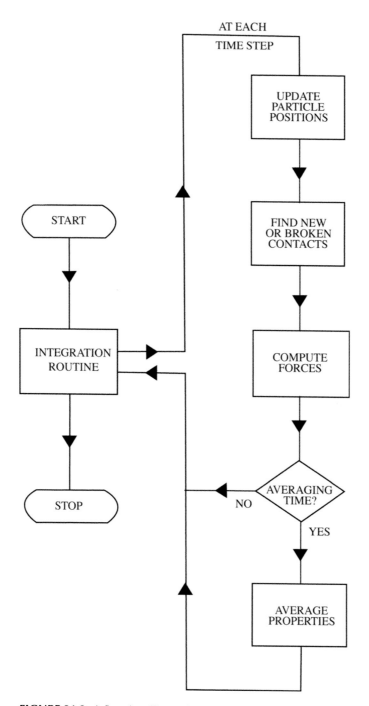

FIGURE 21.2 A flowchart illustrating the soft-particle algorithm.

as the forces on each particle and consequently the differential equations they obey will vary drastically as contacts with other particles are made or broken. The soft-particle algorithm is shown in Figure 21.2. In essence it is simply a numerical integration; however, at each time step, it is necessary to check for new and broken contacts and to compute the interparticle forces. Furthermore the stimulation properties are sampled and averaged at each time step to yield whatever quantities are of interest. The size of the time increment by which the integration routine proceeds should be chosen so that a typical binary collision would occur in 30 to 50 time steps. (On top of pure numerical inaccuracies, there are other sources of inaccurate collision results if very large time steps are chosen. For example, if two particles come into contact with a large relative velocity, too large a time step may allow erroneously large initial overlaps to occur within that first time step, which results in an exaggerated particle response and a net addition of energy to the granular system.) But on the positive side, the soft-particle algorithm also permits contacts to exist for arbitrarily long duration, allowing the simulation of static situations; little computational effect is expended in the actual integration of the equations of motion, hence there is little to be gained in using elaborate integration techniques. (Typically second-order algorithms are used.) Most of the time is taken up in updating the positions and forces on the particles at each time step and, in particular, checking for new or broken contacts between particles.

In binary collisions of linear spring contacts, the collision time will vary as the square root of the ratio of spring constant to particle mass (and is independent of the impact velocity). As mentioned in the last paragraph, the size of the time step is a specified fraction of the binary collision time and is thus controlled by the smallest mass and stiffest spring in the granular system. The computational time required to run a given simulation is not dependent on the collision rate (as it is for rigid-particle models) and is therefore independent of the particle packing. Hence, soft-particle models are computationally inefficient at small concentrations where collisions are infrequent, and most of the computer time will be spent in updating the particle positions as they move unimpeded along their kinematic trajectories. But the simulations become more efficient at larger concentrations when collisions are frequent.

Ironically, it is also at the largest densities that the soft-particle models may run into trouble. This is because an improved computational efficiency is often obtained by choosing an extremely soft spring in the interparticle force model (one that is much softer and permits larger overlaps than would be found in common granular materials). This lengthens the average collision time and with it the time step required to accurately represent a given collision. At small particle concentrations, this should not be a problem, but at larger concentrations, the "excluded volume" taken up by the particles causes a singular behavior in many of the material transport properties within an actual granular material. If the simulated particles are inordinately soft, the overlaps will be large, causing a reduction in the excluded volume and a dampening of the singular behavior, resulting in significant reductions in the measured values of the transport properties. Furthermore, the extended contact time permits a greater number of simultaneous multiparticle contacts to occur than would be encountered in a system with a more realistic particle stiffness. Typically, the stiffness is made large enough so that the particle overlaps do not exceed a specified value (usually 1% of a particle diameter).

Recently, another limit on the stiffness has become apparent. Campbell[24,25] has shown that in shear flows, materials can undergo a regime transition to elastic-inertial at either large shear rates γ, or small stiffness, k. The transition occurs at small values of a parameter k/pd^3y^2 (where d is the particle diameter and p the solid density). Thus, inordinately soft particles may operate in a different flow regime than particles with more realistic stiffnesses. However, flow maps given by Campbell[24,25] indicate the regime transition point so that they may be avoided. Shear flows at fixed concentration were studied by Campbell.[24] Realizing that the majority of granular flows occur in situations where the stress and not the volume is fixed (for example, by a material overburden)[25] repeats those studies for fixed-stress cases where the volume is allowed to change slightly to keep the stress fixed. The results show that fixed-stress flows were shown to behave very differently from fixed-concentration flows and follow a different set of flowmaps. Thus simulators must be careful to use the appropriate flowmaps to avoid unintentional flow regime changes.

One advantage of the soft-particle method is that there is no limitation on how the particle's material is modeled or in the nature of the forces to which the particle is subject. This may be very important in future simulations of systems with very soft plastic materials or for fine powders where long-range electrostatic or van der Waals forces are important. Langston et al.[30] describe a model with a continuously varying normal force that actually extends slightly beyond the outer edges of the particles; but this is done mostly to enhance the computational efficiency of the simulations.

Rigid- vs. Soft-Particle Models

It has long ago become apparent to those that work in the field that soft-particle simulations are preferential to rigid-particle simulations. The major reason is not just because soft-particle codes are capable of handling static assemblies of particles and of incorporating more elaborate collision models, but rather a result of the way that the execution time scales with the number of simulated particles. Time in a rigid-particle simulation updates from collision to collision, and the rate of time progression is dependent on the collision rate. However, in soft-particle simulations, time progresses with the integration time step, which is determined by the particle stiffness, making the time progression independent of the collision rate (although that time step is a fraction of the particle contact time). Rigid-particle models had definite advantages when computers were slow and the number of particles that could be simulated were small. In such cases the time between collisions is large (especially when the concentration is small), and a rigid-particle simulation will progress much more rapidly than a soft-particle simulation. However, as the number of particles becomes large, the time between successive collisions goes to zero. In contrast, the progress of a soft-particle simulation depends on the integration time step and is independent of the number of particles in the simulation (although the computational demands that need be performed at each time step rise roughly proportionally to the number of particles). The break-even point between the two simulation types occurs when approximately one collision will occur in the entire simulated system in one soft-particle integration time step. It is clear that at the large concentrations that are typically encountered in industrial simulations (usually more than 50% by volume), the break-even point will be less than a thousand particles. As the future of granular simulations lies in simulating large systems of particles (a million and more), the soft-particle method is clearly preferential.

However, the work by Campbell[24,25] has added another reason to abandon rigid-particle models, at least for flows at concentrations above 50% by volume, where most common granular flows occur. At such concentrations stress is transmitted through the material through "force chains," highly loaded structures that form internal to the flow and carry the majority of the stress (see, e.g., work by Cundall and Strack,[22] Drescher and De Josselin de Jong,[37] Mueth et al.,[38] Howell et al.,[39] and Howell et al.[40]). Force is carried along the chain by elastic deformation of the particle contacts, and the force generated is proportional to the stiffness, k, of the contacts. Such behavior cannot be modeled by rigid-particle models, as they implicitly assume that $k = \infty$ and thus would always predict infinite stresses in any elastically deformed force chains. Campbell[24,25] has shown that the stiffness, k, is an essential rheological parameter, thus putting solid properties into the rheology of granular solids. Such rheological effects can only be handled by soft-particle models. As a result, soft-particle simulations are required to accurately model the vast majority of common granular flows.

Approximate Simulation Techniques

There have been several approximate algorithms (ones that do not exactly follow the motion of the particles and/or the interparticle forces) employed to study powder shows, usually with the goal of improving the efficiency of the simulation.

Hopkins[19,41] developed a hybrid technique that attempts to exploit the advantages of the both rigid- and soft-particle models. In essence, it employs a rigid-particle interaction model but marches time with a fixed time step, allowing collisions to occur whenever overlaps are detected at the end of a time step. This increases efficiency, as the time step is not limited by material properties and can be much larger than for a soft-particle simulation. However, only one collision between any two particles is permitted at each time step, and thus the time step effectively sets an upper limit on the collision rate. As a result, such a model is only advantageous at small solids concentrations. Where the collision rates are small a large time step is permissible.

The earliest molecular computer simulations (e.g., work by Metropolis et al.[42]) were not true dynamical simulations but made use of the apparently random nature of the molecular motion in gases to approach the problem from a statistical point of view. The goal was to derive distribution functions from which transport properties could be determined from integral relations derived for the theory of nonequilibrium gases (see, e.g., work by Chapman and Cowling[43]). Implicit in this method is Boltzmann's "Stosszahlansatz" or assumption of molecular chaos. Briefly this is the assumption that the velocities of individual particles are independently distributed within a velocity distribution function without regard to history or to the behavior of neighboring particles. This will clearly not be the case in dense concentrations where particles will experience long-duration contacts with their nearest neighbors, and there will be strong correlations between the positions and velocities of neighboring particles. Thus Monte Carlo methods have limited usefulness in granular flow studies. Nonetheless, molecular chaos is an essential assumption in the theory of rapid granular flows,[8] so that Monte Carlo simulations do provide a forum[44,45] for evaluating rapid-flow theories within the context of their basic assumptions.

Lattice gas models have been a popular computational technique for studying gas flows, and it should not be strange that they should be adapted for granular flows. There have been several realizations of this technique (e.g., work by Baxter and Behringer,[46] Gutt,[47] and Désérable[48]), so that only a generalized picture will be given here. Basically the particles in a "lattice grain" model are confined to fixed and regularly spaced lattice points in space. Only one particle may occupy any given point. At each "time" step, particles may move into any of the neighboring lattice points, provided that they are vacant. However, the probability of a given movement is limited by potentials acting on the systems; for example, under gravity a particle is much more likely to move to a lattice point located below it, than to ones located to the side or upward. Walls and flow obstructions can be inserted by simply eliminating lattice points along the boundaries and thus prohibiting the motion of particles in those directions. The best that can be said about this type of model is that it is kinematically correct. There are no dynamics in the simulation scheme, no forces act on the particles (as described above, gravity acts only to affect the directional probability of motion and is not an acceleration), and the particles exert no forces on each other. Consequently, no information about forces can be obtained, and it is not clear that the simulation is responding in an accurate manner to the interaction with other particles or the bounding walls of the system. In addition, velocities are limited by the lattice spacing and the time step in the sense that the fastest a particle can move is one lattice spacing in one time step. In that sense, the velocities of the particles are really meaningless. Thus the information that can be obtained from these simulations is very limited.

Nonround Particles

Most of these simulations have been performed for round particles. This is largely because it is easy to detect contacts between round particles, as contact occurs when the particle centers are the sum of their radii apart. When particles are not round, the orientation and local shape of the particle become important in the contact decision, and that is computationally difficult to handle.

Yet particle shape certainly has a strong effect on granular flows, both due to the ability of angular particles to interact and form strong internal structures and simply by the added resistance to rolling. Convincing evidence can be easily seen in the angle of repose. For round particles, it is difficult to obtain angles of repose of more than a few degrees and much less than the 30° angles of repose typical of granular materials. One solution to this problem is simply to prevent the particles from rolling so that they only interact frictionally in the direction tangential to the point of contact. This allows one to build sand piles with realistic angles of repose but also can produce significant nonphysical effects such as stable columns of particles that would clearly collapse were the particles capable of rotation. Also the surface friction values required to produce realistic angles of repose are, themselves, unrealistically small. For example, it requires a surface friction coefficient of only about 0.2 to produce a 30° angle of repose. Thus, even though the angle of repose may be realistic, the internal particles interactions are not.

The most natural solution for this problem is to use particles that are not round. Several such models have been proposed. In general they can be divided into two classes. The most general technique is to choose particles that are polygonal (e.g., work by Hopkins,[49] Hogue and Newland,[50] and Kohring et al.[51]; those utilized in fracture simulation simulations of Potapov et al.[52–54]; and for three-dimensional polyhedral particles, Cundall,[55] Ghaboussi and Barbosa,[56] and Potapov and Campbell[57]). Fairly realistic looking particles may be created in this manner, but at tremendous computational costs. Polygonal or polyhedral simulations can take orders of magnitude more computer time than their round counterparts. The majority of the computer time is spent in determining contact between particles and in calculating intersections of the sides of contacting polygons, which is a necessary part of the overlap determination.

A computationally more efficient technique is to use shapes for which an analytic expression exists. For example, algorithms have been developed for particles with elliptic or ellipsoidal shapes.[58–60] These can be generalized to a large variety of shapes known as superquadrics (e.g., work by Mustoe and DePooter[61]). These are described by equations similar to ellipsoidal particles, except with exponents larger than 2, and can have shapes that approach polygons, although the corners stay rounded. Due to their angular nature, such particles can interlock, but they do not resemble natural materials. Also, aspect ratios for elliptical particles must be significantly different from unity in order to obtain realistic angles of repose which may present other problems, as the majority of natural granular materials have aspect ratios that are close to one. For ellipsoidal or superquadric particle shapes, the majority of the computer time is spent in iterative solutions of the nonlinear equations used to calculate the particle overlap.

More computationally efficient techniques have recently been developed. Potapov and Campbell[62] built two-dimensional particles out of circular sections. Particles of arbitrary convex shapes can be constructed in this way, and the particle overlap and contact determination with an edge is essentially the same as for circular particles. Additional complications arise from determining which circular segment is involved in a contact, but the execution times are only about twice as long as for their round counterparts. Unfortunately the technique could not be generalized to three-dimensional particles. Ellipsoidal particles were built[31] out of four overlapping spherical particles. As in Potapov and Campbell's[62] work, the contact and overlap determinations are essentially the same as for spherical particles, although one must potentially check four sphere contacts for each quasi-ellipsoidal particle. Also there are complications in determining the moment of inertia of the composite particles, since the spheres overlap and thus their properties are not the same as four separate spheres. Another possibility is to create agglomerates[63,64] of circular or spherical particles "glued" together at the contacts of their surfaces. This gives less choice in particle shape but is easy to implement and requires no special calculation of moment of inertia and so forth; the intraparticle force transmission is accomplished by the forces transmitted across the glued joint between the particles.

It is unlikely that a completely accurate model of particle shape could ever be incorporated into computer simulations. Not only is it difficult to model the many complex shapes of real granules, but one would also have to take into account the way that the shapes change due to handling (e.g., as angular corners are broken off). So one should attempt to create an adequate rather than an exact model, one that captures the essential physics and is easy to implement. This is still an open question, but hopefully, some of the similar, more computationally efficient models will surface in most cases.

Particle Interaction Models

From the above discussion, it is clear that the soft-particle models are by far the most useful, and as the power of computers continually improves, most of the advantages of the rigid-particle and other approximate models will disappear. All soft-particle models are built upon assumptions about how particles interact with one another and with solid boundaries of the system, and how they react to any driving force such as gravity. All the approximations in the models lie in these basic assumptions, and thus interaction models deserve a place in any discussion of granular simulations.

All particle interaction models must contain three basic components: (1) a mechanism for generating a normal force at the contact point that works to separate the particle surfaces, (2) some mechanism for dissipating the collisional energy, and (3) some frictional interaction that acts tangential to the particle surfaces. But the appropriate way of modeling these forces still involves a great source of controversy (see, e.g., work by Walton,[18] Thornton and Randall,[65] and Tüzün and Walton[66]).

As mentioned previously, the forces applied in the normal direction of soft-particle models are usually assumed to act as linear springs and generate a force proportional to the overlap between the particles. Generally, that spring is connected in parallel to a linear dashpot to provide the energy dissipation. Constant spring and dashpot coefficients can be shown to correspond to constant coefficients of restitution. Nonlinear dashpots have also been proposed. Walton and Braun[34,35] used a "latched spring" model that loads with one spring constant and unloads with another as a way of incorporating the energy dissipation; they found this to be closer to the results of elastic–plastic modeling of the impact of round particles.[18,67]

Hertz[68] solved for the interaction between linearly elastic particles with arbitrary radii of curvature and found a nonlinear response that stiffens as the particle deforms. Because Hertz produced a nearly exact solution, one is tempted to believe that it should be *the* correct model for interparticle contacts. But there is a great deal of evidence that this is not the case for many granular solids. One such source lies in the elastic-particles analyses[18,67] mentioned in the last paragraph. The interaction of the elasticity and the plasticity generated a nearly linear response to displacement, at least when the particle's yield strength had been exceeded. But there is some indication that the Hertz theory may fail in granular materials even when the particles are behaving elastically. The evidence comes from measurements of the contact stiffness[69] and of sound speed in granular materials (see the discussion by Goddard[70]). From a Hertzian point of view, the bulk elastic modules of a granular material should increase with the applied pressure. This occurs because the higher the contact force, the more the particle is deformed, and the stiffer the contacts between particles. Hertz predicts a repulsive force that varies as the displacement to the three-halves power. That corresponds to a bulk elastic modulus that varies as the cube root of the applied pressure. But sound speed measurements indicate that the modulus varies more like the square root of the pressure, indicating non-Hertzian behavior. In other words, the experimental results are not in agreement with the theoretical, although the reason for this discrepancy is not certain. Goddard[70] has attributed this to the fact that the applied forces are often too small to deform the macroscopic shape of the particle (which is assumed to be the case in Hertzian theory), but instead interact across and only deform the asperities that project from the surface. Only for heavily loaded contacts will the asperities be flattened so that the macroscopic shape of the particles deforms and Hertz theory becomes valid. From Goddard's observations, one would anticipate using a spring where the generated force is quadratically dependent on the displacement for small displacements but demonstrate Hertzian behavior at large displacements. Direct measurements[69] support this general notion, although they demonstrate a more complicated behavior at small loadings, but also approach Hertzian behavior at large loadings. Some of the tangential force measurements support the idea of contact through asperities. To further confuse the issue, Drake and Walton[71] (Walton is one of the authors of the article by Mullier et al.[69]) reinterpret the large stress data in the article by Mullier et al.[69] and argue that it plots better as a linear spring. It should be noted that it is not difficult to place a nonlinear spring in any existing model, the only difficulty is deciding which nonlinear spring to use and whether a nonlinear spring is justified.

In many cases, however, the choice of a normal force model is moot. For example, in a true rapid granular flow, only binary collisions occur, and the collision solution can be completely determined from momentum and energy considerations (as is done with a rigid-particle simulation). Thus, the collision result is independent of any assumptions about the nature of the particle stiffness. This seems to be validated by flow simulations[34] which found little difference in the behavior of simulations with linear and Hertzian contacts. On the other hand, if the concern is the propagation of elastic waves through a granular material, then the stiffness model is critical in correctly modeling the wavespeed and, given the discrepancy between the theoretical and experimental results alluded to above, would probably have to be determined experimentally. Campbell[24] showed that the contact model was important in his force-chain-dominated "elastic" flows and suggested a preliminary scaling behavior for nonlinear springs. How important the contact model is for the areas lying between these extremes has yet to be determined. To sum up, we have three conflicting models for the particle interactions. Hertz's model, which may only be good for large deformations if that, Goddard's model, which predicts a different behavior at small deformations, and Walton's elastic–plastic model, which shows linear behavior. On top of all that, it is not clear under what flow conditions the differences between these models are significant. With the current confusing state of affairs, it appears that the usual linear spring is as good as any for studying flowing systems unless there are careful measurements that indicate the contrary for the particular material involved.

In much the same way, the linear dashpot commonly employed to introduce inelasticity into the simulation is clearly not a representation of the true physics of any contact. It has been used largely because it is a simple way to add inelasticity into the contact model, and a linear dashpot coupled with a linear spring produces a constant coefficient of restitution which can be easily compared to theoretical rapid-flow models which make similar assumptions. But it has also been shown that the coefficient of restitution is far from a constant but, instead, is a steadily decreasing function of impact velocity.[72,73] This reflects the increasingly larger plastic damage and plastic energy dissipation that accompanies larger impact velocities. Walton and Braun[34,35] utilized an approximation to this inelastic behavior in their computer model. While it is clear that this effect may be important, especially for rapid granular flows in which large relative velocities might be encountered, it is also clear that assuming an invariant from for the energy loss as a function of velocity introduces its own source of error. A clue to this can be found in an article by Raman,[72] who states that in order to obtain repeatable results it was necessary to polish his test spheres between measurements. The implication of this remark is that the plastic damage from earlier tests alters the surface properties and changes the particle's response to subsequent impacts. If such effects were to be incorporated into computer models it would be necessary to follow the impact history of every location on the particle surface—and even then one would have to have a good model for characterizing how each impact affects the local surface properties.

The case is similar, but even more difficult, for incorporating frictional models and tangential compliance into the computer models. The frictional equivalent to Hertzian contact theory would be the very detailed work of Mindlin and Dereiewicz,[74] which assumes Hertzian behavior in the normal direction and predicts complex hysteretic behavior in the tangential direction whenever the frictional limit has been exceeded (i.e., when there is slip between the particle surfaces). Thornton and Randall[65] have incorporated an approximation to Mindlin's theory into their model. But even more complicated models have been proposed.[66] However, problems with using Mindlin's theory appear from the very start. Measurements by Sondergaard et al.[75] indicate that the particle surface friction may not be a well-defined property, but vary statistically about some mean value. This is easy to understand if one assumes that the friction is strongly affected by microscopic surface asperities that are not evenly distributed about the particle's surface. Except under very intense loadings, the contact area between two particles will be very small and involve only few asperities; consequently the frictional behavior detected by Sondergaard et al. reflects the significant local frictional variations about the surface of the particle. Conversely, friction measurements on different materials[76] obtained a constant friction coefficient. Along the same lines, measurements by Maw et al.[77] seem to demonstrate a well-defined frictional behavior for steel, but a large degree of scatter for similar experiments performed on rubber. None of this should be surprising, as surface friction is, after all, a material property, and whether it has a constant value or is statistically distributed over the surface also appears to vary from material to material. In any case, it would seem to be fruitless to incorporate a complicated theory like Mindlin's into the model if one cannot even be certain of the value of the surface friction coefficient.

Compounding the confusion are the surface frictional measurements by Mullier et al.[69] These found strong inconsistencies between the experiments and Mindlin's theory. The odd measurements were traced to observations of plastic alteration of the surface characteristics that accompanied frictional yielding; in fact the asperities that strongly affect the frictional behavior were observed to have been sheared off the surface. (This can be seen quite clearly in before and after SEM photos that are presented in the paper.) As a result the frictional behavior has changed between the loading and unloading phases. The same effect may be qualitatively experienced by rubbing large glassbeads in one's hands after they have been run through a flowing experiment a few times and comparing the feel to that of fresh glassbeads; it is clearly apparent that surface properties have changed even long before there is any visible change in the surface characteristics. This would seem to indicate that a truly accurate simulation would somehow have to incorporate these changes and implies that once again it would necessary to follow the detailed loading history of each portion of the particle surfaces and to have some model for how that history affects the surface frictional properties in addition to how it affects the normal response.

The upshot of this whole section is that it is currently impossible to incorporate anything approaching realistic particle interaction into the simulations, especially considering the lack of knowledge of the effects of plastic deformation working on the particle surface properties. Given that, it seems unnecessary to include complex interaction models into these simulations—at least until the problems are better understood, or that detailed experiments exist that show that current simple interaction models lead to inaccurate results.

Conclusions

This chapter has several methods, both approximate and nearly exact, for creating computer simulations of granular flows. These days, the approximate models are of limited usefulness, and only the soft-particle models should be seriously considered as an investigative and design tool. This is for a variety of reasons, including an increase in computational efficiency when modeling dense flows and large systems of particles. However, the most important reason for using soft-particle models is that the interparticle stiffness has recently been shown to be an important rheological parameter, which cannot be modeled by rigid-particle or other approximate techniques.

How good are even these computer simulations? In the section on particle interaction models, I may have left the reader with the impression that these models have inherent inaccuracies arising from approximations in the material modeling, and that it is nearly hopeless to expect the development of contact models that accurately mimic natural particles. That is the best view of the future as seen from the present, but again the same is true of most models and techniques that are commonly employed for engineering analysis and design. Consider for example an industrial material such as coal which usually has particles of sizes ranging from submicron to several centimeters in size, that are angular in shape and are brittle and thus will break into smaller pieces while being handled. The flow of coal is thus very complex, especially as the bulk material properties will change when the particles break. But I think I can say with confidence that the soft-particle computer simulations described herein, even those that make the most gross material approximations such as linear springs and dashpots, describe the flow of coal as accurately as laboratory experiments performed on "nice" materials such as glassbeads that share few of the properties of coal. Furthermore, because these simulations at least approximate the actual mechanics of the system, they must be much more accurate than other commonly used engineering models such as $K–\varepsilon$ models for turbulence, or two-fluid models for multiphase flows. In such company, this type of computer simulation stands out as a most promising technique for guiding industrial design, especially as the rapid increase in computer power allows the simulation of progressively larger systems.

1.21.2   BREAKAGE OF AGGREGATES

When a three-dimensional large aggregate of arbitrary shape composed of N spherical particles of radius a and density ρ_p is placed in a flow, the hydrodynamic drag force and torque act on the outside particles exposed directly to the flow and are propagated into the inside particles through interactions between constituent particles. This will result in the deformation and breakup of the aggregate.

In the present model, the total force and torque on each particle is evaluated at time t, and then the trial displacement at $t + \Delta t$ is estimated using the discrete element method (DEM). Repeating this procedure for all the constitutive particles, the kinetic behavior of the whole aggregate is simulated. The translational and rotational motions of a particle i in the aggregate are expressed by the following equations:

$$m\frac{du_{pi}}{dt} = F_{hi} + \sum_j F_{mij} \tag{21.1}$$

$$I\frac{d\omega_{pi}}{dt} = M_{hi} + a\sum_j F_{mij} \times n_{ij} \tag{21.2}$$

where m $(= (4/3)\pi\alpha^3\rho_p)$ and $I (= (8/15)\pi\alpha^5\rho_p)$ are the mass and moment of inertia of a particle, respectively, u_{pi} and ω_{pi} are the velocity and angular velocity of particle i, F_{hi} and M_{hi} are the hydrodynamic drag force and torque, respectively, F_{mij} is the mutual interaction force imposed on the particle i by the particle j, $n_{ij} = r_{ij}/r_{ij}$, $r_{ij} = (x_i - x_j)$, and x_i is the position vector of the center of particle i.

Because the flow field around aggregates is extremely complicated in general, it is almost impossible to evaluate the values of F_{hi} and M_{hi} rigorously. In conventional models, the drag force on the constituent particle is assumed to be given by the Stokes law for a single particle, neglecting the disturbance due to neighboring particles. This is called "free-draining approximation." In the present model, the drag force is assumed to act only on the particle surface exposed directly to the flow. This exposed area S_i, illustrated schematically in Figure 21.3, is determined as follows. The surface of a particle is divided into 2592 sections such that the angle between the grid lines is $\pi/36$. Then a straight line is drawn to a corner of a given section from the particle center. If all the lines for four corners do not intersect with the surface of the other particles within the distance of $6a$, the section is assumed to be exposed directly to the flow. Repeating this procedure for all the sections, the exposed area S_i for particle i is determined, as illustrated as the dark sections in Figure 21.3.

Supposing that a single particle of velocity u_{pi} and angular velocity ω_{pi} is in an applied homogeneous flow of velocity u_0 and angular velocity ω_0, the fluid velocity $u_f(x)$ and the corresponding pressure field $p(x)$ at an arbitrary position x around the particle is given by the following equation[78,79]:

$$u_f(x) = (E+\Omega)\cdot x - \left(\frac{a}{r_i}\right)^5 E \times r_i - \frac{5}{2}\left(\frac{a}{r_i}\right)^3 \left(1-\frac{a^2}{r_i^2}\right)\frac{r_i r_i \cdot E \cdot r_i}{r_i^2} + \left(\frac{a}{r_i}\right)^3$$
$$(\omega_{pi}-\omega_0)\times r_i + \frac{3a}{4r_i}\left(1+\frac{a^2}{3r_i^2}\right)(u_{pi}-u_0)+$$
$$\frac{3a}{4r_i}\left(1-\frac{a^2}{r_i^2}\right)\frac{r_i(u_{pi}-u_0)\cdot r_i}{r_i^2} \tag{21.3}$$

$$p(x) = -\frac{5\mu_f a^3}{r_i^3}\frac{r_i \cdot E \cdot r_i}{r_i^2} + \frac{3\mu_f a}{2r_i^3}(u_{pi}-u_0)\cdot r_i \tag{21.4}$$

where $r_i = x - x_i$, $r_i = 1/2 r_i 1/2$, μ_f is the fluid viscosity, E is the rate of strain tensor, and Ω is the vorticity tensor. The stress tensor τ for a Newtonian fluid around a particle is given by

$$\tau = -p\mathbf{I} + \mu_f\left(\nabla u_f + \nabla u f^T\right) \tag{21.5}$$

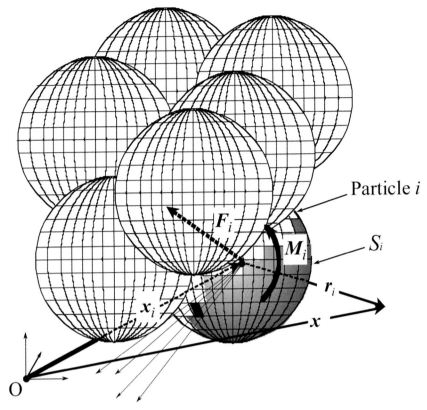

FIGURE 21.3 Schematic drawing of the simulation model of an aggregate and the coordinate system. Dark sections on the particle i indicate sections which are regarded as exposed directly to the flow. Lines are drawn to 4 corners of each section from the particle center in order to determine whether the section is exposed to the fluid or not.

Then the force and torque acting on the area S_i are calculated by the following equations, respectively:

$$F_{hi} = \int_{S_i} \boldsymbol{\tau} \cdot \boldsymbol{n}\big|_{r_i=a} \, dS \tag{21.6}$$

$$M_{hi} = \int_{S_i} \boldsymbol{r}_i \times \boldsymbol{\tau} \cdot \boldsymbol{n}\big|_{r_i=a} \, dS \tag{21.7}$$

where $\boldsymbol{n}$ is the unit vector normal to the surface.

It is known that the velocity field around a particle is influenced by the neighboring particles. This effect is taken into account as follows. The local velocity around a particle in the particle bed of porosity ε_v is given by $\varepsilon_v f(\varepsilon_v)\boldsymbol{u}_0$, where $f(\varepsilon_v)$ is a porosity function and $0 \le f(\varepsilon_v) \le 1$. Hence we assume that the velocity $\boldsymbol{u}_0$ in Equation 21.3 and Equation 21.4 may be replaced by $\varepsilon_v f(\varepsilon_v)\boldsymbol{u}_0$, using the local porosity ε_v around particles. We use the following Steinour's equation for $f(\varepsilon_v)$[80]:

$$f(\varepsilon_v) = 10^{-1.82(1-\varepsilon_v)} \tag{21.8}$$

The local porosity ε_v around a particle is defined as the porosity for the spherical space between the inner radius a and the outer radius $2a$.

TABLE 21.1   Parameters of Aggregates Employed in the Simulation and Estimated Values of c and p

Flow Field	Aggregate	N[–]	Type of Aggregate	D_{fr}[–]	a or $a_{av} \times 10^9$[m]	$\delta \times 10^9$[m]	$c \times 10^{-3}$[–]	P[–]
Shear	I	512	pc	2.47	100	0.4	64.1	0.936
	II	256	pc	2.44	100	0.4	61.6	0.981
	III	1024	pc	2.48	100	0.4	59.1	0.875
	IV	512	pc	2.46	100	2.0	5.70	0.945
	V	512	pc	2.45	500	0.4	12.5	0.946
	VI	512	pc	2.43	500	2.0	0.850	1.04
	VII	512	cc	1.74	100	0.4	7.52	0.725
	VIII	512	pc	2.31	100 ($\sigma_{st} =$ 3.336)	0.4	4.21	0.881
	Exp	–	–	2.2	70	2.58	21.7	0.879
Elongation	IX	512	pc	2.47	100	0.4	111	1.60

The hydrodynamic force and torque given above will be propagated to inside particles through the interparticle interactions. Two kinds of propagation mechanisms are considered. When particles are not contacting, they interact through the interaction forces given by the DLVO theory. Here only the van der Waals force for equal spheres is considered for the sake of simplicity, although the electrostatic repulsive force is able to be taken into account, if needed. On the other hand, when the particle surfaces contact or overlap each other because of the trial displacement by the DEM, a repulsive force acts because of their volume exclusion effect. The interaction due to the volume exclusion is calculated by the conventional method of DEM.[81,82] The quantitative validity of the present model was confirmed by comparing the dynamic shape factor simulated by the present model with the experimental one for well-defined rectangular aggregates composed of chromium spherical particles in quiescent silicon oil.[83] Values simulated by the present model are found to be in a good agreement with the experimental data. Hence, the model is now applied to simulate the behavior of aggregates in flows. All the aggregates I~IX employed are listed in Table 21.1, which may be classified into two kinds in terms of fractal dimension: a particle–cluster (pc) aggregate of rather compact structure whose value of fractal dimension D_{fr} is 2.4, and a cluster–cluster (cc) aggregate of rather loose structure whose value of D_{fr} is 1.7.

Figure 21.4 shows a series of snapshots of the deformation and breakup of the pc-aggregate I composed of monodispersed particles in the shear flow of $\mu_f \gamma_s = 500$ Pa, where γ_s is the shear rate. It is found that the aggregate is rotated, elongated into the flow direction, and then split into smaller fragments, but not eroded one by one to single particles from the aggregate surface as in the case of the breakup by ultrasonication.[84] This splitting breakup is consistent with the photographic observation given by van de Ven,[85] and also with the rupturing process of highly viscous droplets.[86,87] Repeating the similar computation for shear flows of various intensities, the relation between the average number of particles in the final fragments $<i>$ and the shear stress $\mu_f \gamma_s$ can be obtained. It is worth noting that a power-law relation holds between $<i>$ and $\mu_f \gamma_s$, as follows:

$$<i> = c\left(\mu_f \gamma_s\right)^{-P} \tag{21.9}$$

where values of c and P are listed in Table 21.1. It is clear that the value of P is nearly constant irrespectively of the aggregates. This implies that aggregates I~VI and VIII are fragmented essentially in the same fashion. On the other hand, the value of c varies with values of a, the number of particles

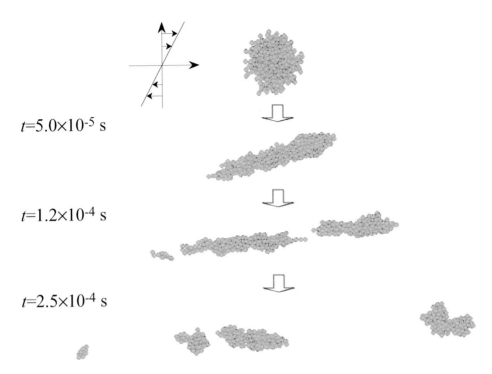

$t=5.0\times10^{-5}$ s

$t=1.2\times10^{-4}$ s

$t=2.5\times10^{-4}$ s

FIGURE 21.4 Snapshots of the fragmentation of an aggregate I of Table 21.1 in simple shear flow.

in an aggregate N, the gap between particles δ, and the standard deviation of size distribution σ_{st}. It is especially sensitive to the value of δ; the final size of fragments is very sensitive to the minimum gap between particles. It is clear that the power-law relation also holds for cc-aggregates, though the value of P is smaller than that of pc-aggregates. These results indicate that P depends mainly on D_{fr}, but not on N, a, δ, and σ_{st}; the aggregates with the same D_{fr} will be broken in a similar fashion. This is consistent with the report by Yeung and Pelton[88] that the strength of aggregates does not vary with the size but rather with the fractal dimension.

It is examined whether the fragmentation process of aggregates may follow any scaling law. It is plausible to assume that the final size of fragments is determined by the balance between the adhesive force between particles and the hydrodynamic drag on particles. The ratio of the magnitudes of these forces, N_{DA}, is defined by the following equation:

$$N_{DA} = 6\pi\mu_f a^2 \gamma_s \left/ \left(\frac{Aa}{12\delta^2} \right) \right. = \frac{72\pi\mu_f a\gamma_s\delta^2}{A} \tag{21.10}$$

where A is the Hamaker constant. All the data of $<i>$ are plotted against N_{DA} in Figure 21.5. It is important to note that almost all data for the aggregates I ~VIII fall around a single line, illustrated by a solid line, although the data for the pc-aggregate IV of large minimum separation and for the cc-aggregate VII of loose structure tend to deviate from the line. This line is expressed by the following equation:

$$<i> = 27.9 \times N_{DA}^{-0.872} \tag{21.11}$$

This equation gives us a good tool to estimate the average size of flocs that exist stably in the shear flow.

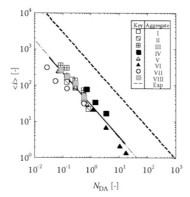

FIGURE 21.5 Dependence of the average number of particles in a fragment on dimensionless parameter NDA (Exp indicates the experimental relation given by Sonntag & Russel, the solid line indicates the best fit for simulated results, and the thin chain line indicates the experimental data for $\delta_{min} = 0.65$ nm.).

The present simulation was compared with quantitative experiments in which effects by reaggregation among broken fragments are carefully avoided. Sonntag and Russel carried out experiments of the breakage of aggregates with $D_{fr} = 2.2$ in the simple shear flows, which are composed of monodispersed latex particles of $a = 7.0 \times 10^{-8}$m.[89] It is found that a power-law relation with $c = 2.11 \times 10^4$ and $P = 0.879$ holds between $<i>$ and $\mu_f \gamma_s$. The comparison with Equation 21.11 is made by the dashed line in Figure 21.5. We consider that the agreement between the simulation and experiments are satisfactory, because the slope is nearly the same, that is, aggregates are fragmented in the similar fashion. As for the absolute magnitude of $<i>$, the experimental value is greater. Sonntag and Russel estimated that δ is 2.58 nm, using their model. This value is much larger than the minimum surface separation widely employed, that is, 0.4 nm. But, if $\delta = 0.65$ nm is assumed, their data are expressed by the thin chain line drawn in the figure and coincide extremely well with Equation 21.11. We consider this value of δ to be much more reasonable.

Simulation by the present model is carried out also for elongational flows. It is found that aggregates do not rotate but are elongated to the flow direction and then split into smaller fragments in the same manner as the shear flow. The value of $<i>$ is also expressed by Equation 21.9, replacing γ_s by the elongation rate γ_e. Values of c and P are listed in Table 21.1. We cannot find any quantitative data to compare. This is probably because the experiment for purely elongational flows is difficult. As for qualitative observation, the breakage process was observed using a four-roller device.[90] It was found that aggregates of irregular shape are split into a few fragments followed by erosion of much smaller fines, but spherical aggregates tend to be broken by erosion. The present simulation indicates that aggregates are broken by splitting, but not by erosion, which is essentially consistent with the above-mentioned observation for aggregates of irregular shape.

It is important to know which flow is more adequate to break up aggregates, shear or elongational flows. Values of $<i>$ for the aggregate I are plotted against the dissipation energy ε_{dis} in Figure 21.6 under the flow conditions that appear in usual industrial processes. The value of $<i>$ for the elongational flow is always smaller than that of the shear flow in this range. This indicates that the elongational flow is more effective and preferable to disperse coagulated particles. This result is consistent with the observation by Kao and Mason,[91] who claimed that the elongational flow is more effective for floc breakup, because the energy of flow is consumed to break up the aggregates but not to rotate them.

1.21.3 PARTICLE MOTION IN FLUIDS

Trajectories of particle motion in fluids are obtained by integrating the Newtonian equation of motion considering the fluid force. There are two approaches concerning how to give fluid force

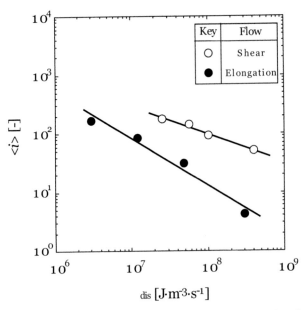

FIGURE 21.6 Comparison between the fragment size of aggregates I in shear flow and that in elongational flow.

to the equation of motion. The first one is to make use of established expressions of the force, which are shown in monographs and handbooks. If the particle Reynolds number is sufficiently small, theoretical expressions such as Stokes drag and Suffman lift forces are applied. If the particle Reynolds number is not sufficiently small, the fluid forces are expressed by the use of the drag and lift coefficients. The Newtonian equation of motion taking into account the fluid forces and other forces acting on a particle is called the Basset–Boussinesq–Oseen equation (BBO equation). The monograph by Crowe et al.[92] describes the details of the BBO equation. The simplified form of the BBO equation is called the Langevin equation. The second approach to give the fluid forces is based on calculated results of flow around moving particles. The second approach has been developed in the last 10 years, while the first one has been used in most simulations in the past. Expressions of drag and lift described in monographs and handbooks are precise when a single particle meets the fluid with constant velocity. If the particle concentration becomes higher, those expressions should be modified. Wen and Yu's modification[93] is well known.

In the second approach, the flow around each particle is solved based on the full Navier–Stokes equation[94–96] or the Lattice–Boltzmann simulation,[97] as shown in Figure 21.7. The fluid forces acting on particles are estimated by integrating stresses on the surface of particles that are the solutions of the basic equation. In this method, particles are regarded as moving boundaries. In principle, the fluid forces acting on the particle can be estimated precisely in any circumstances, but the computation load in this method is much heavier than the first one. Thus, as the number of particles increases, it is difficult to use the second approach. In a recent simulation, cases of more than 2000 particles were calculated by the second approach.

The following part of this section focuses on the first method. If the particle concentration is sufficiently small, particle motion does not affect fluid motion and particle–particle interaction can be neglected. In that case, as long as the flow field of the fluid is given, numerical integration of the BBO equation or Langevin equation is straightforward. The problem is how to give the fluid motion, particularly in the case of turbulent flows. One method is to express instantaneous fluid motion empirically or semiempirically by assuming turbulent intensity and eddy life time. The other method is to use the results of Large Eddy Simulation (LES) or Direct Numeric Simulation (DNS) of the Navier–Stokes equation. Many studies of turbulent diffusion of small particles have been done

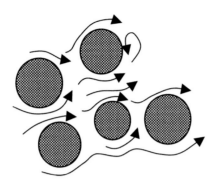

FIGURE 21.7 Flow around particles.

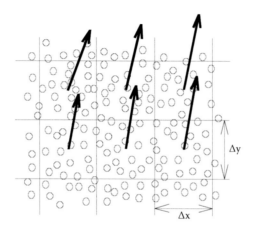

FIGURE 21.8 Cell for local averaging.

along this line. Apart from particle diffusion, some researchers are interested in heterogeneity in particle concentration caused by turbulent vortices.[98,99]

In most industrial processes, particle concentrations are not so low. In such cases, the fluid motion is affected by the presence of particles, and thus not only forces acting on particles but forces acting on fluid due to particles must be taken into account. This interaction of particle with fluid is sometime called two-way coupling. It is necessary to solve fluid motion and particle motion simultaneously.

In numerical analysis of the two-way coupling, a flow field is divided into cells, as shown in Figure 21.8. The size of the cell should be larger than the particle size and smaller than the size of the flow system. The effects of the presence of particles on the fluid are taken into account by the void fraction of each phase and momentum exchange through drag force. This approach has been developed by Anderson and Jackson[100] and called the "local averaging approach."

In addition to two-way coupling, particle–particle interaction should be considered at higher concentrations. Concerning particle–particle interaction, two phenomena are known: collision and contact. If particles are in a dispersed phase, binary collision between particles is dominant. As the particle concentration becomes much higher, particles keep in contact with each other. In the foregoing part of this section where the particle–particle interaction can be neglected, it is not necessary in calculation to treat all individual particles; a certain number of sample particles are enough to represent all actual particles if the number of sample particles is large enough to obtain statistically stable values.

In cases where the particle–particle interaction should be considered, calculation is made for all particles if collision or contact partners are found from trajectories in a deterministic way. The deterministic method of finding partners is inevitable in dense phase flows where particles are in contact.

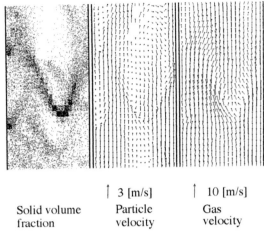

↑ 3 [m/s]   ↑ 10 [m/s]

Solid volume   Particle   Gas
fraction   velocity   velocity

FIGURE 21.9 Configuration of particle clusters in a
vertical duct flow (Ref. 101).

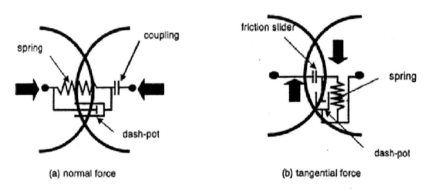

(a) normal force   (b) tangential force

FIGURE 21.10   Model of contact force.

In collision-dominated flows, a method based on ample particles is available: the direct simulation
Monte Carlo method, which makes, use of the collision probability. As an example to which this
method was applied, a gas–solid flow in a vertical channel is shown in Figure 21.9.[101] It is found
that particles form clouds or clusters in the channel by repeated collision.

The contact forces in contact-dominated flows are modeled by using mechanical elements such as
springs, dashpots, and friction sliders, as shown in Figure 21.10.[102]

This model was originally proposed in the field of soil mechanics. Combining particle
motion based on this model with fluid motion, simulation has a wide range of applications such
as dense-phase pneumatic conveying[103] and dense-phase fluidized beds.[104,105] Figure 21.11 shows
a calculated bubble rising in a fluidized bed.

1.21.4 PARTICLE METHODS IN POWDER BEDS

Powder flow controls the performance of many engineering operations and is also important in natural processes. What kind of method can be used to reveal the flow mechanism of innumerable powder
particles? It would be a continuum method for which the constitutive equation for the stress–strain
relation and the equation of the state for the bulk density stress relation are necessary.

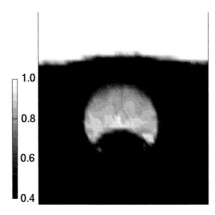

FIGURE 21.11 Bubble rising in a fluidized bed. [From Kawaguchi, T., Yamamoto, Y., Tanaka, T., and Tsuji, Y., *Proc. 2nd Int. Conf. on Multiphase Flow*, Kyoto, FB2–17, 1995. With permission.]

We present numerical simulation results of the velocity and stress fields for a flowing powder using the smoothed particle hydrodynamics (SPH) method[106,107] based on the constitutive equations obtained by DEM. The SPH method, which has been used in astrophysics and hydrodynamics, is a Lagrangian continuum model. In this method the Lagrangian equation of motion is solved for hypothetical particles constituting the continuum. The Lagrangian trajectories of hypothetical imaginary particles are calculated, and therefore it is possible to describe the discrete characteristics of particulate matters.

Numerical Method

The computational domain is divided into many imaginary particles (element pieces) which overlap with each other, and the trajectories of particles with mutual interaction are calculated to describe the motion of the continuum, for example, an approximated powder bed. The physical quantity of the particle is distributed by the kernel function, and the quantity at a fixed point is obtained by integration over all overlapping particles. The particle concentration shows the continuum density which decides the interaction range of the kernel function, since the computational domain at high density is occupied by small particles and that at low density is occupied by large particles.

In the SPH method, the physical property f is determined from the kernel function W. Smoothed $f(r)$ is expressed as

$$\langle f(r) \rangle_i = \int_D f(r_j) W(r_i - r_j, h) \, dr_j \tag{21.12}$$

The integral interpolant is given by

$$\langle f(r) \rangle_i = \sum_{j=1}^{N} m_j \frac{f_j}{r_j} W(r_i - r_j, h) \tag{21.13}$$

where $W(r_i - r_j, h)$ is an interpolating kernel with

$$\int W(r_i - r_j, h) = 1 \tag{21.14}$$

and m_j is the mass of an imaginary particle and ρ_j is bulk density of the powder.

The Gaussian function was used as an interpolating kernel. To satisfy the principle of action and reaction between particles, the kernel function should be symmetric, that is, $W(r_i-r_j)=W(r_j-r_i)$. Then the kernel function takes the form

$$W\left(r_{ij}\right)=\frac{1}{2\pi}\left\{\frac{e^{-(r_i-r_j)^2/h_i^2}}{h_i^2}+\frac{e^{-(r_i-r_j)^2/h_j^2}}{h_j^2}\right\} \tag{21.15}$$

h is given as

$$h_i=\zeta\left(\frac{m_0}{\rho_{b_i}}\right) \qquad h_j=\zeta\left(\frac{m_0}{\rho_{b_j}}\right) \tag{21.16}$$

,

where $\xi = 1.4$ and subscripts i and j show the reference and the surrounding particles. The effects of other particles with centers which are within the radius $3h_i$ are taken into account to obtain the smoothed value.

Substitution of velocity into $v_j = f_j$ into Equation 21.13 gives

$$\langle v\rangle_i=\sum_j\frac{m_j}{r_j}v_j W\left(r_{ij},h\right) \tag{21.17}$$

The gradient of $f(r)$ may be obtained as

$$\langle\nabla f\left(r\right)\rangle_i=\sum_{j=1}^{N}m_j\frac{f_j}{\rho_j}\nabla W\left(r_{ij},h\right) \tag{21.18}$$

The Rankine static stress model was used for the initial stress distribution of the powder in the tank.

To determine the pressure generated by bulk density differences due to motion in powder, the following equation is applied:

$$P=K\rho\, gH\left(\frac{\rho}{\rho_0}-1\right) \tag{21.19}$$

where H is the depth of powder bed. The constant for the powder is not clear but should be chosen so as to be in agreement with experimental results.

The bulk density is calculated using the continuity equation transformed as

$$\frac{d\rho_b}{dt}=\nabla(\rho v)+v\nabla\rho_b \tag{21.20}$$

Yuu et al.[108] have calculated the forces acting on each powder particle in the powder bed using DEM and obtained the shear and the normal stresses by locally averaging these forces on the plane in the powder bed that assumed continuum. This is essentially the same method of molecular dynamics that gives the stress field in the fluid. The experimental relationships were also obtained under the same conditions. The comparison of calculated and experimental stress–stain rate relationships shows good agreement, and both dynamic shear and dynamic normal stresses are expressed as linearly dependent on strain rates over a fairly wide strain rate region. The following equations obtained by Yuu et al. show the stress–strain rate relationships in the particulate matter:

$$\sigma_x = \sigma_{x0} + \sqrt{\sigma_{x0}^2 + \sigma_{y0}^2}\; A1 D_{xx} \tag{21.21}$$

$$\sigma_y = \sigma_{y0} + \sqrt{\sigma_{x0}^2 + \sigma_{y0}^2}\; A1 D_{yy} \tag{21.22}$$

and

$$\tau_{xy} = \tau_{xy0} + \sqrt{\sigma_{x0}^2 + \sigma_{y0}^2}\; A2 D_{xy} \tag{21.23}$$

where D_{xx}, D_{yy} and D_{xy} are the deformation rates.

The coefficient $A2$ is much larger than coefficient $A1$. This means that shear deformation occurs more easily than normal deformation in the particulate matter.

To calculate the stagnant zone on the powder bed, the Bingham-type plastic model is used. Momentum equations for powder beds are expressed as

$$\frac{du}{dt} = -\frac{1}{\rho_b}\frac{\partial P}{\partial x} - \frac{1}{\rho_b}\left(\frac{\partial \sigma_x}{\partial x} + \frac{\partial \tau_{xy}}{\partial y}\right) \tag{21.24}$$

and

$$\frac{dv}{dt} = -\frac{1}{\rho_b}\frac{\partial P}{\partial y} - \frac{1}{\rho_b}\left(\frac{\partial \tau_{xy}}{\partial x} + \frac{\partial \sigma_y}{\partial y}\right) - g \tag{21.25}$$

We have calculated Equation 21.21 through Equation 21.25 for the velocity field of the powder bed using the SPH method.

Figure 21.12(a) shows results obtained with the SPH method using Lagrangian equations by Yoshida.[109] At 0.2 s, the powder bed height has decreased further, and at 1.0 s, the discharge of powder has diminished. The stagnant zone of powder in the corner of the tank was found to be fairly in good agreement with the measured one the same as indicated by experiment. Figure 21.12b shows experimental results for powder in the tank. The calculated flow patterns and shape of the stagnant zone and those measured are basically the same.

1.21.5 TRANSPORT PROPERTIES

To evaluate numerically transport properties such as the diffusion coefficient and shear viscosity for dispersed systems, we have two methods: direct and indirect simulations. In the former, we simulate the experimental situation on the computer and measure the quantity directly, while in the latter, we calculate time-correlation functions in the equilibrium state and interpret them into the transport properties by the Green–Kubo formula, which is widely used for simple liquids.[110] The main difference between simple liquids and dispersed systems such as colloidal suspensions is the existence of

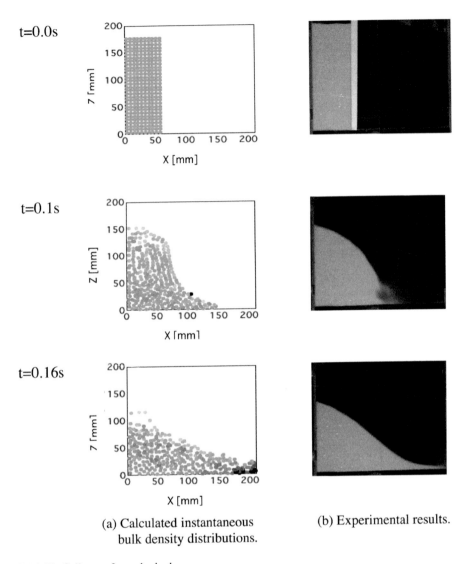

(a) Calculated instantaneous
bulk density distributions.

(b) Experimental results.

FIGURE 21.12 Collapse of powder bed.

hydrodynamic interaction by the fluid surrounding the particles.[111] Here we focus on the contribution of the hydrodynamic interaction.[112] Other contributions such as interparticle forces and Brownian motion are briefly commented on later.

Diffusion Coefficients

The self-diffusion coefficient is defined by the mean-square displacement of a tracer particle as

$$D^s = \frac{1}{6} \frac{d\left\langle \left| x(t) - x(0) \right.^2 \right\rangle}{dt} \tag{21.26}$$

where $x(t)$ is the position of the tracer particle at time t. The brackets $<>$ denotes the average of many time sequences. It has two asymptotes: short- and long-time diffusion coefficients. The splitting timescale is a^2/D_0, where a is the particle radius and D_0 is the Stokes–Einstein diffusion coefficient of a single isolated particle in the fluid defined by

$$D_0 = \frac{k_B T}{6\pi\eta a} \tag{21.27}$$

Here k_B is the Boltzmann constant, T is the temperature, and η is the viscosity of the fluid. The long-time self-diffusion coefficient is also written as

$$D_\infty^s = \lim_{t\to\infty} \frac{\langle |x(t)-x(0)|^2\rangle}{6t} = \frac{1}{3}\int_0^\infty \langle U(0)U(t)> \, dt \tag{21.28}$$

where U is the velocity of the tracer particle. The last form with the time-correlation function of the velocity is called the Green–Kubo formula. As in Eqs. (21.26) and (21.28), the self-diffusion coefficients are calculated from many series of dynamical simulations; this is the indirect way.

The direct way to calculate diffusion coefficients is to calculate the mobility of the tracer particle. The factor $1/6\pi\eta a$ in Eq. (21.27) is the mobility of a single isolated particle. For dispersed systems, on the other hand, the mobility is not just $1/6\pi\eta a$ because of the hydrodynamic interaction among particles. Under the Stokes approximation, the hydrodynamic interaction is expressed by the resistance equation

$$\begin{pmatrix} F \\ S \end{pmatrix} = \begin{pmatrix} R_{FU} & R_{FE} \\ R_{SU} & R_{SE} \end{pmatrix} \begin{pmatrix} U-U^\infty \\ -E^\infty \end{pmatrix} \tag{21.29}$$

where F is the force and torque vector with $6N$ components, S is the stresslet, $U-U^\infty$ is the translational and rotational velocities relative to the imposed flow, and E^∞ is the strain imposed to the system, for N particles in the system. The whole resistance matrix with $11N \times 11N$ elements depends only on the configuration of the particles and is calculated by the Stokesian dynamics method.[113] From Eq. (21.29) the particle velocity U under no external flow is given by

$$U = R_{FU}^{-1} . F \tag{21.30}$$

Then, the Stokes–Einstein relation is extended for dispersed systems as

$$D_0^s = k_B T \langle \left(R_{FU}^{-1} \right)_{\alpha\alpha} \rangle \tag{21.31}$$

where D_0^s is a matrix with 6×6 elements and $R[-1/Fu]\alpha\alpha$ denotes the self part of the inverse of the resistance matrix with 6×6 elements. The diagonal elements of the translational part of D_0^s is the short-time self-diffusion coefficients D_0^s. Figure 21.13 shows the numerical results of D_0^s/D_0 with experimental results.[112]

Shear Viscosity

The shear viscosity is usually calculated by the direct way rather than the indirect way. The reason is that the interparticle force of hydrodynamic interaction needed for the Green–Kubo formula of the

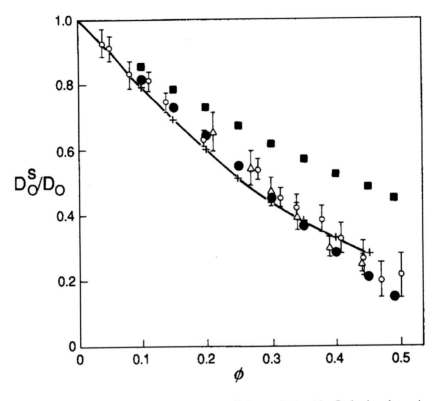

FIGURE 21.13 Short-time self-diffusion coefficients obtained by Stokesian dynamics methods (solid symbols), a theoretical calculation (solid line), and experimental results (open symbols). [From Phillips, R. J., Brady, J. F., and Bossis, G., *Phys. Fluids.*, 31, 3462–3472, 1988. With permission.]

shear viscosity is complicated. In the direct way, the bulk stress p is calculated under the shear flow. Neglecting Brownian motion and interparticle forces, the bulk stress is given by

$$\langle R \rangle = -\langle p \rangle \mathbf{I} + 2\eta \langle \mathbf{E}^{\infty} \rangle - n \langle \mathbf{S} \rangle \tag{21.32}$$

where p is the pressure and n is the number density of particles.[114] From Brady and Bossis,[113] the stresslet for force-free particles is given by

$$\mathbf{S} = \left(\mathbf{R}_{SU} . \mathbf{R}_{FU}^{-1} . \mathbf{R}_{FE} - \mathbf{R}_{SE} \right) . \mathbf{E}^{\infty} \tag{21.33}$$

After taking the average of the stresslet $\mathbf{S}$, the shear viscosity of dispersed system η_r scaled by the fluid viscosity η is obtained as

$$-n \langle \mathbf{S} \rangle = 2\eta (\eta_r(\phi) - 1) \langle \mathbf{E}^{\infty} \rangle \tag{21.34}$$

where ϕ is the volume fraction of particles. Note that in the dilute limit, we have the Einstein's result

$$\eta_r = 1 + \frac{5}{2}\phi \tag{21.35}$$

Figure 21.14 shows the numerical results of η_r with experimental results.[112]

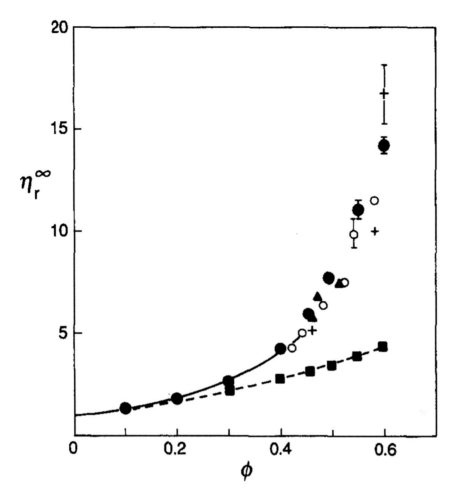

FIGURE 21.14 Shear viscosities obtained by Stokesian dynamics methods (solid symbols), theoretical calculations (lines), and experimental results (open symbols). [From Phillips, R. J., Brady, J. F., and Bossis, G., *Phys. Fluids.*, 31, 3462–3472, 1988. With permission.]

Remarks

Using the Stokesian dynamics method, we can calculate the microstructures of dispersed systems, as shown in Eq. (21.29), under arbitrary circumstances and evaluate any quantities such as sedimentation velocity and normal stresses as well, by the direct way. To do this, however, we need to know the quantity to calculate, as shown by Batchelor.[115] The Brownian contributions are shown for the diffusion coefficient[116] and the bulk stress.[115] The general form of the bulk stress has been shown,[117] and thorough numerical analysis on the Péclet number dependence of diffusion coefficients and shear viscosity has been shown.[118]

1.21.6 ELECTRICAL PROPERTIES OF POWDER BEDS

Percolation Theory

Electrical conductivities are very sensitive to the structure of powder beds. Simulations of electrical conductivity can be utilized for the structural evaluation or analysis of powder beds. For example, a simulation of the electrical conductivity for a mixture of conducting and insulating particles can be performed as an application of percolation concepts. If an arrangement of the particles in the

TABLE 21.2   Percolation Thresholds for Several Lattice Models

	Lattice	Coordination Number	Threshold
Two dimension	Honeycomb	3	0.6962
	Square	4	0.592745
	Triangle	6	0.5
Three dimension	Diamond	4	0.428
	Cubic	6	0.3117
	Body center cubic	8	0.2460
	Face center cubic	12	0.198

mixture is described as a lattice model, the electrical conductivity can be derived as a function of concentration of the conducting particle. In this simulation, a drastic change of the electrical conductivity with changing concentration especially relates to a critical point or percolation threshold.[119–121] Percolation thresholds of some lattice models are tabulated in Table 21.2. The lattice model should be selected in taking into account the packing density of the powder bed.

Calculation by Applying Kirchhoff's Law

There is another approach to analysis of powder bed structures by simulation. The simulation of electrical resistance for binary systems leads to understanding the change of particle arrangements or the degree of mixing. Particle contact resistances were calculated by using an electrical equivalent circuit based on Kirchhoff's law. Fujihara and coworkers[122,123] performed the simulation of a binary particles system consisting of stainless steel and lead particles and compared the simulated results with experimental ones in a milling process. In the simulation, the randomness or degree of mixing was estimated from a standard deviation σ concerning the concentration σ can be expressed as

$$\sigma = \frac{\sqrt{\sum_{i=1}^{N}(C_i - C_0)}}{N}$$

where N is the number of samples, C_i concentration of sampled particles in i specimen, and C_0 is the average concentration in the mill. Then, they discussed their simulation model to reduce the calculation load.

DEM Simulation of Charged Particles

Flow dynamics of charged particles is very important for the formation of powder beds. Yoshida and his coworkers performed a DEM simulation of polymer and toner particles in terms of a charge site model 124. In this model, a particle was divided into several independent parts, as shown in Figure 21.15. Charge transfer occurs from one site to another site at intersite contact in a probability accompanied by the charge difference between the sites. The simulation of polymer and toner particles was compared with a multiple-impact charging experiment in a metal container. In these simulations and experiments, the particles collided many times in a vibration field. Figure 21.16 shows the time dependence of the total charge of the polypropylene particles. In this comparison, the space charge effect has a large contribution in the powder charging by multiple contacts. Charged particles need to be controlled for required formation of powder beds.

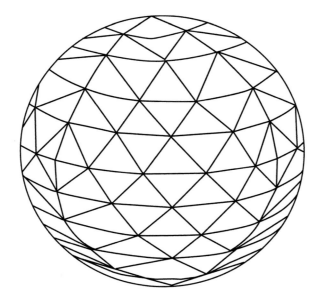

FIGURE 21.15 Illustration of the simulation model for charged particles. This particle is divided by 200 charged sites.

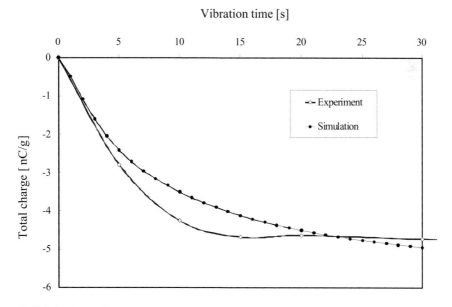

FIGURE 21.16 Time dependence of the total charge of polypropylene particles.

REFERENCES

1. Alder, B. J. and Wainwright, T. *J. Chem. Phys.*, 31, 459–466, 1959.
2. Cundall, P. A., *A Computer Model for Rock-Mass Behavior Using Interactive Graphics for Input and Output of Geometrical Data*, U.S. Army Corps of Engineers (Missouri River Division), Tech. Rep. No. MRD-2074, 1974.
3. Tsuji, Y., Kawaguichi, T., and Tanaka, T., *Powder Technol.*, 77, 79–87, 1993.

4. Haff, P. K., Jiang, Z., and Forrest, S. B., in *Advances in the Mechanics of Granular Materials,* Shen, H. H., Satake, M., Mehrabadi, M., Chang, C. S., and Campbell, C. S., Eds., Elsevier, Amsterdam, 1992, pp. 373–380.

5. Potapov, A. V., Hunt, M. L., and Campbell, C. S., *Powder Technol.,* 116, 204–213, 2001.

6. Campbell, C. S., in *Proceedings of the Tenth U.S. National Congress of Applied Mechanics,* Austin Texas, June 1986, ASME, New York, 1986, pp. 327–338.

7. Herrmann, H. J. and Luding, S., *Cont. Mech. Thermo.,* 10. 189–231, 1998.

8. Campbell, C. S., *Annu. Rev. Fluid Mech.,* 22, 57–92, 1990.

9. Campbell, C. S., Ph.D. Thesis, California Institute of Technology, Pasadena, CA, 1982.

10. Campbell, C. S. and Brennen, C. E., *J. App. Mech.,* 52, 172–178, 1985.

11. Campbell, C. S. and Brennen, C. E., *J. Fluid Mech.,* 151, 167–188, 1995.

12. Campbell, C. S., *J. Fluid Mech.,* 203, 449–473, 1989.

13. Campbell, C. S., *J. Fluid Mech.,* 247, 111–136, 1993.

14. Campbell, C. S., *J. Fluid Mech.,* 247, 137–156, 1993.

15. Campbell, C. S., *J. Fluid Mech.,* 348, 85–104, 1997.

16. Hawkins, G. W., in *Mechanics of Granular Materials, New Models and Constitutive Relations,* Jenkins, J. T. and Satake, M., Eds., Elsevier, Amsterdam, 1983, pp. 305–312.

17. Walton, O. R., Kim, H., and Rosato, A. D., in *Mechanics Computing in the 1990's and Beyond,* Vol. 2, Adeli, H. and Sierakowski, R. L., Eds., ASCE, New York, 1991, pp. 1249–1261.

18. Walton, O. R., in *Particulare Two-Phase Flow,* Roco, M. C., Ed., Butterworth-Heinemann, Boston, 1993, pp. 884–911.

19. Hopkins, M. A. and Logue, M. Y., *Phys. Fluids,* A, 3, 47–57, 1991.

20. McNamara, S. and Young, W. R., *Phys. Fluids,* A, 5, 34–45, 1992.

21. Luding, S. and McNamara, S., *Granular Matter,* 1, 113–128, 1998.

22. Cundall, P. A. and Strack, O. D. L., *Geotechnique,* 29, 47–65, 1979.

23. Babicacute, M., Shen, H. H., and Shen, H. T., *J. Fluid Mech.,* 219, 81–118, 1990.

24. Campbell, C. S., *J. Fluid Mech.,* 465, 261–291, 2002.

25. Campbell, C. S., *J. Fluid Mech.,* submitted.

26. Campbell, C. S., Cleary, P., and Hopkins, M. A., *J. Geophys. Res.,* 100, 8267–8283, 1995.

27. Campbell, C. S. and Zhang, Y., in *Advances in Micromechanics of Granular Materials,* Shen, H. H., Satake, M., Mehrabadi, M., Chang, C. S., and Campbell, C. S., Eds., Elsevier, Amsterdam, 1992, pp. 261–270.

28. Haff, P. K. and Werner, B. T., *Powder Technol.,* 48, 239–245, 1986.

29. Herrmann, H. J., *Physica A,* 191, 263–276, 1992.

30. Langston, P. A., Tüzün, U., and Heyes, D. M., in *Powders and Grains 93,* Thornton, C., Ed., A. A. Balkema, Rotterdam, 1993, pp. 357–363.

31. Vu-Quoc, L., Zhang, X., and Walton, O. R., *Comp. Meth. App. Mech. Eng.,* 187, 483–528, 2000.

32. Walton, O. R., *Energy and Technology Review, Lawrence Livermore Lab.,* May, 24–36, 1984.

33. Walton, O. R., *Mech. Mater.,* 16, 239–247, 1993.

34. Walton, O. R. and Braun, R. L., *J. Rheol.,* 30, 949, 1986.

35. Walton, O. R. and Braun, R. L., *Acta Mech.,* 63,: 73–86, 1986.

36. Zhang, Y. and Campbell, C. S., *J. Fluid Mech.,* 237, 541–568, 1992.

37. Drescher, A. and De Josselin de Jong, G., *J. Mech. Phys. Solids,* 20, 337, 1972.

38. Mueth, D. M., Jaeger, H. M., and Nagel, S. R., *Phys. Rev. E.,* 57, 3164–3169, 1998.

39. Howell, D. W., Behringer, R. P., and Veje, C. T., *Chaos,* 9, 559–572, 1999.

40. Howell, D. W., Behringer, R. P., and Veje, C. T., *Phys. Rev. Lett.,* 26, 5241–5244, 1999.

41. Hopkins, M. A., *Particle Simulation,* Vol. 1, Rep. No. 87-7, Department of Civil Engineering, Clarkson University, Potsdam, NY, 1987.

42. Metropolis, N., Rosenbluth, M., Teller, A., and Teller, E., *J. Chem. Phys.,* 21, 1087–1092, 1953.

43. Chapman, S. and Cowling, T. G., *The Mathematical Theory of Non-Uniform Gases,* 3rd Ed., Cambridge University Press, 1970.

44. Hopkins, M. A., M.S. thesis, Clarkson University, Potsdam, NY, 1985.

45. Hopkins, M. A. and Shen, H. H., *J. Fluid Mech.,* 244, 477–491, 1992.

46. Baxter, G. W. and Behringer, R. P., *Phys. Rev. A,* 42, 1017–1020, 1990.

47. Gutt, G. M., Ph.D. thesis, California Institute of Technology, Pasadena, CA, 1989.

48. Désérable, D. and Martinez, J., in *Powders and Grains 93,* Thornton, C., Ed., A. A. Balkema, Rotterdam, 1993, pp. 345–350.
49. Hopkins, M. A., *The Numerical Simulation of Systems of Multitudinous Polygonal Blocks,* U.S. Army Cold Regional Research Engineering Laboratory, USACRREL Rep. CR 99-22, 1992.
50. Hogue, C. and Newland, D. E., in *Powders and Grains 93,* Thornton, C., Ed., Balkema, Rotterdam, 1993, pp. 413–420.
51. Kohring, G. A., Melin, S., Puhl, H., Tillemans, H. J., and Vermohlen, W., *Comput. Methods Appl. Mech. Eng.,* 124, 273–281, 1995.
52. Potapov, A. V., Hopkins, M. A., and Campbell, C. S., *Int. J. Mod. Phys.,* C6, 371–398, 1995.
53. Potapov, A. V., Campbell, C. S., and Hopkins, M. A., *Int. J. Mod. Phys.,* C6, 399–425, 1995.
54. Potapov, A. V. and Campbell, C. S., *Int. J. Mod. Phys.,* C7, 155–180, 1996.
55. Cundall, P. A., *Int. J. Rock Mech. Mining Sci.,* 25, 107–116, 1988.
56. Ghaboussi, J. and Barbosa, R., *Int. J. Num. Anal. Methods Geomech.,* 14, 451–472, 1990.
57. Potapov, A. V. and Campbell, C. S., *Int. J. Mod. Phys.,* C7, 717–730, 1996.
58. Rothenburg, L. and Bathurst, R. J., *Geotechnique,* 1, 79–85, 1992.
59. Rothenburg, L. and Bathurst, R. J., *Polym.-Plast. Tech.,* 35, 605–648, 1996.
60. Ting, J. M., *Comp. Geotech.,* 13, 175–186, 1992.
61. Mustoe, G. G. W. and DePooter, G., in *Powders and Grains 93,* Thornton, C., Ed., A. A. Balkema, Rotterdam, 1993, pp. 421–427.
62. Potapov, A. V. and Campbell, C. S., *Granular Matter,* 1, 9–14, 1998.
63. Thornton, C. and Kafui, K. D., *Powder Technol.,* 20, 109–124, 2000.
64. Kafui, K. D. and Thornton, C., in *Powders and Grains 93,* Thornton, C., Ed., A. A. Balkema, Rotterdam, 1993, pp. 401–406.
65. Thornton, C. and Randall, C. W., in *Micromechanics of Granular Materials,* Satake, M. and Jenkins, J. T., Eds., Elsevier, Amsterdam, 1988, pp. 133–142.
66. Tüzün, U. and Walton, O. R., *J. Phys. D,* 25, A44–A52, 1992.
67. Thornton, C., *J. Appl. Mech.,* 64, 383–386, 1997.
68. Hertz, H., *J. Math Crelles J.,* 92, 156–171, 1882.
69. Mullier, M., Tüzün, U., and Walton, O. R., *Powder Technol.,* 65, 61, 1991.
70. Goddard, J., *Proc. Roy. Soc.,* 430, 105–131, 1990.
71. Drake, T. G. and Walton, O. R., *J. Appl. Mech.,* 62, 131–135, 1995.
72. Raman, C. V., *Phys. Rev.,* 12, 442–447, 1918.
73. Goldsmith, W., *Impact,* Edward Arnold, London, 1960.
74. Mindlin, R. D. and Deresiewicz, H., *J. App. Mech. Trans. ASME,* 20, 327–344, 1953.
75. Sondergaard, R., Chaney, K., and Brennen, C. E., *J. Appl. Mech.,* 57, 694–699, 1990.
76. Foerster, S. F., Logue, M. Y., Chang, H., and Allia, K., *Phys. Fluids,* 6, 1108–1115, 1994.
77. Maw, N., Barber, J. R., and Fawcett, J. N., *J. Lub. Technol.,* 103, 74–80, 1981.
78. Russel, W. B., Saville, D. A., and Schowalter, W. R., *Colloidal Dispersions,* Cambridge University Press, 1989.
79. Dhont, J. K. G., *An Introduction to Dynamics of Colloids,* Elsevier, Amsterdam, 1996.
80. Steinour, H. H., *IEC,* 36, 618–624, 1944.
81. Cundall, P. A. and Strack, O. D. L., *Geotechnique,* 29, 47–65, 1979.
82. Society for Powder Technology, Japan, Ed., *Funtai Simulation Nyumon,* Sangyo-Tosho, Tokyo, 1998.
83. Niida, T. and Ohtsuka, S., *KONA,* 15, 202–211, 1997.
84. Higashitani, K., Yoshida, K., Tanise, N., and Murata, H., *Colloids Surf. A,* 81, 167–175, 1993.
85. van de Ven, T. G. M., in *Colloidal Hydrodynamics,* Academic Press, London, 1989, p. 532.
86. Torza, S., Cox, R. G., and Mason, S. G., *J. Colloid Interface Sci.,* 38, 395–411, 1972.
87. Williams, A., Janssen, J. J. M, and Prins, A., *Colloids Surf. A,* 125, 189–200, 1997.
88. Yeung, A. K. C. and Pelton, R., *J. Colloid Interface Sci.,* 184, 579–585, 1996.
89. Sonntag, R. C. and Russel, W. B., *J. Colloid Interface Sci.,* 113, 399–413, 1986.
90. Pandya, J. D. and Spielman, L. A., *J. Colloid Interface Sci.,* 90, 517–531, 1982.
91. Kao, S. V. and Mason, S. G., *Nature,* 253, 619–621, 1972.
92. Crowe, C., Sommerfeld, M., and Tsuji, Y., in *Multiphase Flows with Droplets and Particles,* CRC Press, 1997, pp. 81–88.
93. Wen, C. Y. and Yu, Y. H., *Chem. Eng. Prog. Symp. Ser.,* 62, 100–111, 1966.

94. Hu, H. H., *Int. J. Multiphase Flow,* 22, 335–352, 1996.
95. Takiguchi, S., Kajishima, T., and Miyake, Y., *JSME Int. J. Ser. B,* 42, 411–418, 1999.
96. Pan, T. W., Joseph, D. D., Bai, R., Glowinski, R., and Sarin, V. J., *J. Fluid Mech.,* 451, 169–191, 2002.
97. Qi, D., *Int. J. Multiphase Flow,* 26, 421–433, 2000.
98. Crowe, C. T., Chung, J. N., and Troutt, T. R., *Prog. Energy Combust. Sci.,* 14, 171–194, 1988.
99. Squire, K. D. and Eaton, J. K., *Phys. Fluids,* A3 5, 1169–1178, 1991.
100. Anderson, T. B. and Jackson, R., *Ind. Eng. Chem. Fundam.,* 6, 527–539, 1967.
101. Yonemura, S., Tanaka, T., and Tsuji, Y., *ASME/FED Gas-Solid Flows,* 166, 303–309, 1993.
102. Cundall, P. A. and Strack, O. D. L., *Geotechnique,* 29, 47–65, 1979.
103. Tsuji, Y., Tanaka, T., and Ishida, T., *Powder Technol.,* 71, 239–250, 1992.
104. Tsuji, Y., Kawaguchi, T., and T. Tanaka, *Powder Technol.,* 77, 79–87, 1993.
105. Kawaguchi, T., Yamamoto, Y., Tanaka, T., and Tsuji, Y., in *Proceedings of the Second International Conference on Multiphase Flow,* Kyoto, 1995, pp. FB2–17.
106. Monaghan, J., *J. Comput. Phys.,* 110, 399–406, 1994.
107. Monaghan, J., *Phys. D,* 98, 523–533, 1996.
108. Yuu, S., Hayashi, M., Waki, T., and Umekage, A., *J. Soc. Powder. Technol. Jpn.,* 34, 212–220, 1997.
109. Yoshida, K., Masters thesis of Kynshu Inst. Tech., p. 44, 2003.
110. Hansen, J. P. and McDonald, I. R., *Theory of Simple Liquids,* Academic Press, London, 1986.
111. Russel, W. B., Saville, D. A., and Schowalter, W. R., *Colloidal Dispersions,* Cambridge University Press, 1989.
112. Phillips, R. J., Brady, J. F., and Bossis, G., *Phys. Fluids.,* 31, 3462–3472, 1988.
113. Brady, J. F. and Bossis, G., *Annu. Rev. Fluid Mech.,* 20, 111–157, 1988.
114. Batchelor, G. K., *J. Fluid Mech.,* 41, 545–570, 1970.
115. Batchelor, G. K., *J. Fluid Mech.,* 83, 97–117, 1977.
116. Batchelor, G. K., *J. Fluid Mech.,* 74, 1–29, 1976.
117. Brady, J. F., *J. Chem. Phys.,* 98, 3335–3341, 1993.
118. Foss, D. R. and Brady, J. F., *J. Fluid Mech.,* 407, 167–200, 2000.
119. Ottavi, H., Clerc, J., Giraud, G., Roussenq, J., Guyon, E., and Mitescu, C. D., *J. Phys. C Solid Phys.,* 11, 1311–1328, 1978.
120. Stauffer, D., *Introduction to Percolation Theory,* Taylor & Francis, London, 1988.
121. Odagaki, T., *Introduction to Percolation Physics,* 2nd Ed., Shokabo, Tokyo, 1995.
122. Fujihara, Y. and Yoshimura, Y., *J. Soc. Mater. Sci. Jpn.,* 36, 1198–1204, 1987.
123. Yoshioka, T., Kaneko, T., Fujihara, Y., and Yoshimura, Y., *J. Soc. Mater. Sci. Jpn.,* 34, 1255–1259, 1985.
124. Yoshida, M., Shimosaka, A., Shirakawa, Y., Hidaka, J., Matsuyama, T., and Yamamoto, H., *Powder Technol.,* 135–136, 23–34, 2003.

Part II

Process Instrumentation

2.1 Powder Sampling

Hiroaki Masuda
Kyoto University, Katsura, Kyoto, Japan

2.1.1 SAMPLING EQUIPMENT

When powder is poured into a heap, size segregation will occur: fine particles tend to remain at the center of the heap and coarse ones congregate at the periphery. Even when the powder is slowly transported by a belt conveyor, coarse particles tend to float on the surface of the powder bed. It is known that size segregation occurs more frequently for powders having higher flowability.

The following are said to be the golden rules of sampling:

1. A powder should be sampled while in motion.
2. The entire stream of powder should be taken out for a long series of short time intervals.

The first rule recommends sampling from flowing powder, such as a discharging flow from a belt conveyor or feeding flow from one storage vessel to another. Even in these cases, size segregation may occur. Therefore, the entire stream of powder should be sampled by traversing the stream, and the sampling should continue for a long series of short time intervals. If the sampling speed v (m/s) is low, the mass of the sampled powder may be too large. The sampled mass w (kg) is calculated from the relation

$$w = \frac{L}{v}\frac{W}{L}b = \frac{Wb}{v} \tag{1.1}$$

where L (m) is the width of the powder stream, W (kg/s) the powder flow rate, and b (m) is the width of the sampler's mouth (i.e., sampling width or cutter width).

Samplers obeying the rules above are called full-stream samplers. One of the full-stream samplers, named a cutter sampler, is shown schematically in Figure 1.1. There are several types of full-stream samplers, such as car, belt, and rotary. The sampling width (or cutter width) and the sampling speed are adjustable in these samplers. Therefore, it is possible to take out the desired mass of powder according to Equation 1.1. Random sampling is also possible by randomly changing the starting time of the sampler.

Other types of samplers are the snap sampler and the slide spoon sampler. These samplers take a small amount of powder out of several positions of a powder stream or powder bed using a specially designed spoon. The mass of the sampled powder, called the increment, is much less than that of the full-stream sampler, and the final sample collected of all increments can be treated easily in a laboratory measurement.

A scoop is used to take a sample from powder loaded on a truck. In this case it is recommended that the sample be dug out from 30 cm below the surface, because the surface region is usually affected by segregation. Sampling positions are determined so as to divide the powder bed into equal areas. Sampling spears can be used to take samples from deep inside the powder bed.

A sampling device of the air-suction type (Figure 1.2) is also useful to get vertical samples of the powder bed in which particles are mixed with air at the sampler nose and the resulting gas-solid mixture is sucked into a gas–solid separator such as a cyclone or bag filter.

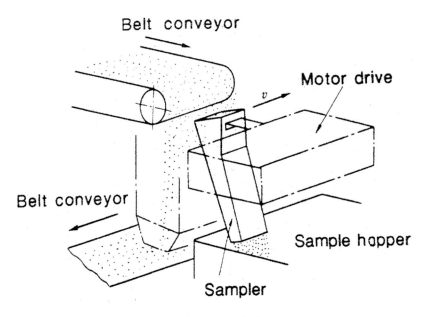

FIGURE 1.1   Full-stream sampler (cutter sampler).

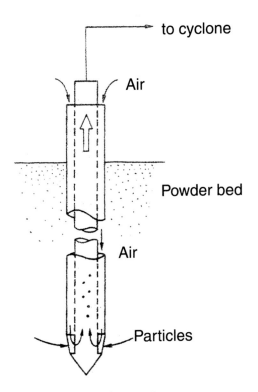

FIGURE 1.2   Suction sampler.

When continuous sampling from a storage vessel is necessary, a screw conveyor may be used. A table feeder with multiscrapers developed in the author's laboratory is useful to take small samples during continuous feeding of powder to a process.[1] Particles flowing through the table feeder do not suffer from breakage, whereas a screw conveyor often causes particle breakage. A special constant-volume sampler or slide-valve type of sampler may be adopted in the sampling of powder sliding on an inclined chute. These samplers are inserted into the powder stream very rapidly, so that mechanical parts of the samplers could cause problems. Moreover, if the powder stream is only partly sampled, the effect of size segregation is unavoidable.

The powder assembly of all increments is called the gross sample. The total mass of the gross sample is usually too much to be used as a sample for laboratory measurements of chemical composition, particle size distribution, and so on. It is necessary to reduce the mass of the sample before use in the laboratory. The procedure used to reduce a sample mass is called sample division.

Sample division should be done without segregation. The principles involved here are the same as those already discussed regarding collection of the gross sample. Figure 1.3 shows a sample divider called a spinning riffler, which satisfies the golden rules of sampling. This spinning riffler consists of a turntable with many chutes and bottle-receivers under the table. The performance of the sample divider depends on the rotational speed of the table. A higher speed (as high as 300 rpm) gives a better result.

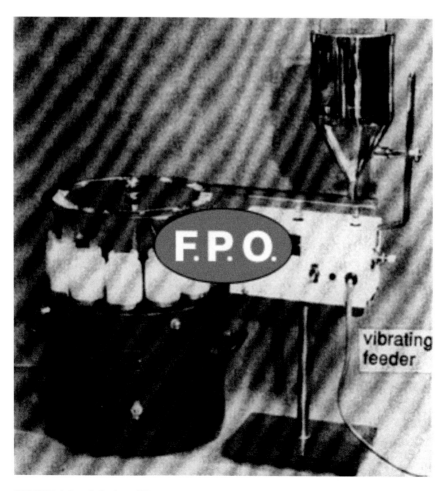

FIGURE 1.3 Spinning riffler.

There is another type of divider which consists of a rotary feeder and several receivers. A sample divider of this type is called a mechanical distributor or sample splitter. Similar but simpler are an oscillating hopper sample divider and an oscillating paddle sample divider.[2] These two sample dividers are not as effective as the spinning riffler or mechanical distributor in reducing the sample mass.

All of the sample dividers mentioned above have moving parts, which are apt to cause mechanical problems in a dusty atmosphere. On the other hand, a sample divider called a chute riffler or chute splitter has no moving parts. It has a series of two chutes directing right and left, one after another. The number of chutes on the right is equal to that on the left. Powder poured onto a chute riffler is divided by the series of the two chutes. Sample dividing by the chute riffler is affected by particle segregation, although the effect becomes insignificant as the number of chutes is increased. The segregation effect may be diminished by use of a procedure called Carpenter's method.

A chute riffler requires repeated dividing to reduce the sample size (mass). A sample divider called a table sampler or multicone divider can sufficiently reduce the sample size by only one trial. The table sampler has a series of holes and prisms on an inclined table. Powder is fed to the top of the sampler (divider). Prisms placed across the path of the powder stream divide it into many fractions. Some particles fall through the holes and are discarded, while the powder remaining on the table proceeds to the next row of prisms and holes, some powder is removed, and so on. The powder reaching the bottom of the table is the final sample. Sample dividers similar to the table sampler but with no holes are called a 1/16 divider, 1/32 divider, and so on, according to the number of final samples obtained.

2.1.2 ANALYSIS OF SAMPLING

Various sorts of data can be obtained by analyzing the sampled powder. The mass median diameter (MMD) of the sample, for example, is obtainable from the particle size measurement. The value of MMD will change for every sample. An entire set of these values is called a population. Let the value x_1 be a first sample extracted from the population with a certain probability. Such a variable (e.g., the MMD) is called a random variable. The random x variable is characterized by a probability density function or frequency distribution $f(x)$. The probability that the sampled value is found in the range $x \sim x + dx$ is given by $f(x)dx$. The following relation is applicable for various types of random variables in natural phenomena and industrial processes:

$$f(x)dx = \frac{1}{\sqrt{2\pi}\sigma}\exp\left(-\frac{(x-\mu)^2}{2\sigma^2}\right)dx \qquad (1.2)$$

where μ is the population mean and σ^2 is the population variance. The function $f(x)$ is called the normal distribution or Gaussian distribution.

The population mean μ and the population variance σ^2 are unknown parameters in Equation 1.2, which should be estimated from the following sample mean $\bar{x}$ and sample variance s^2:

$$\bar{x} = \frac{1}{n}\sum_{j=1}^{m} x_j f_j \qquad (1.3)$$

$$s^2 = \frac{1}{n-1}\sum_{j=1}^{m}\left(x_j - \bar{x}\right)^2 f_j \qquad (1.4a)$$

$$= \frac{1}{n-1}\sum_{j=1}^{m}\left[x_j^2 f_j - \frac{1}{n}\left(\sum_{j=1}^{m} x_j f_j\right)^2\right] \qquad (1.4b)$$

where x_j is the representative value of the jth class, f_j is the frequency of x_j ($j=1$ to m), and n is the sample size $\left(n = \sum\limits_{j=1}^{m} f_j\right)$. The denominator of Equation 1.4a is $(n-1)$ in place of n to make the variance unbiased.

The confidence interval of the population mean with the confidence level is given by the following relation:

$$\bar{x} - t(v,p)\frac{s}{\sqrt{n}} < \mu < \bar{x} + t(v,p)\frac{s}{\sqrt{n}} \tag{1.5}$$

Some of the numerical values of $t(v,p)$ are listed in Table 1.1.[3] The definition of t is

$$t = \frac{\bar{x} - \mu}{s}\sqrt{n} \tag{1.6}$$

This is also a random variable and follows Student's t–distribution with degrees of freedom $v=n-1$. If the population mean is known, one can obtain the bias $d=x-\mu$.

On the other hand, the confidence interval of the population variance with confidence level $(1-p)$ is given by

$$\frac{ns^2}{\chi^2\left(v,\dfrac{1}{2}p\right)} < \sigma^2 < \frac{ns^2}{\chi^2\left(v,1-\dfrac{1}{2}p\right)} \tag{1.7}$$

The definition of χ^2 is

$$\chi^2 = \frac{ns^2}{\sigma^2} \tag{1.8}$$

This is also a random variable obeying the χ^2 distribution.

The precision of the sampling is usually represented by 2σ. The standard deviation σ is called the average error and $(2/3)\sigma$ is the probable error. The quantity CV (coefficient of variation) is more often utilized in expressing the precision of the sampling.

$$CV = \frac{\sigma}{\mu} \times 100\,(\%) \tag{1.9}$$

TABLE 1.1 Student's t-Distribution $t(v,p)$

v/p	0.25	0.05	0.01
3	1.4226	3.1825	5.8409
4	1.3444	2.7764	4.6041
5	1.3009	2.5706	4.0321
6	1.2733	2.4469	3.7074
7	1.2543	2.3646	3.4995
8	1.2403	2.3060	3.3554
9	1.2297	2.2622	3.2498

On the other hand, the accuracy of the sampling is represented by the bias d. Now we will consider, as an example, whether a snap sampling is accurate or not with confidence level of 95% by assuming that the full-stream sampling is correct. Let the data obtained from the full-stream sampling be denoted by x_i and that from the snap sampling be denoted by x_i'. The procedure is as follows:

1. Calculate $d_i = x_i - x_i'$ for $i = 1, 2, \ldots n$
2. Calculate its variance from Equation 1.4.
3. Calculate t from Equation 1.6 with $\bar{d} = \bar{x} - \mu$.
4. Find the value of $t(v, 0.05)$ from Table 1.1.
5. Compare t with $t(v, 0.05)$. If $t > t(v, 0.05)$, the snap sampling has a significant bias with confidence level of 95%. The bias is given by $\bar{d}$.

The next problem is how many particles should be sampled so as to obtain, for example, the particle size distribution by sizing and counting. In case that the log-normal size distribution is applicable, distribution of sample mean particle size can be analytically obtained.[4] Based on the sample mean distribution, the minimum number of particles $n*$ can be estimated by the following equation:

$$n* = \frac{\omega}{\delta^2} \tag{1.10}$$

where δ is an error admissible in the evaluation of particle size analysis, and ω is a parameter defined by

$$\omega = z_c^2 m^2 s^2 \left(\frac{1}{2} m^2 s^2 + 1 \right) \tag{1.11}$$

where z_c is the standardized normal variable for the stated level of confidence which is determined by Table 1.2, m is a parameter in the evaluation function $\bar{x}_{m,0}$ given by Equation 1.13 below, and s is a standard deviation of log-normal particle size distribution which is given by the natural logarithm of geometric standard deviation s_g as follows.

$$s = \ln s_g \tag{1.12}$$

The evaluation function $\bar{x}_{m,0}$ is

$$\bar{x}_{m,0} = \bar{x}_{1,0} \exp\left(\frac{1}{2} m s^2 \right) \tag{1.13}$$

TABLE 1.2   Parameter z_c in Equation 1.11

Confidence level (%)	90	95	97.5	99	99.5
	1.64	1.96	2.24	2.57	2.81

For particle size measurements, the following evaluation functions are recommended.

Mass median diameter:
$$\overline{x_{6,0}} = \overline{x_{1,0}} \exp\left(3s^2\right) \qquad (1.14)$$

Sauter diameter:
$$\overline{x_{5,0}} = \overline{x_{1,0}} \exp\left(\frac{5}{2}s^2\right) \qquad (1.15)$$

Mean volume diameter:
$$\overline{x_{3,0}} = \overline{x_{1,0}} \exp\left(\frac{3}{2}s^2\right) \qquad (1.16)$$

The minimum number of particles can be estimated if one of these evaluation functions is adopted.[5] The procedure under the mass median diameter as the evaluation function is as follows:

1. Assume the estimation probability (confidence level). If it is assumed as 95%, then Table 1.2 gives $u = 1.96$.
2. Comparing Equation (1.13) with Equation (1.14), the parameter $m = 6$.
3. Calculate the parameter ω by Equation (1.11).

$$\omega = 36u^2s^2\left(18s^2 + 1\right) = 138.3s^2\left(18s^2 + 1\right) \qquad (1.17)$$

4. Set the admissible error δ. If it is assumed as $\delta = +-0.05$ (within $+-5\%$ error), Equation (1.10) gives

$$n* = 138.3s^2\left(18s^2 + 1\right)/\delta^2 = 55320s^2\left(18s^2 + 1\right) \qquad (1.18)$$

5. Determine the geometric standard deviation s_g of the measured particle size distribution. If s_g for the sample powder is 2.0, for example, then $s = \ln s_g = \ln 2.0 = 0.693$. Therefore, Equation 1.18 gives $n* = 256228$. This is the minimum number of particles which should be sampled so as to obtain the particle size distribution whose mass median diameter is within 5% of the corresponding population value with 95% confidence level.

The above procedure is applicable to other cases such as the analysis of specific surface area, chemical element of particles, and so on, as far as these variables depend on particle size. The evaluation function should be determined appropriately based on the mean particle diameter suitable for each variable.

REFERENCES

1. Masuda, H., Kurahashi, H., Hirota, M., and Iinoya, K., *Kagaku Kogaku Ronbunshu*, 2, 286–290, 1976.
2. Allen, T., in *Particle Size Measurement*, 3rd Ed., Chapman & Hall, London, 1981, chap. 1.
3. Hoel, P. G., *Elementary Statistics*, Wiley, New York, 1966, appendix.
4. Masuda, H. and Iinoya, K., *J. Chem. Eng. Jpn.*, 4, 60–66, 1971.
5. Masuda, H. and Gotoh, K., *Adv. Powder Technol.*, 10, 159–173, 1999.

2.2 Particle Sampling in Gas Flow

Hideto Yoshida

Hiroshima University, Higashi-Hiroshima, Japan

Hisao Makino

Central Research Institute of Electric Power Industry, Yokosuka, Kanagawa, Japan

Some problems related to the sampling of particles are reviewed and discussed here. In the particle sampling processes, a so-called sampling loss occurs at the entrance and inner wall of the sampling probe, so that the measured value can be incorrect. The subjects to be discussed here are (1) anisokinetic sampling errors, (2) sampling in stationary air, and (3) practical applications of particle sampling.

2.2.1 ANISOKINETIC SAMPLING ERROR

If the sampling velocity u is different from that of the main stream velocity u_o, the particle concentration in the sampling tube becomes different from that in the main stream. This type of discrepancy in particle concentration is referred to as anisokinetic sampling error. Figure 2.1 shows a computer simulation of particle trajectories near the sampling probe under anisokinetic sampling conditions. When the sampling velocity u is higher than that of the main stream velocity u_o, the measured concentration C_o becomes lower than the main stream concentration C_0. Conversely, when $u < u_o$, C becomes greater than C_o.

Table 2.1 lists some typical proposed expressions for anisokinetic sampling errors. Figure 2.2 shows the relation between the concentration ratio and the velocity ratio. The use of Davies' equation[1] is in good agreement with the numerical results.[2,3] Watson's equation[4] underestimates the sampling error.

Figure 2.3 shows the comparison between several empirical or semitheoretical equations listed in Table 2.1 and the results obtained by the numerical calculations for both potential and viscous flow. The error estimated by a numerical calculation for potential flow is the highest among them, whereas C/C_o, as estimated by Davies' equation, is intermediate between the numerical results of potential flow and viscous flow. From Figure 2.2, it can be seen that the concentration ratio C/C_o approaches unity when the velocity ratio u_o/u is less than 0.25. Figure 2.4 shows the relation between the concentration ratio and Levin's parameter k for a velocity ratio u_o/u of less than 0.25. The experimental data of Gibson and Ogden[7] and Yoshida et al.[8] are compared with Levin's equation[9]:

$$\frac{C}{C_o} = 1 - 0.8k + 1 - 0.8k^2, k = 2P\sqrt{\frac{u_o}{u}}$$

(2.1)

The experimental data agree well with Levin's equation and with the numerical calculation for point sink flow.

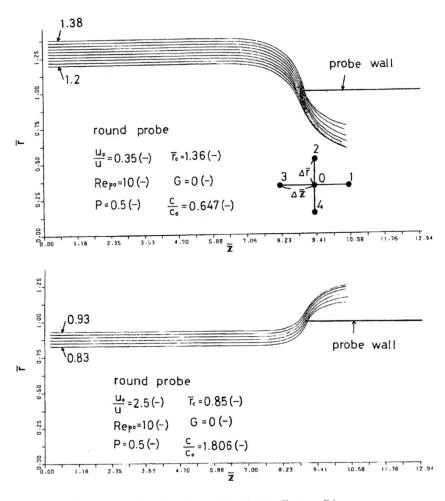

FIGURE 2.1   Particle trajectories for anisokinetic sampling conditions.

2.2.2   SAMPLING IN STATIONARY AIR

It is impossible to carry out isokinetic sampling in stationary air. Therefore it is important to estimate the error of anisokinetic sampling. This sampling error has been investigated by Levin,[9] Davies,[1] Kaslow and Emrich,[10] Breslin and Stein,[11] Gibson and Ogden,[7] Yoshida et al.,[8] and Agarwal and Liu.[12] Figure 2.5 shows the calculated fluid stream lines near the sampling probe for the case of viscous flow and potential flow. The fluid stream lines for both cases are nearly the same except for the probe inlet.

Figure 2.6 shows one of the calculated trajectories for particles in the viscous flow field. Sampling efficiency was also calculated using the point sink approximation. The analytical solution of Levin[9] agrees with the numerical calculations for the point sink flow up to Levin's parameter $k = 0.5$. For larger values of k, some particles orbit about the point sink, as shown in Figure 2.7. Experimental data by Yoshida et al.[8] agree well with their theoretical calculations based on the viscous flow. Figure 2.8 shows the calculated C/C_o as a function of inertia parameter P and gravitational parameter G. The solid lines represent upward sampling and the dashed lines show downward sampling. The results obtained by Agarwal and Liu[12] and Davies'[1] region of perfect sampling are also shown in Figure 2.8. When the probe is facing downward, the concentration ratio C/C_o is less than unity. However, when the probe is facing upward, the ratio C/C_o can become larger than unity because of

TABLE 2.1 Expressions for Anisokinetic Sampling Error

Researchers	Empirical or Semiempirical Equation
Watson (1954)	$\dfrac{u_0}{u}\left[1 + F\left(\sqrt{u/u_0} - 1\right)\right]^2$, F is a function of P
Badzioch (1959)	$1 + \alpha\left(\dfrac{u_0}{u} - 1\right)$, $\alpha = \dfrac{PD}{2L}\left[1 - \exp\dfrac{(-2L)}{PD}\right]$, $L = 6 - 1.6D$
Davies (1968)	$\dfrac{u_0}{u}\left(\dfrac{P}{P+0.5}\right) + \left(\dfrac{0.5}{P+0.5}\right)$
Zenker (1971)	$\dfrac{u_0}{u} + K\left(1 - \dfrac{u_0}{u}\right)$, $K\{1 + \exp(1.04 + 2.06\log(P/2))\}^{-1}$
Belyaev and Levin (1974)	$1 + \dfrac{P(1 + 0.31\,u/u_0)}{1 + P(1 + 0.31\,u/u_0)}\left(\dfrac{u_0}{u} - 1\right)$
	$0.18 < \dfrac{u}{u_0} < 6.0, \quad 0.36 < P < 4.06$

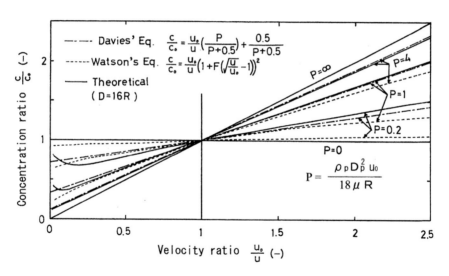

FIGURE 2.2 Calculated anisokinetic sampling errors.

the direct settling of larger particles. The effect of gravity is clearly shown in Figure 2.9, where the sampling efficiency C_m/C_o is represented by

Upward sampling:

$$\frac{C_m}{C_o} = \frac{C}{C_o}\frac{1}{1+G} \tag{2.2}$$

Downward sampling:

$$\frac{C_m}{C_o} = \frac{C}{C_o}\frac{1}{1-G} \tag{2.3}$$

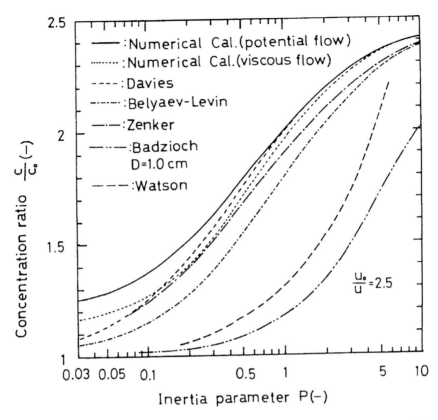

FIGURE 2.3   Relationship between concentration ratio and inertia parameter (numerical results compared with other equations).

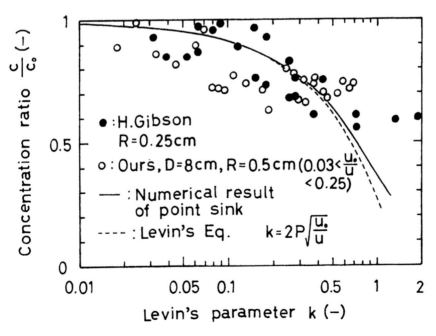

FIGURE 2.4   Experimental results of high-speed sampling errors compared with theoretical ones.

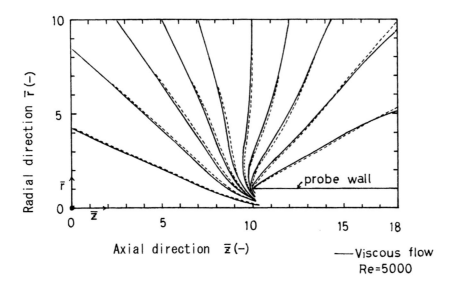

FIGURE 2.5   Calculated stream lines near sampling probe.

The maximum value of C_m/C_o is unity, as shown by the particle mass balance. A rapid fall in the efficiency of the downward sampling as the inertia parameter increases is found.

Particles with a diameter of less than about 10 μm are regarded as harmful to humans because small particles are easily deposited in the human lung. In Japan, particle concentration measurements with a diameter of less than 10μm are regulated as a 10 μm cut. The particle deposition efficiency on the lung surface increases sharply for particle diameters of less than about 2 μm. The concentration measurement of particles with a diameter less than 2.5 μm is referred to as PM-2.5 sampling. Figure 2.10 shows U.S. regulations regarding a PM-2.5 sampler. In this case, the sampling flow rate and particle density are regulated to 16.7 l/min and 1,000 kg/m³, respectively.

Classification performance of the sharp-cut cyclone (SCC) inlet cyclone does not satisfy this criterion and further research will be needed to determine the sharp cut PM-2.5 sampler.

2.2.3   PRACTICAL APPLICATIONS OF PARTICLE SAMPLING

Sampling Probe and System Used in Practical Application

Figure 2.5 shows the sampling probes that are commonly used for measurement of dust concentration. They are referred to as static-pressure-balance-type sampling probes and standard-type sampling probes. The inner static pressure must be equal to the outer static pressure for isokinetic sampling in the case of static-pressure-balance-type sampling probe. However, if the inside cross-sectional area of the sampling probe is constant, the sampling velocity is lower than the main flow velocity because of the pressure drop from the nozzle inlet to the static-pressure detecting hole in the sampling probe. Tamori[13] has proposed an advanced static-pressure-balance-type sampling probe, in which the inner diameter expands.

Whitely and Reed[14] investigated the effect of the shape of the sampling probe on measurement of dust concentration. They measured dust concentration using the simplified probe and the standard probe. For the simplified probe, the distance between the probe tip and the bend is short. They concluded that the difference in measured concentration between these two sampling probes is small.

Measured concentration also varies with the angle between the sampling probe and main flow. Watson[4] and Raynor[15] examined the sampling error by changing the angle between the sampling probe and main flow. They found no significant difference at an angle less than ±5°. Belyaev and

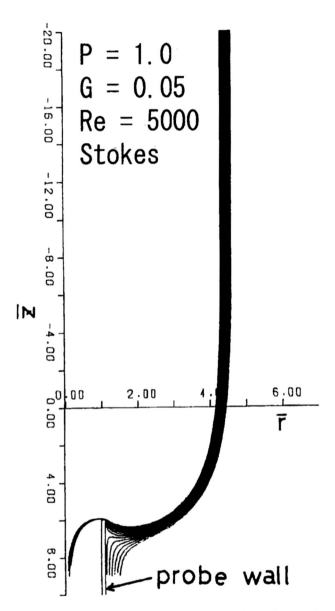

FIGURE 2.6   Simulated particle trajectories in stationary air (viscous flow).

Levin[16] investigated the effect of the thickness of the probe tip and concluded that the sampling error increases when the probe diameter is decreased.

Measuring Method for Particle Size Distribution by Anisokinetic Sampling

The sampling efficiency during anisokinetic sampling is dependent on the particle size when the main flow velocity and the diameter of the sampling probe are fixed. Consequently, it should be noted that the anisokinetic sampling performs the classification of the particles. Particle size distributions can be measured by using this "classification effect" during anisokinetic sampling.[17]

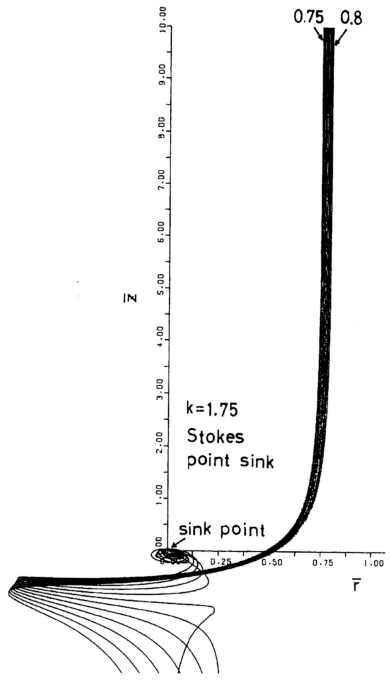

FIGURE 2.7 Calculated particle trajectories in stationary air (point sink approximation).

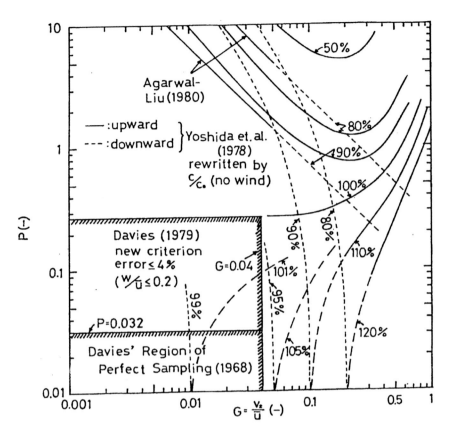

FIGURE 2.8   Constant sampling efficiency curves in stationary air.

The sampling efficiency of monodisperse particles has a linear relation with the velocity ratio, according to Davies' equation, so that the anisokinetic sampling efficiency can be easily estimated. The slope a of this linear line varies with the particle size, as shown in Figure 2.12. In the case of polydisperse particles, the concentration ratio also has a linear relation with the velocity ratio. The slope A of this relation for polydisperse particles is obtained by integrating the product of the slope a and the particle size distribution function $f(x)$:

$$A = \int_0^\infty af(x)\,dx = \int_0^\infty \left(\frac{P(x)}{P(x)+0.5}\right) f(x)\,dx \tag{2.4}$$

Equation 2.4 can be used to determine $f(x)$ when the slope A is experimentally obtained because the inertial parameter can be easily calculated. However, when a large number of variables in $f(x)$ exist, a solution of the above equation cannot be obtained. Some assumptions are necessary to calculate the particle size distribution. For example, the particle size distribution function is required. Figure 2.13 shows the particle size distribution of fly-ash particles calculated by this method. Fly-ash particles properly follow a log-normal distribution, and the geometric standard deviations of these particles are nearly 2.5, so that the median diameter x_{50} can be calculated. In this figure, the results measured by the Andersen stack sampler (ASS), which is one of the most common methods used in particle size distribution measurements, are also shown. It is concluded that the measured values by two different methods fully coincide with each other is impossible, because the actual

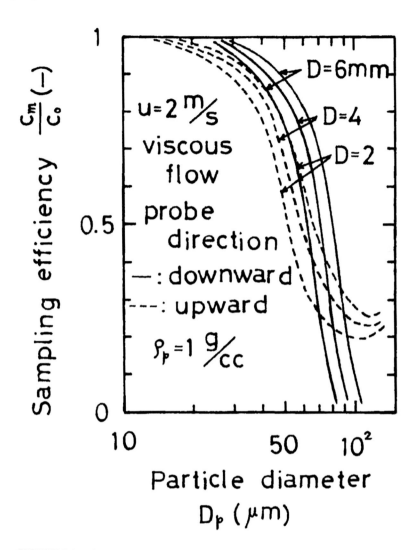

FIGURE 2.9 Sampling efficiency curves for downward and upward probe directions.

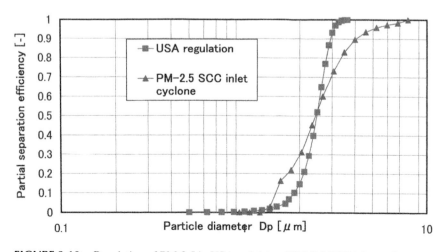

FIGURE 2.10 Regulation of PM-2.5 in USA and data of PM-2.5 SCC inlet cyclone.

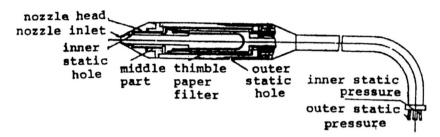

Static pressure balance type sampling probe

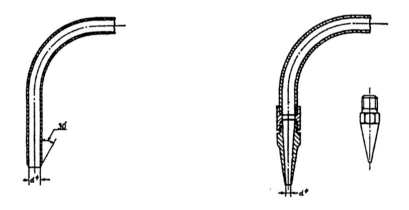

Standard type sampling probe

FIGURE 2.11   Sampling probes for dust measurement.

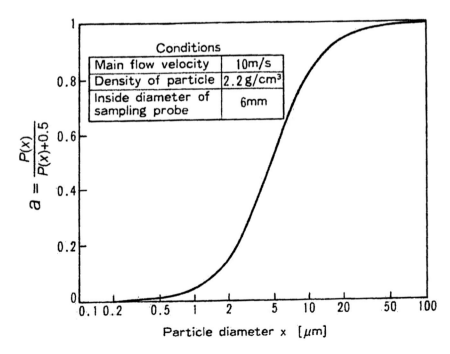

FIGURE 2.12   Change of the slope "a" in Davies' equation.

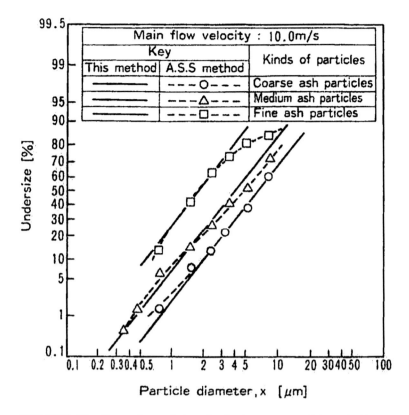

FIGURE 2.13   Comparison of size distributions measured by anisokinetic sampling and ASS.

particle size distribution does not follow the ideal log-normal distribution. However, the differences between this method and a conventional method are very small.

On the other hand, to measure complicated particle size distributions, the new method proposed by Tsuji et al.[18,19] uses plural sampling probes of different diameters. The inertia parameter can be changed by altering the diameter of the sampling probe. It, therefore, is possible to obtain plural equations such as Equation 2.4, which show the relationship between slope A and $f(x)$ when plural sampling probes of different diameters are used. If more than two sampling probes are used, these equations can be solved by use of Twomey's method.[20] Figure 2.14 shows the particle size distributions of fly-ash measured by this method (the three-probe method) and ASS. While there are small differences in the measured values by two methods, these methods are roughly in agreement.

Since the size distributions of droplets and melted particles may change after sampling and these particles flow on the plates in ASS, measurements of these particles by use of conventional methods becomes very difficult. On the other hand, the method shown in this section is suitable for these particles, because the result is obtained from only the measurement of particle concentration.

Measuring Method for Particle Size Distribution by Backward Sampling

The method using anisokinetic sampling is capable of measuring the size distribution of particles whose size is in the order of 10 μm. the accuracy of this method becomes lower, because the slope a in Figure 2.12 is small and does not change significantly with particle size. The reason for this is that the fine particles easily follow the changes of the stream lines near the probe during the anisokinetic

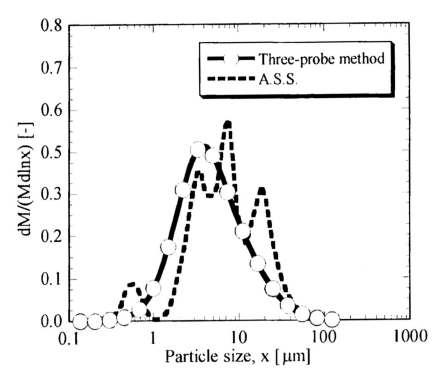

FIGURE 2.14   Particle size distribution of fly-ash measured by the three-probe method and ASS.

sampling due to their small inertia forces. The backward sampling, in which the probe is aligned at 180° to the flow stream, is effective for the classification of fine particles.[21,22]

In backward sampling, the stream line near the sampling probe changes substantially compared to anisokinetic sampling. Large particles cannot follow these stream lines and are not aspirated into the probe. Fine particles, which easily follow the stream line, enter the probe.

The next equation was proposed to express the classification efficiency during backward sampling.[21,22]

$$\frac{C_i}{C_0} = \exp\left\{-5.09\frac{U_0}{U_i}P(x)\right\}$$

(2.5)

In order to measure particle size distribution by use of classification during backward sampling, the backward sampling method and the combined method are proposed. The backward sampling method involves classification during backward sampling, and the combined method involves classification during both backward sampling and anisokinetic sampling.

The accuracy of these methods was compared using a computer simulation.[23] Figure 2.15 shows one of the results. In this figure, it is assumed that the true distribution in the main flow is the bimodal distribution whose peaks are 1 μm and 10 μm. The value by the three-probe method is also shown in this figure. While the three-probe method cannot estimate the two peaks, the backward sampling method is able to estimate these peaks. However, there are small differences between the true distribution and the value by the backward sampling in the 10 μm region. The combined method is the most accurate of these methods. The reason for this is that the combined method utilizes both back-

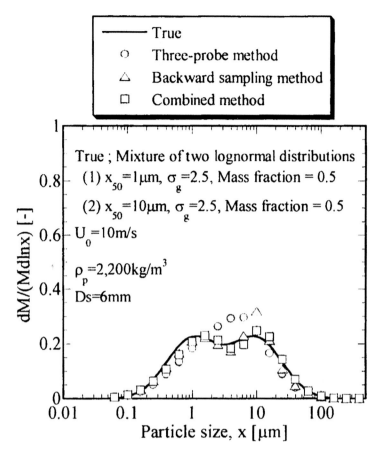

FIGURE 2.15 Comparison of the accuracy of the three-probe method, backward sampling method, and the combined method.

ward sampling, which is effective for fine particles less than several μm, and anisokinetic sampling, which is effective for particles of the order of 10 μm.

The particle size distribution of fly-ash in a coal combustion exhaust gas measured by the backward sampling method and the combined method are shown in Figure 2.16. This fly-ash has a typical size distribution whose peak is at 5 μm. The distributions measured by these two methods nearly agree with the result by ASS. The accuracy of the methods by use of the backward sampling is similar to ASS.

Notation

a	Slope of linear relation between concentration ratio and velocity ratio for monodisperse particles
A	Slope of linear relation between concentration ratio and velocity ratio for polydisperse particles
C/C_o	Particle concentration in a sampling probe and main stream (kg/m³) Sampling efficiency in stationary air
D, D_1	Probe diameter and duct diameter (m)
$F(x)$	Particle size distribution function (μm⁻¹)
G	Gravitational parameter in duct flow and in stationary air ($= \rho_p x^2 g/18\mu u_o$, $= \rho_p x^2 g/18\mu u$)

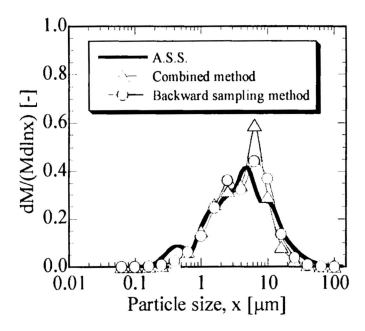

FIGURE 2.16   Particle size distribution of fly-ash measured by backward sampling and the combined method.

k	Levin's parameter
p	Inertia parameter in duct flow and in stationary air ($= \rho_P x^2 u_o/18\mu R$, $= \rho_P x^2 u/18\mu R$)
R	Probe radius (m)
Re	Flow Reynolds number
Re_{po}	Free-stream particle Reynolds number ($= x u_o \rho/\mu$)
$r(=r/R)$, $z(=z/R)$	Dimensionless coordinates
$r_c(=r_c/R)$	Radial coordinate of critical particle trajectory
u, u	Sampling velocity of duct flow and in stationary air (m/s)
u_o	Main flow velocity (m/s)
v_t	Particle settling velocity (m/s)
μ	Viscosity of fluid (Pa s)
	Density of fluid (kg/m³)
P_p	Particle density (kg/m³)
x	Particle diameter (μm)
x_{50}	Mass median particle diameter (μm)

REFERENCES

1. Davies, C. N., *J. Appl. Phys.*, 1, 921–929, 1968.
2. Yoshida, H., Ohsugi, T., Masuda, H., Yuu, S., and Iinoya, K., *Kagaku Kogaku Ronbunshu Japan*, 2, 336–340, 1976.
3. Masuda, H., Yoshida, H., and Iinoya, K., *J. Powder Technol. Jpn.*, 18, 177–183, 1981.
4. Watson, H. H., *AIHA Quart.*, 15, 21–29, 1954.
5. Badzioch, S., *Br. J. Appl. Phys.*, 10, 26–29, 1959.
6. Zenker, P., *Staub. Rein Luft*, 31, 30–38, 1971.

7. Gibson, H., and Ogden, T. L., *J. Aerosol Sci.*, 8, 361–369, 1977.
8. Yoshida, H., Uragami, M., Masuda, H., and Iinoya, K., *Kagaku Kogaku Ronbunshu Japan*, 4, 123–128, 1978.
9. Levin, L. M., *Bull. Acad. Sci. USSR Geophys. Ser.*, 7, 87–94, 1957.
10. Kaslow, D. E. and Emrich, R. J., *Technical Report 23*, Lehigh University, 1974.
11. Breslin, J. A. and Stein, R. L., *AIHA J.*, 36, 576–586, 1975.
12. Agarwal, J. K. and Liu, B. Y. H., *AIHA J.*, 41, 191–199, 1980.
13. Tamori, Y., *Funtai Kogaku Kenkyu Kaishi Japan*, 11, 3–13, 1974.
14. Whitely, A. B. and Reed, L. E., *J. Inst. Fuel*, 32, 316–325, 1959.
15. Raynor, G. S., *Am. Ind. Hyg. Assoc. J.*, 31, 294–299, 1970.
16. Belyaev, S. P. and Levin, L. M., *J. Aerosol Sci.*, 3, 127–136, 1972.
17. Makino, H., Tsuji, H., Kimoto, M., Yoshida, H., and Iinoya, K., *Kagaku Kogaku Ronbunshu Japan*, 21, 896–903, 1995.
18. Tsuji, H., Makino, H., Kimoto, M., Yoshida, H., and Iinoya, K., *J. Aerosol Sci.*, 26 (Suppl. 1), S113–S114, 1995.
19. Tsuji, H., Makino, H., Yoshida, H., and Iinoya, K., *J. Aerosol Sci.*, 28 (Suppl. 1), S681–S682, 1997.
20. Twomey, S., *J. Comput. Phys.*, 18, 188–200, 1975.
21. Tsuji, H., Makino, H., Yoshida, H., Ogino, F., Inamuro, T., and Fujita, I., *Kagaku Kogaku Ronbunshu Japan*, 25, 780–788, 1999.
22. Tsuji, H., Makino, H., and Yoshida, H., *Powder Technol.*, 118, 45–52, 2001
23. Tsuji, H., Makino, H., and Yoshida, H., *J. Powder Technol. Jpn.*, 36, 810–818, 1999.

2.3 Concentration and Flow Rate Measurement

Hiroaki Masuda and Shuji Matsusaka
Kyoto University, Katsura, Kyoto, Japan

2.3.1 PARTICLE CONCENTRATION IN SUSPENSIONS

The concentration of particles suspended in gas or liquid is expressed in several ways. The various definitions of particle concentration are summarized as follows:

1. Mass flow ratio m (–):

$$m = \frac{W}{Q} \tag{3.1}$$

where W and Q are the mass flow rates of particles and fluid, respectively.

2. Flow concentration c (kg/m³):

$$c = \frac{W}{Q_v} = \rho_a m \tag{3.2}$$

where Q_v is the volumetric fluid flow rate and ρ_a the density of fluid. If the concentration c is divided by the particle density ρ_p, it becomes dimensionless and is called the volumetric concentration.

3. Holdup concentration c_h (kg/m³):

$$c_h = \frac{dw}{V} = \frac{Wdl}{vV} = \frac{W}{vA} \tag{3.3}$$

where dw is the mass of suspension, V the volume occupied by the suspension, dl the length, v the mean velocity of particles, and A the cross section.

The above-mentioned three types of concentration are often called the particle concentration for simplicity. The holdup concentration c_h is related to the flow concentration c by the following equation:

$$c_h = \frac{u}{v}c \tag{3.4}$$

where u ($= Q_v/A$) is the velocity of fluid. In gas–solids suspension flow, the particle velocity v is usually smaller than the gas velocity u, so that the holdup concentration c_h is higher than the flow concentration c. The number concentration is also utilized instead of the mass or volumetric concentration. The optical particle counter, for example, measures the number concentration.

Table 3.1 lists the methods for the measurement of relatively high particle concentration in powder-handling processes. A gamma-ray densitometer or density gauge is used especially for the continuous measurement of slurries with a high solids concentration. Figure 3.1 shows this instrument, which

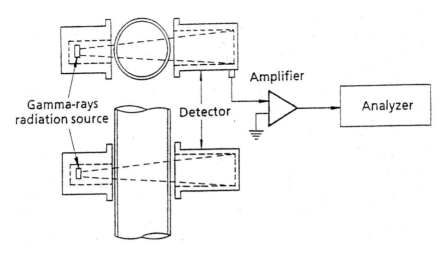

FIGURE 3.1   Gamma-ray density gauge.

TABLE 3.1   Powder Concentration Meter

Principle	Notes
Electric capacitance increase	Electric condenser, parallel-plate condenser
Gamma-ray attenuation	Gamma-ray densitometer, Lambert-Beer law
Statistical correlation of various noise	Autocorrelation
Microwave reasonance	Resonant frequency decrease

TABLE 3.2   Dust Concentration Meter

Principle	Notes
Photo extinction	Photometer, transmissometer, Lambert-Beer law
Light scattering	Photometer, nephelometer, Optical particle counter
Laser radar	Lidar (long-range measurement)
Nucleated condensation	Condensation nuclei counter (ultrafine particles)
Electrostatic charging	Charge transfer by impact

consists of a radioactive source, radiation detector, amplifier, and analyzer. For a radioactive source, ^{137}Cs ($E = 662$ keV) and ^{241}Am ($E = 60$ keV) are usually utilized. The gamma rays produced by these sources pass through the pipe line, and the output is manipulated to get the slurry concentration. The slurry or pulp density can be measured by various methods, including the above-mentioned nuclear density meters, hydrometers, balanced flow vessels, differential air bubblers, ultrasonic density meters, and so on.[1-4]

The methods listed in Table 3.2 are applicable to fine particles. Table 3.3 lists the methods for detecting the mass of fine particles sampled on a filter or sampled in the sensing zone of an instrument.

TABLE 3.3   Dust Monitor (Sampled Particles)

Principle	Notes
Beta-ray attenuation	Beta gauge, Lambert-Beer law
Light attenuation	Lambert-Beer law
Electrically forced vibration	Mechanical resonance (frequency decrease)
Piezoelectric vibration	Piezobalance[a] (frequency decrease)
Electrostatic charging	Charge measurement after unipolar charging

[a] Trade name.

The various electrical phenomena and the attenuation technique can be used in measurement as follows.

Electric Capacitance Change

Electric capacitance of a parallel-plate capacitor increases with increasing particle concentration in a gas–solids suspension. The measured concentration is the volumetric concentration. As the area of the parallel-plate capacitor is made as small as 1 cm^2, the sensor is utilized to measure the local concentration of particles in powder-handling equipment such as a fluidized bed.[5]

Electrostatic Charging

The electric current generated by particle impact is proportional to the powder flow rate. Therefore, the flow concentration c of a gas–solids suspension will be obtained if the suspension is sampled with a constant velocity. Particles in sampled gas impact on the sensor target placed in the sampled flow or impact on an inside wall of the sensor tube. Any sensitive materials can be used as the sensor.[6]

Induction Charging

The electroconductive particles will be polarized in an electric field. If the polarized particle contacts with a target electrode, negative or positive charge will be neutralized by the opposite charge supplied from the electrode; the particle obtains net charge after the contact. Therefore, the current required for the charging will be found between the electrodes as a voltage pulse.[7] The number of pulses found per unit time can give the number concentration.

Piezoelectric Effect

A piezoelectric sensor can be applied to the measurement of particle concentration in a flowing suspension. The sensor detects the number frequency of impact. At the same time, the particle mass is measured so that one can transform the frequency into the corresponding mass concentration. In this kind of sensor, the impact efficiency should be correctly taken into consideration.[6,8]

Electrically Forced Vibration (Resonance Technique)

This method is based on the electrically forced vibration of collected particle mass. Suspended particles are sampled isokinetically onto a fibrous filter that is connected to a spring. The sensing element is electrically vibrated. If the frequency of the forced vibration coincides with the natural frequency, resonance will be found. Therefore, if the resonance frequency is detected, the mass m will be calculated. As the particles deposit on the filter, the mass m increases from m to $m + \Delta m$, and the resonance frequency changes. If the frequency shift is measured, the mass of the particles will be obtained.[9] Then the flow concentration c is calculated by

$$c = \frac{\Delta m}{Q_v \Delta t} \tag{3.5}$$

where Δt is the duration of sampling and Q_v is the sampled flow rate.

Quartz Crystal Microbalance

A mass sensor utilizing a quartz oscillator is based on the same principle as discussed above. The sensor is called the quartz crystal microbalance (QCM). Suspended fine particles are sampled by use of a suction pump and are collected on the quartz crystal.[10]

The resonance frequency decreases as discussed above. The frequency shift Δf (Hz) for the quartz crystal with the natural frequency of 5 MHz is given by

$$\Delta f = -5.65 \times 10^6 \frac{\Delta m}{S} \tag{3.6}$$

where S is the particle-collection area (m^2). The mass sensitivity ($C_f = 5.65 \times 10^6$) is independent of the physical properties of the deposited particles. A frequency shift as small as 0.1 Hz is measured by use of a simple frequency counter. Therefore, Equation 3.6 shows that the QCM can detect a small amount of particles of 1.8×10^{-8} kg/m^2 ($= 1.8$ ng/cm^2).

Attenuation Technique

The energy of waves such as light (electromagnetic wave), sound, or radioactive rays will be dissipated through suspended particles or a powder bed because of wave absorption or scattering. If a light beam passes through a uniformly suspended aerosol layer, the intensity of light will decrease as follows:

$$\frac{I}{I_0} = \exp(-\eta_m C_h L) \tag{3.7}$$

where I_0 is the initial incident intensity of light (i.e., intensity of the transmitted light without particles), C_h is the holdup concentration, η_m is the mass absorption coefficient, and L is the total thickness of the aerosol layer. Equation 3.7 implies that the intensity of the light beam decreases exponentially with the thickness of the layer. This fact is known as the Lambert–Beer law. The maximum attenuation will be found for a certain particle size between 0.1 and 2 µm. On the other hand, the mass absorption coefficient for very short wavelengths such as UV light (0.01 to 0.4 µm) or X-rays (less than 0.01 µm) is approximately independent of the particle size. For the mass measurement of particles sampled on dust filters, gamma rays used in the density gauge are too strong, and beta rays from ^{14}C,^{147} Pm (promethium), or ^{63}Ni are utilized. The concentration meter based on the beta rays is applicable to the measurement of higher concentration than the QCM. The mass absorption coefficient η_m is a function of the maximum energy E_{max} of the beta rays, and it is almost independent of the kinds of materials deposited. Therefore, the meter is easily calibrated by use of a plastic film such as a Mylar film.[11]

2.3.2 POWDER FLOW RATE

There are two types of powder flow: bulk solids flow and suspension flow. The principles of powder flowmeters for bulk solids flow and suspension flow are listed in Table 3.4 and Table 3.5, respectively. The weighing method in Table 3.4 belongs to direct measurement, but others belong to inferential measurements.

The powder flow rate W (kg/s) in suspension flow is the product cvA (kg/s) of particle concentration c (kg/m^3), particle velocity v (m/s), and cross-sectional area A (m^2). Therefore, the measurement of the powder flow rate requires a combined measurement of particle concentration and particle velocity. Table 3.6 lists the principles of the particle velocity measurement. The mean velocity of powder flow across the cross-sectional area should be measured to obtain the powder flow rate from the combined concentration-velocity measurement.

Weighing Method

The weighing method is the most reliable and accurate in the measurement of the powder flow rate in the bulk solids flow.[1] The hopper scale (weigh hopper) and the belt scale (belt weigher or conveyor scale) are the typical flowmeters of this kind.

The hopper scale receives powder in the hopper, weighs it, and then discharges. The frequency of receiving and discharging is 100–400 cycles per hour, which can be changed according to the process requirement. A combination of high speed in the early stage of the feeding and low speed in the final stage allows a short cycle time with high accuracy.[6] A hopper scale which weighs the scale hopper after the feeding is perfectly stopped is called the net-weight mode hopper scale. For continuous powder feeding, the hopper scale is operated in the loss-in-weight mode, in which the powder is added before the hopper becomes empty and the discharge flow is kept continuous. The powder adhered on the hopper wall does not cause a weigh error of the hopper scale in this mode.

The belt scale is, however, more suitable than the hopper scale for the continuous feeding of powder, where the flow rate is obtained as a product of the weight of powder on the belt of unit

---

TABLE 3.4   Powder Flowmeter for Bulk Solids Flow

Principle	Notes
Weighing	Belt scale, hopper scale
Impulsive force	Impact flow meter
Coriolis force	Massometer,[a] etc.
Back pressure of air jet	Gas purge type of flowmeter
Volume displacement	Turbine type of flow meter

[a] Trade name.

TABLE 3.5   Powder Flowmeter for Suspension Flow

Principle	Notes
Pressure drop	Venturi-type flowmeter, Venturi-orifice flowmeter
Electrostatic charging	Particle electrification by impact
Microwave	Microwave-type flowmeter (microwave resonance)
Statistical phenomena	Correlation-type flowmeter (concentration × velocity)
Heat absorption	Metallic particles

TABLE 3.6   Particle Velocity Meter

Principle	Notes
Statistical Phenomena	Correlation technique (electrostatic current, electric capacitance, light extinction
Spatial filtering	Microwave type, laser velocimeter (lattice window)
Piezoelectric phenomena	Piezocrystal (momentum detector)
Doppler effect Time of flight	Laser Doppler velocimeter Double-beam laser

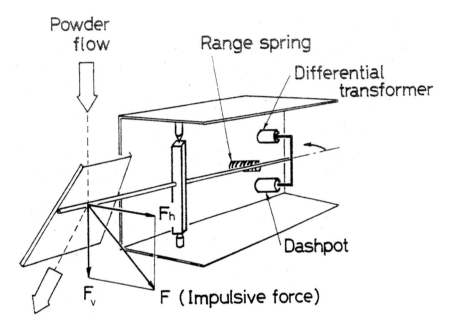

FIGURE 3.2   Impact flowmeter.

length and the belt speed. When the flow rate is controlled so as to feed a constant powder mass, the belt scale is called a constant-feed weigher. Belt speed control is more preferable than gate opening control in the operation of the constant-feed weigher.[12] A mechanical unit or load cell is used to weigh the powder.

Impulsive Force Flowmeter

The flowmeter shown in Figure 3.2, which is based on the impulsive force, is called an impact flowmeter. Powder falls down by gravity and impacts on an inclined plate. The force acting on the plate is calculated from the impulse versus momentum relation, and it is expressed by the horizontal and vertical components. The impact flowmeter has been able to detect the vertical component since the early stage of its development,[13] which is the sum of the impulsive force and the weight of the powder adhered on the plate. On the other hand, a recently developed one detects the horizontal component and it is not affected by the adhered powder.[6] The horizontal component of the impulsive force is given by

$$F_h = \frac{1}{2}(e + \mu) K \sin 2\theta \sqrt{2gh}\, W \qquad (3.8)$$

where e is the coefficient of restitution of particles, μ is the coefficient of friction, K is the air-drag coefficient, h is the free-falling height of the powder, θ is the inclination of the plate, and W is the powder mass flow rate.

The flowmeter is very compact and its dynamic response is very fast. Further, dust intrusion into the mechanical parts is easily shut out. The principle has also been applied to gas–solids suspension flow.[14] In this case, the gas and particle velocities should be measured by a suitable method. Also, the particle inertia must be large enough so that all particles impact on the detector.

Coriolis Force Flowmeter

When the powder flow is received on a rotating disk and thrown off from the periphery of the disk, the powder suffers the Coriolis force in the direction counter to the rotation. The resulting total torque T on the rotating axis is given by

$$T = \omega r^2 W \tag{3.9}$$

where ω is the angular velocity of the table, r is the table radius, and W is the powder mass flow rate. Equation 3.9 contains no calibration constant, which is the most important feature of the Coriolis force flowmeter.

Differential Pressure Method

Mass flowmeters based on the differential pressure or pressure drop are applied mainly to suspension flow. However, the principle is also utilized in the measurement of bulk powder flow.[15] One feature of the method is that the electric parts of the flowmeter can be isolated from the main process where there might be radioactive rays. When particles are transported by a pneumatic conveyor, the pressure drop along the pipeline usually becomes larger than the corresponding airflow only. The additional pressure drop depends on the mass flow ratio m. In the measurement of powder flow rate, it is necessary to know the relationship between the mass flow ratio and the pressure drop, besides the gas flow rate in the gas–solids mixture flow. The venturi meter[16] or the orifice meter is applicable for this purpose.

The powder flow rate W is calculated from

$$W = \mathrm{K}\sqrt{\rho}\,\frac{\Delta P - k\Delta P_a}{\sqrt{\Delta P_a}} \tag{3.10}$$

where ΔP is the pressure drop at the straight pipe section, ΔP_a is the pressure recovery at the diffuser, ρ is the air density, and K and k are the calibration constants. The calibration constant k is adjustable to unity if the distance between pressure taps in the straight pipe section is designed so that the pressure drop for airflow only is equal to the pressure recovery at the diffuser.

The pressure drop of a venturi nozzle is utilized in the venturi-orifice flowmeter.[17] A venturi with a long throat is also used to increase the pressure drop and attain higher meter sensitivity.[18] Another version of the flowmeter is composed of a straight pipe with a core inserted so as to make an annular flow between them. The particles will be accelerated by high-speed annular flow, resulting in a pressure drop. The flowmeter is insensitive to upstream disturbances and does not require a long straight inlet section as in the case of the venturi flowmeter.

The above-mentioned flowmeters can be used in gas–solids mixture flow with mass flow ratio m below 10. For a higher mass flow ratio, the relation between the pressure drop ratio ($\Delta P/\Delta P_a$) and the mass flow ratio m becomes nonlinear.

Electrical Method

The electric currents generated by particle impact can be applied to the measurement of the powder flow rate in gas–solids mixture flow. However, the concentration should be rather small, because the relation between the current and the flow rate becomes nonlinear for a higher concentration.[19] The sensitivity depends on the material of the detector.[20] If the detector is covered with particles, impact electrification becomes ineffective; therefore, the air velocity should be higher to prevent the particles from adhering. Also, there is a measuring system that uses two detectors that can measure both the powder flow rate and the average electrical charge of particles in aerosol pipe flow.[21,22]

The electromagnetic flowmeter is based on Faraday's law of electromagnetic induction. Figure 3.3 shows an electromagnetic flowmeter, which is truly obstructionless and is widely used for slurry flow measurement. The pipe is made from a nonmagnetic material such as stainless steel, through which the magnetic field can pass. The inside wall is covered by a nonconductive plastic or ceramic. A pair of electrodes is set on the wall perpendicularly to the magnetic field. The potential difference induced between the electrodes as the slurry moves across the magnetic field is given by

$$V = kBDu \qquad (3.11)$$

where k is the proportional constant (unitless), B is the magnetic flux density (T), D is the inside diameter of the pipe (m), and u is the average flow velocity (m/s). The volumetric flow rate Q_v (m³/s) is given by

$$Q = \frac{\pi D}{4kB} V \qquad (3.12)$$

The electromagnetic flowmeter is not affected by variations in density, viscosity, pH, pressure, or temperature. If the concentration is higher than 10%, the proportional constant can become larger because of the increase in magnetic flux density, and recalibration of the flowmeter will be necessary. Also magnetic solids such as magnetite can cause an erroneous reading, and particle deposition on the wall will affect the pickup of potential difference.

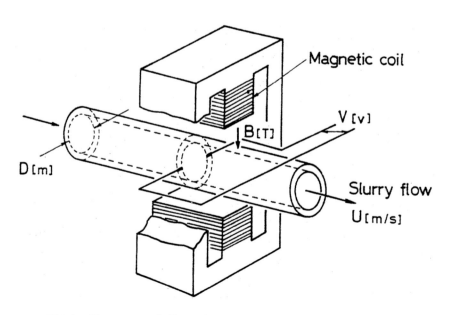

FIGURE 3.3   Electromagnetic flowmeter.

There is another method based on electrostatic induction. By monitoring the electrostatic flow noise generated while charged particles are conveyed through the pipeline, the mass flow rate could be calculated.[23]

The piezoelectric method is used only to detect local flow.[24] The sensor is made of a piezoelectric crystal or ceramics. If the dielectric crystal is deformed by an impulsive force, displacement of ions in the crystal takes place (polarization), which induces an electric pulse between the upper and lower surfaces of the crystal. The pulse can be detected if the surfaces are coated with metal and connected to a pulse analyzer. The particle mass should be known to measure the local mass flux or concentration by use of the piezoelectric sensor. A strain gauge can be used as a sensor of the impact force instead of the piezoelectric crystal.[25]

The microwave (i.e., electromagnetic wave with frequency range 10^9–10^{12} Hz) method utilizes a circular tube resonator lined with a dielectric material.[26] If the gas–solids mixture flows through the resonator, the resonance frequency changes so that the volumetric concentration c_v can be obtained. On the other hand, there exists a standing wave in the resonator, and it works as a lattice window for the velocity measurement (i.e., spatial filtering). The powder flow rate is obtainable as a product of concentration and velocity.

Statistical Method

A pair of sensors is set along a conveyor line a distance L apart. The signals detected by the sensors will fluctuate because of the flow turbulence. If the autocorrelation is calculated for various values of the delay-time parameter τ, it takes a maximum at a certain value of $\tau = \tau_0$. τ_0 gives the time interval for the particle cloud to travel between two sensors. Hence the particle flow velocity v is determined as

$$v = \frac{L}{\tau_0} \tag{3.13}$$

The powder flow rate W will be determined as the product of the particle velocity by particle concentration and the cross-sectional area of the flow.[27]

The statistical method has no obstruction to the particle flow and hence is preferable for process instrumentation. The method can be applied to the bulk solids flow in hoppers or chutes by use of a sound noise pickup (i.e., microphone). Fiber optics can be used in the measuring system for particles flowing on a chute or moving in a vessel.

REFERENCES

1. Perry, R. H. and Green. D. W., in *Perry's Chemical Engineers' Handbook,* 6th Ed., McGraw-Hill, New York, 1984, p. (5) 8–20.
2. Mitchel, J. W., in *Coal Preparation,* American Institute of Mining, Metallurgical, and Petroleum Engineers, New York, 1979, p. (19) 42.
3. Wills, B. A., in *Mineral Processing Technology,* Pergamon Press, Oxford, 1979, p. 45.
4. Weiss, N. L., in *Mineral Processing Handbook,* American Institute of Mining, Metallurgical, and Petroleum Engineers, New York, 1985, p. (10) 173.
5. Bakker, P. J. and Heertjes, P. M., *Br. Chem. Eng.,* 3, 240, 1958.
6. Iinoya, K., Masuda, H., and Watanabe, K., in *Powder and Bulk Solids Handling Processes,* Marcel Dekker, New York, 1988, pp. 65–130.
7. Keily, D. P. and Millen, S. G., *J. Meteorol.,* 17, 349, 1960.
8. Mann, U. and Crosby, E. J., *Ind. Eng. Chem. Process Des. Dev.,* 16, 9, 1977.
9. Wang, J. C. F., Kee, B. F., Linkins, D. W., and Lynch, E. Q., *Powder Technol.,* 40, 343–351, 1984.
10. Lu, C. and Czanderna, A. W., *Application of Piezoelectric Quartz Crystal Microbalances,* Elsevier, New York, 1984.

11. Sem, G. J. and Borgos, J. A., *Staub Reinhalt. Luft,* 35, 5, 1975.
12. Grader, J. E., *Control Eng.,* 15 (3), 60, 1968.
13. Iinoya, K., Yoneda, T., Kimura, N., Watanabe, K., and Shimizu, T., *J. Res. Assoc. Powder Technol. Jpn.,* 3, 424–431, 1966.
14. Barth, W., *Chem. Ing. Technol.,* 29, 599, 1957.
15. Kagami, H., Maeda, M., and Yagi, E., *Kagaku Kogaku Ronbunshu,* 1, 327–329, 1975.
16. Masuda, H., Ito, Y., and Iinoya, K., *J. Chem. Eng. Jpn.,* 6, 278–282, 1973.
17. Farber, L., *Trans. ASME,* 75, 943, 1953.
18. Iinoya, K. and Gotoh, K., *Kagaku Kogaku,* 27, 80, 1963.
19. Masuda, H., Mitsui, N., and Iinoya, K., *Kagaku Kogaku Ronbunshu* 3, 508–509, 1977.
20. Matsusaka, S., Nishida, T., Gotoh, Y., and Masuda, H., *Adv. Powder Technol.,* 14, 127–138, 2003.
21. Masuda, H., Matsusaka, S., and Nagatani, S., *Adv. Powder Technol.,* 5, 241–254, 1994.
22. Masuda, H., Matsusaka, S., and Shimomura, H., *Adv. Powder Technol.,* 9, 169–179, 1998.
23. Gajewski, J. B., *J. Electrostat.,* 37, 261–276, 1996.
24. Heertjes, P. M., Verloop, J., and Willems, R., *Powder Technol.,* 4, 38–40, 1970.
25. Raso, G., Tirabasso, G., and Donsi, G., *Powder Technol.,* 34, 151–159, 1983.
26. Kobayashi, S. and Miyahara, S., *Keisoku Jido Seigyo Gakkai Ronbunshu,* 20, 529, 1984.
27. Beck, M. S., Drane, J., Plaskowski, A., and Wainwright, N., *Powder Technol.,* 2, 269–277, 1969.

2.4 Level Measurement of a Powder Bed

Hiroaki Masuda and Shuji Matsusaka
Kyoto University, Katsura, Kyoto, Japan

2.4.1 LEVEL METERS AND LEVEL SWITCHES

Level meters are classified into two types: a continuous type and a discrete type. Discrete level meters are also called level switches and are usually used to maintain the powder level between the prescribed upper and lower limits by feeding or discharging powder.[1] Table 4.1 illustrates the principles of continuous level meters. The electric capacitance method is also utilized in measurement of the powder level. Table 4.2 lists the principles of various level switches.

Level meters or level switches should be located at suitable positions because the powder bed does not take a horizontal surface. Further, the profile of the surface will gradually change during feeding or discharging. For free-flowing particles, the profile may be estimated from the angle of repose in general. If the estimation is difficult, experiments should be carried out to determine suitable locations for level meters.

TABLE 4.1 Powder Level Meter (Continuous Type)

Principle	Notes
Weighing	Load cell
Electric capacitance	
Electric resistance	
Sounding	Wire rope with a sinker
Attenuation	Radiation, ultrasonic wave
Reflection time (reflectometry)	Ultrasonic wave, electromagnetic pulse, pulsed microwave

TABLE 4.2 Powder Level Switch

Principle	Notes
Back pressure increase	Gas purge or gas injection
Elastic deformation (powder pressure)	Diaphragm with mercury switch
Mechanical blockage of motion by powder	Reciprocation, rotation, swinging motion
Electrically forced vibration	Piezoelectric vibration, tuning fork
Attenuation	Microwave, electromagnetic wave

2.4.2   MECHANICAL METHOD

Sounding Level Meters

A sounding level meter mounted at the top of a storage vessel is shown in Figure 4.1. A weight connected with wire rope is lowered slowly from a wire drum of the level meter. When the weight reaches the surface of the powder bed, the wire rope becomes tensionless. Hence the torque acting on the wire drum decreases. The change of torque is detected mechanically, and the wire rope is rolled up by reversing the motor rotation. The powder bed level is measured as the length of rope is released, which can be determined from the number of rotations of the drum.

Level meters of this type can measure up to a depth of 50 m from the top with 0.1 m accuracy. The meter is applicable to a high-temperature vessel such as a blast furnace. The level of solid materials immersed in liquid can also be measured.

Piston Level Switches

Figure 4.2 is a schematic illustration of a piston level switch. A shaft having a disk at the end is moved reciprocally by a motor and crank mechanism. If the disk is blocked by particles from the left-hand side in the figure, the spring is compressed by motor action, and the link changes to a V shape. Then the link pushes a microswitch and the motor is switched off. If the disk is released from the particles, the link is returned to the original shape by spring action. Then the motor rotates again.

Bearings are isolated by bellows so that dust does not intrude into the mechanical part. However, adhesion of particles on the disk, bellows, and the shaft can cause problems.

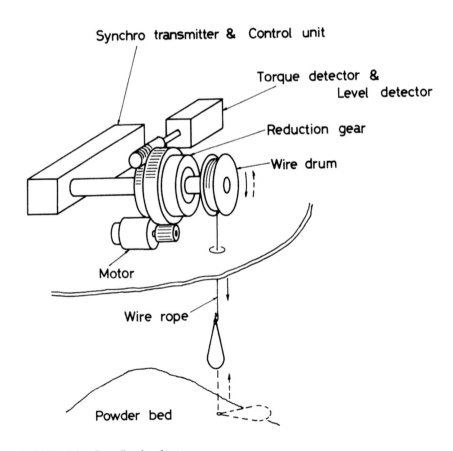

FIGURE 4.1   Sounding level meter.

Rotating Blade Level Switches

A level switch of this type has a blade at the end of a shaft, which is rotated slowly (1–3 rpm) by a synchronous motor. Mechanical blockage of the rotating blade by particles is effectively detected by a spring and two microswitches. If the blade is blocked by particles, the motor housing rotates in the counter direction and pushes a microswitch. Then the motor stops and generates an electrical signal showing the powder level.

Rotating blade level switches are widely used because of their simplicity. The blade can, however, be broken by a firm blockage of powder. If the blade is buried in powder flow, it can be forced to rotate, resulting in an erroneous signal.

Swing Level Switches

Figure 4.3 shows a swing level switch. A blade swings from side to side through a motor rotation and mechanical connectors. Leaf springs are inserted between the blade shaft and the motor. Mechanical parts are isolated from particles by a diaphragm. If the blade is blocked by particles, the leaf springs will bend, and the motor rather than the blade is moved along a guide. Then the motor pushes a microswitch, as in the case of the piston level switch and the rotating level switch. If the blade is released from the particles, the motor returns to the normal position by spring action.

The blade is set vertically in a storage vessel. Therefore, the drag force caused by powder flow is not as large as in the case of rotating blade level switches. Thus, the above-mentioned erroneous signal caused by powder flow can be avoided in swing level switches.

Diaphragm Level Switches

Level switches of the diaphragm type have no motors. A diaphragm is connected with a microswitch or a mercury switch. If powder pressure acts on the level switch, the diaphragm deforms elastically and turns the switch on. After the powder level goes down, the diaphragm returns to the normal position and the switch goes off.

This type of level switch has no parts protruding into storage vessels, in contrast to the other level switches mentioned above. Diaphragm level switches can be strong against mechanical damage in this sense.

Level Switches Based on Vibration

A weak vibration of a rod or a tuning fork causes deformation of a piezoelectric crystal, giving an electric output. The output is amplified and is supplied to the other piezoelectric crystal as its input.

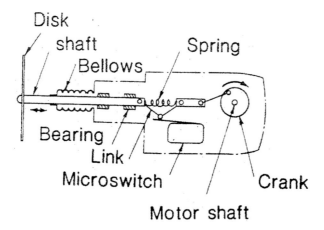

FIGURE 4.2 Piston level switch.

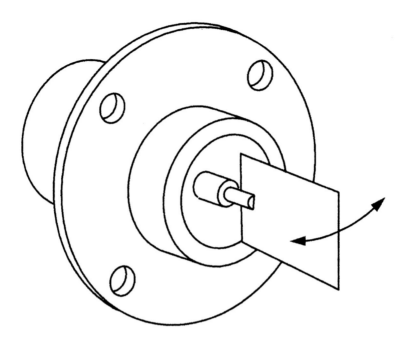

FIGURE 4.3   Swing level switch.

Then the latter crystal is electrically excited and, finally, vibrates at the resonance frequency. The resonance frequency decreases when the rod or the tuning fork is constrained by particles. Thus, the level of powder is detected.

This type of level switch is not affected by the electrical properties of particles (although the vibration is electrically forced, the vibration itself is purely mechanical). The level of powder immersed in liquid can also be detected. However, adhesion of fine particles to the rod or the tuning fork causes a frequency decrease, resulting in erroneous signals.

2.4.3   ELECTRICAL METHOD

Electric Capacitance Level Meters

Electric capacitance increases with particles existing between two electrodes. For continuous measurement of the powder level, a wire electrode is vertically stretched in a storage vessel. The counter electrode is the vessel itself.

Electric capacitance level switches, which have a rod- or bar-type electrode, are also available. Some of them have an earth electrode at their base. Level switches of this type have no moving parts and are easily adapted to high-pressure or high-temperature conditions. It is, however, necessary to adjust their sensitivity at the installation site, because the sensitivity also depends on the shape of the storage vessel. The sensitivity also depends on the dielectric constant of the powder materials. Therefore, readjustment is necessary when the kind of powder changes. Fine powders adhered to the electrodes can increase the electric capacitance and cause erroneous signals.

Electric Resistance Level Meters

Constant voltage is applied between a rod electrode inserted in a storage vessel and the vessel wall. As the specific resistance of particles is smaller than that of air, electric current through the powder layer becomes larger as the powder level increases. The current also depends on the shape of the vessel.

The specific resistance of the powder layer depends on the packing density, types of material, moisture content, and temperature. Therefore, the sensitivity is affected by these variables.

Microwave Level Meters

Microwave level meters can detect a 20-m distance with a resolution of 1 cm. If a microwave is emitted to a powder bed, a part of the wave is reflected on the surface of the powder bed and detected by a sensor of the level meter. The reflection time of a pulsed microwave is utilized to measure the powder level based on the velocity of light.

Emission and detection of the wave are carried out by a single antenna. The microwave can pass through plastics, and the powder level in a vessel closed by plastics can also be measured.

Microwave level meters need no mechanical contact with the powder, and their installation is easy. Their sensitivity is less affected by suspended dust or water vapor than ultrasonic level meters.

2.4.4   ULTRASONIC WAVE LEVEL METERS

An ultrasonic wave is a sound wave with a frequency higher than 20 kHz, and ultrasonic level meters can detect a 50-m distance with 1–2% accuracy. If the ultrasonic wave is emitted from the level meter to a powder bed, a part of the wave is reflected on the surface of the powder bed and detected by a sensor of the level meter. The reflection time of a pulsed ultrasonic wave is utilized to measure the powder level based on the velocity of the wave in air, which is given by

$$u = 331 + 0.6T \tag{4.1}$$

where u is the sound velocity (m/s), and T is the temperature of the air (°C).

Emission and detection of the wave are carried out by a single element such as a piezoelectric vibrator. The element is excited periodically by a pulsed electric power supply. Then the surrounding air vibrates and an ultrasonic pulse wave is emitted. The wave reflected on the powder bed surface is detected by the same element and is transformed into an electric signal.

Ultrasonic level meters need no mechanical contact with powder, and their installation is easy. Their sensitivity is, however, affected by suspended dust or water vapor in storage vessels because of the sound attenuation.

2.4.5   RADIOMETRIC METHOD

Radiation from high-energy gamma-ray sources such as cesium-137 or cobalt-60 can be used. In most cases, detectors are placed opposite the radiation source in order to perceive a change in the intensity of transmitted radiation.

A Geiger-type detector is usually used for simple on-off level switches. For continuous level measurement, more stable scintillation counters are preferable. Random decay of radioisotopes can cause short periodic fluctuations in output signals. Therefore, the output signals will be averaged for a certain period of time (which can be selected by varying the time constant of the detector) in order to avoid the error caused by the fluctuations.

Usually, a larger radioisotope source is necessary for larger vessels or thicker-walled vessels. Special windows for sources and detectors installed in the wall with dead space will substantially reduce source size. However, the alignment of the windows is critical. The use of these level meters on empty vessels must be carefully considered to ensure that the operating and maintenance personnel will not receive excessive radiation. Human tolerance for safe exposure is about 100 mrem/week. A reflectance level switch should be considered, instead of the ordinary transmission unit, for large-scale vessels.[2]

2.4.6 PNEUMATIC METHOD AND OTHERS

Pneumatic Method

Pneumatic level switches utilize air (or gas) nozzles. Air is injected into storage vessels through a nozzle equipped at the side walls. The air flow rate is held constant by a flow regulator. If the air jet is blocked by the powder bed, the back pressure in an air nozzle increases, and that is detected by a pressure transducer.

Air must be injected continuously even if the level measurements are not required, because the nozzles can be blocked by particle intrusion. For adhesive particles, a cave can be formed in the powder bed after the particles near the nozzle are blown off. Hence, no increase in the back pressure can be obtained, and the level switch fails to detect the powder bed.

Weighing Method

If the purpose of level measurements is to determine the content of powder in a storage vessel, a weighing method is preferable. Load cells are applied in weighing level meters as in the case of a hopper scale.[1] It should be noticed that the weighing will be affected if air or gas flows through the vessel.

Numerical calculation must be done based on the bulk density of powder and the geometrical shape of the vessel in order to estimate the level of powder bed according to the weighing method. Powder bridging or rat holing can also cause errors in the estimation.

Optical Method

A high-power laser is applied to obtain the surface profile of a powder bed in a storage vessel. The laser beam scans the surface of the powder bed, and the trajectory of light on the surface is recorded by a highly sensitive TV camera, as shown in Figure 4.4. Data obtained are analyzed by a microcomputer based on the triangulation. The surface profile can be estimated by the optical method within ±50 mm. The measuring time is about 10 s.[3]

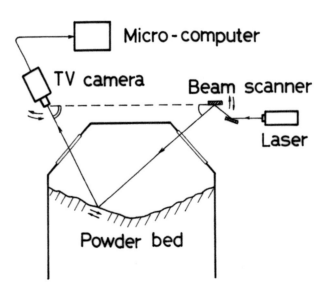

FIGURE 4.4   Laser level meter.

REFERENCES

1. Iinoya, K., Masuda, H., and Watanabe, K., in *Powder and Bulk Solids Handling Processes,* Marcel Dekker, New York, 1988, pp. 131–149.
2. Rowe S. and Cook, H. L., Jr., *Chem. Eng.,* 76 (2), 159–166, 1969.
3. Yamasaki, H., in *Jido Seigyo Handbook-Kiki Ouyou Hen* (Handbook for Automatic Control Instruments and Applications), Ohmsha, Tokyo, pp. 842–843.

2.5 Temperature Measurement of Powder

Hiroaki Masuda
Kyoto University, Katsura, Kyoto, Japan

Kuniaki Gotoh
Okayama University, Okayama, Japan

2.5.1 THERMAL CONTACT THERMOMETERS

Thermocouples

Thermocouples are composed of two different kinds of metal wires that are connected at both ends, making a loop. One of the ends is called the hot junction (i.e., sensing point) and the other the reference junction (i.e., cold junction). The hot junction is usually welded. If there is a temperature difference between these two junctions, electromotive force (or thermo-electromotive force) will be generated in the loop, which is almost directly proportional to the temperature difference. Table 5.1 illustrates common thermocouples, their composition, and their operating temperature range. The numerical values of the electromotive force are found in standard calibration tables.[1] The reference junction is usually kept 0°C in an ice bath, or an electrical cooler (i.e., electrical compensator) can be used for this purpose. The sensing parts of the thermocouples are protected against corrosion and mechanical damage by use of ceramic or metal tubes filled with alumina or magnesium oxide powder.

Resistance Thermometers

Electric resistance thermometers utilize the fact that the electric resistance of conductors (i.e., metals) increases with increasing ambient temperature. The electric resistance of semiconducting materials also changes with the ambient temperature. In the latter case, the concentration and mobility of the charge carrier in the materials increase with increasing temperature, and therefore the resistance decreases. The temperature dependence of the electric resistance of semiconducting materials is expressed by

$$R = aT^{-b} \exp\left(\frac{c}{T}\right) \tag{5.1}$$

where R is the resistance (Ω), T is the temperature (K), and a, b, and c are calibration constants. Resistance thermometers constructed of semiconducting materials are known as thermistors (i.e., thermally sensitive resistors). Thermistors are less stable and accurate than metal thermoresistors but offer the advantages of a lower manufacturing cost and a much higher resistivity than that of a metal thermoresistor.[2] Table 5.2 lists the types of resistance thermometers and their operating temperature ranges.

Industrial resistance thermometers are usually constructed of platinum. The platinum wire (diameter 0.03 to 0.05 mm and length 1 m) is wound on a mica plate or glass rod and is then inserted into a protecting tube or enclosed in tempered glass. The electric resistance of platinum resistance

TABLE 5.1   Thermocouples

Material	Abbreviation	Temperature Range (°C)	EMF (μV/°C)	Accuracy (°C)
Copper-constantan	CC	−180–300	50	2–5
Iron-constantan	IC	0–500	60	3–10
Chromel-alumel	CA	0–1000	40	2–10
Platinum-platinum 10% rhodium	PR	100–1400	10	0.5–5

TABLE 5.2   Resistance Thermometers

Material	Temperature Range (°C)	Accuracy (%)
Platinum	−180–500	0.5–2
Nickel	−50–120	0.5–2
Copper	0–120	0.5–2
Thermistor	−50–200	0.5–2

thermometers changes 0.4% for each temperature change of 1°C (temperature coefficient = 0.4%/°C). The temperature coefficient of thermistors is about 10 times larger than that of the platinum resistance thermometer. The temperature versus resistance relation of thermistors is, however, nonlinear, as one can see from Equation 5.1; and the calibration constants a, b, and c will be different for each thermistor. Although thermistors have this defect, their sensing elements are very small (0.3 to 2 mm diameter), and the dynamic response can be made faster than in platinum resistance thermometers. The time constant is 0.3 to 2 s.

2.5.2   RADIATION THERMOMETERS

Every object emits thermal radiation energy, depending on its temperature. Therefore, the temperature of particles can be estimated by detecting their thermal radiation energy. Radiation thermometers need no thermal contact with the particles. The thermal radiation energy of a perfect emitter (i.e., blackbody) is given by Planck's law. The radiation energy is concentrated mainly in the visible and infrared regions with wavelengths below 10 μm.

There are three definitions of the temperature of solid materials:

1. *Radiation temperature, T_R.* If the emissive power of an object is equal to that of a perfect emitter of temperature T_R, T_R is called the radiation temperature of the object.
2. *Luminance temperature, T_S.* If the monochromatic luminance of an object is equal to that of a perfect emitter of temperature T_S, T_S is called the luminance temperature of the object.
3. *Color temperature, T_F.* If the color of an object is the same as that of a perfect emitter of temperature T_F, T_F is called the color temperature of the object.

Following is the relationship between the luminance temperature T_S and the true temperature T:

$$\frac{1}{T} = \frac{1}{T_s} - \frac{\lambda}{c_2} \ln \varepsilon_1 \qquad (5.2)$$

where ε_λ is the monochromatic emissivity (i.e., blackness) of the object. Further, the relationship between the color temperature T_F and the true temperature T is given by

$$\frac{1}{T} = \frac{1}{T_F} + \frac{1}{c_2} \frac{\ln(\varepsilon_{\lambda 1} / \varepsilon_{\lambda 2})}{(1/\lambda_2) - (1/\lambda_1)}$$

(5.3)

where $\varepsilon_{\lambda 1}$ and $\varepsilon_{\lambda 2}$ are the emissivities at wavelength of λ_1 and λ_2, respectively. From Equation 5.2 and Equation 5.3, the following relationship can be derived:

$$T_S < T < T_F$$

(5.4)

The emissivity of a powder bed is usually approximated by unity (i.e., blackbody), and therefore, $T \cong T_S \cong T_F$. This is incorrect for a particle and suspended particles.

Table 5.3 illustrates the ordinary thermosensor (i.e., detector), measuring temperature ranges, and accuracy of radiation thermometers. Radiation pyrometers measure the total energy $E(T)$, collecting the radiation by lenses and focusing it on a detector such as a thermopile, silicon cell, thermistor bolometer, or photocell (PbSe, PbS). Thermopiles are composed of thermocouples connected in series so as to increase the sensitivity. Accuracy and stability are, however, not as good as in other detectors. Portable radiation pyrometers are also available.

Two-color pyrometers are more expensive than radiation pyrometers. The radiation from particles is collected by lenses and projected to an interference filter made from, for example, indium phosphate (InP). The radiation reflected by the filter and that transmitted through the filter are measured by two different photocells, respectively, and the ratio of the output signals of each photocell is calculated. As the wavelength of reflected radiation is different from that of transmitted radiation, luminance temperatures at two different wavelengths (i.e., two colors) are obtained. The measured temperature is insensitive to variations in the emissivity of particles and is less affected by several disturbances caused by aerosols such as water vapor or fumes. Two different optical filters may also be utilized to get radiation intensities at two different wavelengths. Temperature measurement in gas–solids suspension flow can be done by use of a two-color pyrometer.[3]

A more economical thermometer for the measurement of temperature ranging from 700 to 3000°C is an optical pyrometer, which is a type of brightness pyrometer. The pyrometer determines the temperature of an object by comparing the luminance of the object with that of a reference filament of a standard electric bulb. The accuracy of an optical pyrometer is about ±5°C at 1000°C. It is easy to control the electric current supplied to the filament so as to get the same luminance as the object. Such a pyrometer is called an automatic optical pyrometer.

TABLE 5.3   Radiation Thermometers

Thermometer	Detector	Temperature Range (°C)	Accuracy (%)
Radiation pyrometer	Thermopile	200–1000	±1–2
	Silicon cell	600–3000	±5–1
	Thermistor bolometer	0–500	±0.5–1
	PbSe, PbS	100–1000	±0.5–1
Two-color pyrometer	PbS	300–1000	±0.5–1
	Silicon cell	700–2000	±0.5–1
	Photomultiplier	1000–3500	±0.5–1

The measuring points of radiation thermometers can be changed by use of a rotating or vibrating mirror. These instruments, called scanning thermometers, are applicable to obtain the temperature profile of large equipment such as a cement kiln. Commercially available infrared scanning thermometers can be applied in measurement of the temperature range of 10 to 1000°C with sensitivity 0.5°C at 25°C and resolving power 3% of full scale. The focal length of the instruments is between 1.2 m and 600 m.

REFERENCES

1. West, R. C. and Astle, M. J., *CRC Handbook of Chemistry and Physics*, 61st Ed., CRC Press, Boca Raton, Fla., 1980, p. E-108.
2. Gardner, J. W., *Microsensors—Principle and Application,*: John Wiley, New York, 1994.
3. Themelis, N. J. and Gauvin, W. H., *Can. J. Chem. Eng.*, 40, 157–161, 1962.

2.6　On-Line Measurement of Moisture Content

Satoru Watano
Osaka Prefecture University, Sakai, Osaka, Japan

2.6.1　INTRODUCTION

Powder-handling processes are mainly categorized into wet and dry processes.[1,2] In wet processes, the degree of wetness (moisture content) seriously affects physical properties (size, shape, density, etc.) of the final products. In order to produce the desired product efficiently and continuously without difficulty, monitoring and control of moisture content are required. Even in dry processes, moisture content sometimes causes problems. Generally, powder materials contain moisture, to some extent as water of hydration, bound water, and so forth, which exists on the powder surface or forms liquid bridges between powders. Moisture on the powder surface greatly changes the powder's properties such as flowability and internal friction, causing handling problems. The liquid bridges generate agglomeration of powders and adhesion to the vessel wall. In all cases, control of moisture content is necessary in order to avoid problems.

2.6.2　ELECTRICAL METHODS

Table 6.1 lists moisture sensors and their measuring principles. In powder-handling processes, there are no sensors which can measure moisture content directly. Therefore moisture content is determined by using the correlation between indirect parameters (absorbance of spectrum, dielectric constant, etc.) and moisture content that have been previously measured by a direct method (such as the drying method).

So far, chemical, electrical, and optical methods, as well as methods applying radioactive rays and nuclear magnetic resonance spectrometry, have been available for powder-handling processes. Among them, electrical and optical methods have been widely used in industrial practical application.

In the following, electrical and optical methods for measuring moisture content of powder materials are briefly explained.

Figure 6.1 illustrates a schematic diagram of an electrical resistance-type moisture sensor. In this sensor, moisture content is determined by measuring the electrical resistance change due to moisture increase/decrease. As seen in the figure, powder samples are packed between two electrodes, and apparent specific resistance R is determined by measuring the applied voltage V and the electric current I as

$$R = \frac{VS}{Id} \tag{6.1}$$

where S and d indicate areas of electrode and distance between the two electrode, respectively. The apparent specific resistance R varies with moisture content W as

$$\log R = -a \log W + b \tag{6.2}$$

317

TABLE 6.1   List of Moisture Sensors

Methods		Principle of measurement	Continuous measurement	Detach/contact to powder bed	Wide application	Range
Drying method	Normal drying Infrared heating High frequency induction heating	Measuring weight loss due to drying of wet samples.	×	Contact	○	0.1~100%
Chemical method	Distillation method	Perform distillation and measure water volume.	×	Contact	○	0.01~20%
	Karl Fischer (KF) titration method	Water is titrated with KF reagent containing iodine, sulfur dioxide and pyridine in methanol. Infinitesimal moisture content can be determined.	×	Contact	○	0.001~100%
	Electric resistance method (DC)	Measure electric resistance change (direct current) due to moisture increase/decrease.	○	Contact	○	3~50%
Electric method	Method applying high frequency	Measure electric current (high frequency) change due to moisture increase/decrease. Smaller quantity of water can be detected than the DC method.	○	Contact	○	0.1~60%
	Method applying dielectric constant (electrostatic capacity)	Measure dielectric constant change due to moisture increase/decrease.	○	Contact	○	2~25%
Optical method	Method applying infrared ray	Measure absorbance of infrared –ray at absorption band (1.43, 1.94μm) of water near infrared.	○	Detach	○	1~90%
	Method applying microwave	Measure attenuation of microwave due to water.	○	Detach	○	1~90%
Nuclear Magnetic Resonance (NMR) Spectrometer		Sample is dissolved in a solvent and its NMR is measured.	×	Contact	×	0.1~90%

Method applying radioactive rays	Method measuring moderation/scattering	Measure moderation/scattering of neutron/γ-ray due to water.	○			2–80%
	Method measuring attenuation strength	Measure attenuation strength of β or γ-rays due to water.		Detach	×	
Other method	Balanced relative humidity	Calculate moisture content by relative humidity, which is balanced with circumstances.	○	Detach	○	1–80%

□: Possible ×: Impossible A: Very good B: good C: Poor

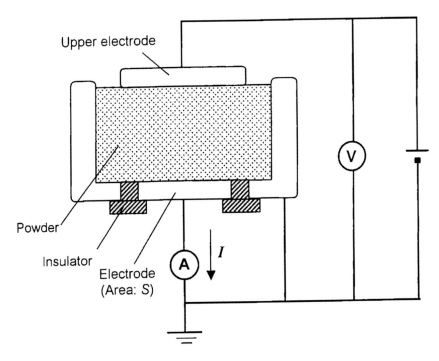

FIGURE 6.1   Schematic diagram of electrical resistance-type moisture sensor.

where a and b are constants depending on the shape of electrode, properties of powder materials, and circumstance conditions (humidity and temperature). By using Equation 6.2, moisture content can be determined. For the electric current, direct current (DC) or high-frequency alternating current is used.

The electrical resistance-type moisture sensor is widely used since its electrical circuit is simple, the measurement principle is easy, and the measurement time is short. However, it has some draw-backs: (1) temperature correlation is required since the electrical resistance generates heat, (2) it cannot be used for a powder bed that moves during the measurement since powder packing conditions affect the resistance, and (3) measurement error may occur if the powder sample contains electrolyte.

Figure 6.2 shows a schematic diagram of a dielectric constant-type moisture sensor. Assuming that the electrical capacity when powder materials exist inside between two electrodes or inside a cylindrical electrode is C_1, and the one without any powder materials is C_0, the apparent dielectric constant ε_a is obtained as

$$\varepsilon_a = \varepsilon_0 \frac{C_1}{C_0} \tag{6.3}$$

By using the apparent dielectric constant, moisture content can be determined, since the apparent dielectric constant ε_a shows an approximately linear relationship to moisture content as

$$\varepsilon_a = j + k\bar{w} \tag{6.4}$$

where j and k are constants. These constants require previous calibration, since they vary with different powder materials. The dielectric constant-type moisture sensor has merits in that it is less affected by temperature and packing conditions than the electrical resistance-type moisture sensor. However, its circuit is complicated, and it cannot measure moisture content of powder materials containing electrical conductors, such as metal and coal powders.

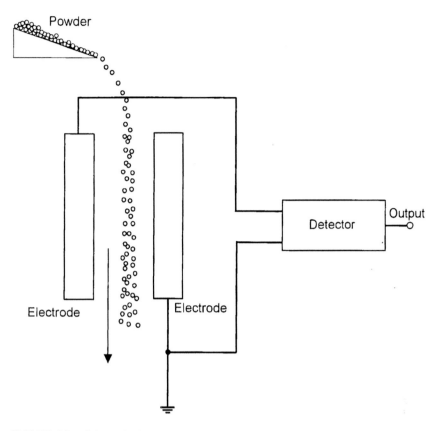

FIGURE 6.2   Schematic diagram of a dielectric constant-type moisture sensor.

A microwave-type moisture sensor is able to measure moisture content continuously without touching the objective material. Figure 6.3 investigates absorbance characteristics of microwaves in the case of flour and cornstarch powders. When the microwaves are transmitted to the powder material containing water, water absorbs the microwaves and the microwaves attenuate. Since the degree of attenuation has a correlation to moisture content, as shown in Figure 6.3, moisture content of powder materials can be determined. However, the system is complicated to avoid the microwave leakage. In addition, the measured value may vary to some extent, depending on the leakage/refraction, powder surface condition, and temperature.

2.6.3   INFRARED MOISTURE SENSOR

In general, an optical sensor utilizes the principle that physical values (concentration, distance, etc.) are indirectly detected by using the absorbed/reflected energy of spectra when the spectra are radiated to the object. The so-called spectra are categorized into ultraviolet, optical, and infrared rays, depending on the wavelength, and each sensor uses a specific wavelength, which fits characteristics of the measured object.

Figure 6.4 shows infrared (IR) absorption characteristics.[3,4] Water absorbs infrared spectra markedly at wavelengths of 1.43, 1.94, and 3.0 μm, due to the resonance with atomic vibration between oxygen and hydrogen in water molecules. Since energy of IR spectrum is absorbed remarkably at these wavelengths in proportion to moisture content, moisture content can be detected by measuring the difference between the irradiated energy and the reflected one. Since the absorption spectrum measurement is easily disturbed by granule surface conditions and fluctuation in beam

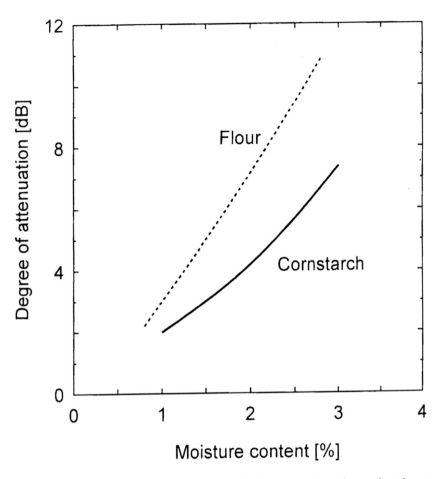

FIGURE 6.3 Relationship between degree of microwave attenuation and moisture content.

length, a reference method which calculates the reflected energy ratio of the absorbed spectrum (1.94 µm) and of the bilateral unabsorbed spectrum (reference spectrum of 1.78 and 2.14 µm) has been adopted. This cancels out any disturbance, since the disturbance affects the absorbed and unabsorbed spectra equally.

Details of the IR moisture sensor are illustrated in Figure 6.5. The continuous light from the source is condensed by a condenser, then changed to a chopper beam by passing through a rotating sector of four optical filters composed of the different selective spectra of 1.94 µm for absorption, the bilateral 1.78 and 2.14 µm for unabsorbed references, and a visible spectrum for targeting. When these portions of the spectra are irradiated to the object, the absorption spectrum is absorbed in proportion to the moisture content of the object, while the reference spectra are fully reflected.[4] These spectra are transformed into electrical signals, and the degree of absorptivity, defined as the ratio of energy absorbed to the supplied energy of the spectrum, is calculated. Based on the degree of absorptivity, the absorbance X, the output from the moisture sensor, is calculated using the following equation:

$$X = k_0 \cdot \ln\left\{\frac{(k_2 \cdot V_2 + V_3)/2}{k_1 \cdot V_1}\right\} \tag{6.5}$$

where, V_1, V_2, and V_3 show absorption at 1.94, 1.78, and 2.14 µm in the IR spectra after temperature compensation, respectively, and k_0, k_1, and k_2 indicate the calibration coefficients, respectively.

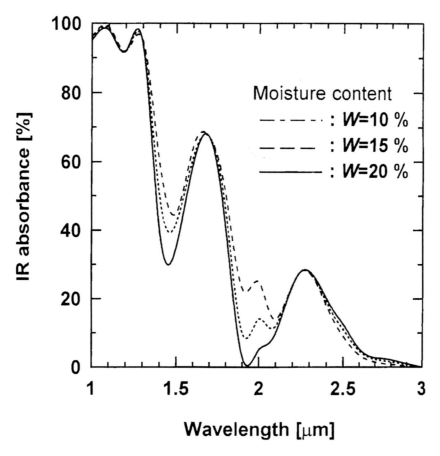

FIGURE 6.4   IR absorption characteristics of powder material (lactose and cornstarch mixed 7:3 by weight) under various moisture contents.

Although the infrared moisture sensor is able to measure moisture content continuously without touching to the object and is not influenced by powder density and packing conditions very much, it measures moisture content on the surface or just below the surface of the object due to the long wavelength. Normally, moisture is distributed inside the powder material, and there are variations in moisture content between the surface and the core. Therefore, in order to measure moisture content by using the infrared sensor, a calibration curve between sensor output (surface moisture content detected by IR sensor) and total moisture content measured by a drying method is used.

Figure 6.6 indicates the calibration curves when the lactose and cornstarch mixing ratio was varied.[3] The calibration curves are different depending on the powder materials.[3] It is because the powder's physical properties such as water-absorbing potential, surface conditions, IR refraction characteristics, and moisture distribution are different.

2.6.4   APPLICATION OF MOISTURE CONTROL TO POWDER-HANDLING PROCESSES

Wet granulation is one of the most sensitive operations to moisture content. Wet granulation is defined as a size enlargement that adheres or sticks powders together by using liquid binders to produce granules having the desired size, shape, and density. The properties of the final product are significantly influenced by the moisture content.[5,6,7] Therefore, monitoring and control of moisture content are required to produce the desired product continuously and with high accuracy.

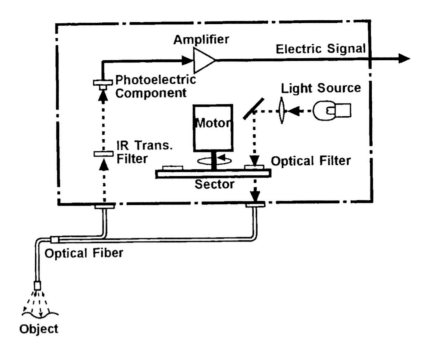

FIGURE 6.5   Details of IR moisture sensor.

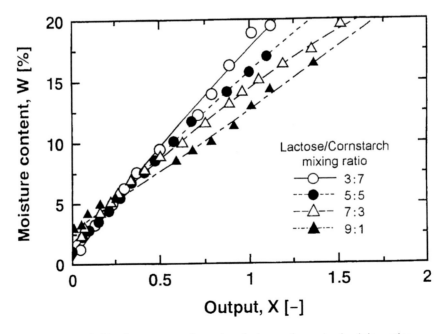

FIGURE 6.6   Calibration curves under various lactose and cornstarch mixing ratios.

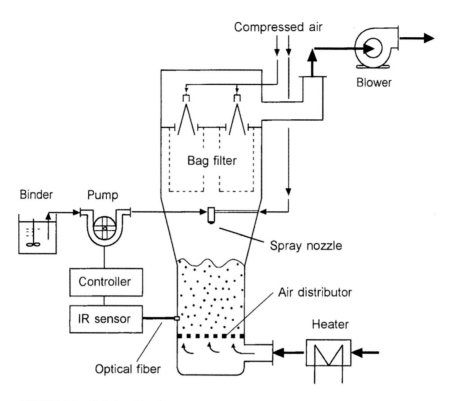

FIGURE 6.7   Relationship between granule mass median diameter and moisture content.

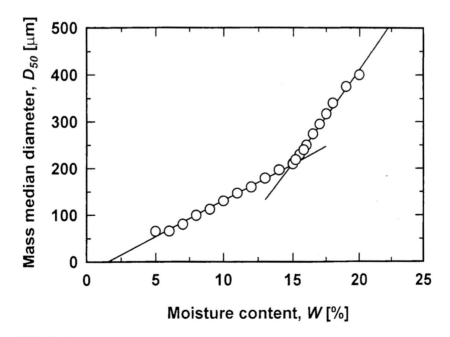

FIGURE 6.8   Moisture control system in fluidized-bed granulation.

Figure 6.7 illustrates a system which measures and controls moisture content by using an IR moisture sensor in a fluidized-bed granulation.[4,8,9] The IR detector and fluidized-bed vessel are connected by optical fibers, and heated purge air is blown at the extremity of the sensor (the place where the optical fibers are connected at the vessel wall) to prevent powder sticking.

Figure 6.8 indicates the relationship between granule mass median diameter and moisture content when powder samples composed of lactose, cornstarch, and hydroxypropylcellulose (HPC) are granulated by a purified water (binder).[3] Figure 6.8 shows the granule mass median diameter has a linear relationship to the moisture content, implying that granule growth can be controlled by measuring and controlling the moisture content.

In the Japanese pharmaceutical industry, automatic manufacturing of drugs (granules) is commonly conducted by fluidized-bed granulation with moisture control, and uniform properties can be maintained easily.[10]

REFERENCES

1. Gotoh, K., Masuda, H., and Higashitani, K., Eds., *Powder Technology Handbook,* 2nd Ed., Marcel Dekker, New York, 1997.
2. Society of Powder Technology Japan, Ed., *Huntaikogaku binran,* 2nd Ed., Nikkan Kogyo Shinbun, Tokyo, 1998.
3. Watano, S., Hideo, H., Sato, Y., Miyanami, K., and Yasutomo, T., *Adv. Powder Technol.,* 7, 279–289, 1996.
4. Watano, S., Sato, Y., and Miyanami, K., *J. Chem. Eng. Jpn.,* 28, 282–287, 1995.
5. Watano, S., Morikawa, T., and Miyanami, K., *J. Chem. Eng. Jpn.,* 28, 171–178, 1995.
6. Watano, S., Morikawa, T., and Miyanami, K., *Chem. Pharm. Bull.,* 44, 409–415, 1996.
7. Schaefer, T. and Worts, O., *Arch. Pharm. Chem. Sci.,* 5, 178–193, 1977.
8. Watano, S., Yamamoto, A., and Miyanami, K., *Chem. Pharm. Bull.,* 42, 133–137, 1994.
9. Watano, S., Takashima, H., and Miyanami, K., *J. Chem. Eng. Jpn.,* 30, 223–229, 1997.
10. Watano, S., Takashima, H., Miyanami, K., Murakami, T., and Sato, T., *Chem. Pharm. Bull.,* 42, 1302–1307, 1994.

2.7 Tomography

Richard A. Williams
University of Leeds, West Yorkshire, United Kingdom

Recent developments in the design and application of noninvasive tomographic sensors now allow direct observation of the behavior of powders and particulate suspensions inside process equipment. This offers the possibility of acquiring measurements *in situ*. The methodology can be used for several purposes: process auditing, fault detection, process control; development of advanced models; and to enable verification of computational fluid dynamics predictions. The main sensing methods are summarized and illustrated with selected application examples.

2.7.1 INTRODUCTION

It is widely recognized that little is known about the behavior of particles during their passage through process equipment. To date, the lack of fundamental data has hindered the development of generic and reliable models. Tomographic measurements seek to address the widespread need for the direct analysis of the internal characteristics of process equipment. Ideally, the measuring instruments for such applications must use robust, noninvasive sensors that can operate in the proximity of aggressive and fast-moving flowing powders and multiphase mixtures.

Tomographic technology involves the acquisition of measurement signals from sensors located on the periphery of a process vessel or the pipeline through which it is conveyed. This reveals information of the nature and distribution of components within the sensing zone. Most tomographic techniques are concerned with abstracting information to form a cross-sectional image. This can, for example, utilize a parallel array of sensors place so that their sensing field interrogates a projection of a suitable radiation across the subject. In order to explore the entire cross section, it is necessary to obtain other projections by rotating the subject in the direction normal to the direction of the sensor field or, preferably, to rotate the measurement sensors around the subject. It might not always be possible to adopt such a methodology because it might be impractical to physically rotate the subject, or the sensors, and the time required to rotate the assembly might take too long compared to changes occurring within the subject under investigation. Process tomography instruments have been devised to overcome these limitations.

The basic components of process tomographic instruments are shown in Figure 7.1,[1] embodied in hardware (sensors, signal/data control) and software (signal reconstruction, display and interpretation facilities, generation of output control signals to process hardware). In this figure sensors are mounted around the exterior of the (nonmetallic) pneumatic conveying pipe. Data are collected and processed to yield a reconstructed image of the profile of the particulate bed at a given instant of time. The image can be analyzed, for instance, to provide measurements of the local solid concentration $C(x,y)$ at any desired point. The procedure can be repeated by acquiring additional frames with time.

It is, however, necessary to emphasize that obtaining a good-quality computed "image" may only represent the first stage of information input for the powder technologist. Sometimes an image is not even required since measurement data can be fitted directly to a model of the process, such as reported by West et al.[2] The ultimate goal is quantitative (numerical) interpretation of an image or, more likely, many hundreds of data sets and/or images corresponding to different spatial and temporal conditions. In most circumstances, a visual diagnosis based on glancing at the images will be

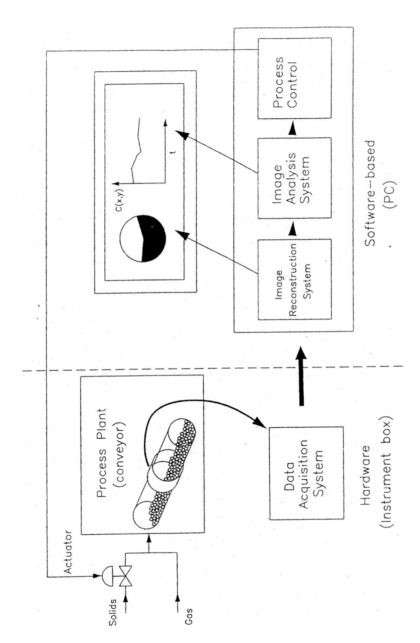

FIGURE 7.1   Components of a process tomographic sensing system, showing here an electrical capacitance system controlling a pneumatic conveyer.

insufficient, except perhaps to diagnose gross malfunction occurring in a process. Therefore, there is a major and fundamental difference between the general philosophy of medical tomography and that of process tomography.

2.7.2 SENSOR SELECTION AND SPECIFICATION

The heart of any tomographic technique is the sensor system that is deployed. The basis of any measurement is to *exploit differences in the properties of the process being examined.* A variety of sensing methods can be employed based on measurements of transmission, diffraction, or electrical phenomena.[3] Table 7.1 shows a selection of the most widely used sensing methods and their principal attributes.

Whereas most devices employ a single type of sensor, there are a number of opportunities for multimode systems using two (or more) different sensing principles. The choice of sensing system will be determined largely by the following:

- The nature of components contained in the pipeline, vessel, reactor, or material being examined (principally whether they exist as a solid, liquid, gas, or a multiphase mixture, and if so in what proportions)
- The information sought from the process (steady state, dynamic, resolution, and sensitivity required) and its intended purpose (laboratory investigation, optimization of existing equipment, process monitoring or control)
- The process environment (ambient operation conditions, safety implications, ease of maintenance, etc.)
- The size of the process equipment and the length scale of the process phenomena being investigated.

High-resolution tomography can be expensive, rather slow, and, in some cases, impose special requirements for safe operation. Nevertheless, such methods can be extremely valuable for detailed laboratory studies.[4] In contrast to this, the lower-resolution methods are less expensive, offer a fast dynamic

TABLE 7.1 Examples of Sensors for Process Tomography

Principle	Typical Spatial Resolution (percentage of diameter of cross section)%	Practical Realization	Comment
Electromagnetic radiation	1	Optical	Fast, optical access required
		X-ray and gamma ray	Slow, radiation containment
		Positron emission	Labelled particle, not on-line
		Magnetic resonance	Fast, expensive for large vessels
Acoustic	3	Ultrasonic	Sonic speed limitation complex to use
Measurement of electrical properties	10	Capacitive Conductive Inductive	Fast, low cost, suitable for small or large vessels

response, and generally impose fewer safety constraints. Therefore, the process need for image resolution must be carefully considered. For the particular case of electrical tomography, considered later, the spatial resolution is determined by the nature of the measurement and associated reconstruction method. Imaging frequency (frame rates) from a few up to several hundred frames per second is attainable. To date, frame rates up to 1000 Hz have been demonstrated.[5] The use of more than one plane of image sensors offers the possibility of three-dimensional imaging. This has led to the adoption of voxel–voxel correlation methods from which velocity information can now be routinely deduced.[6] Related methods such as electrical resistance tomography can be used to image solid concentration profiles in aqueous slurries (e.g., in mixers, separators, etc.), as described in detail elsewhere.[7]

2.7.3   EXAMPLES OF POWDER-PROCESSING APPLICATIONS

Pneumatic Conveying

The transport of dry and damp minerals via hoppers and in conveying lines is of importance, as plant operability is often poor due to pluggage of conveying lines due to subtle changes in the feedstock (e.g., moisture content, size distribution, particle shape). Electrical capacitance tomography (ECT) can be used to image the dielectric constant inside conveying lines for dilute- and dense-phase processes, as reported in detail elsewhere.[8-10] Analyses of the changes in the image with time can be used to identify the prevailing flow regime. For instance, Figure 7.2 shows a real-time visualization of powder flow for two conditions in dense-phase conveying. Analysis of the data also reveals characteristics such as slug length and velocity that can be computed from the images, and also slugging characteristics can be recovered from statistical analysis of void age fluctuation data (Figure 7.3).

These methods have been used in a variety of conveying systems, and in some cases schemes for on-line control have been proposed.[11,12] For example, in an earlier work on coal conveying estimation of mass flow rate using a simplified fuzzy logic approach, each membership function was adapted by means of a genetic algorithm.[13] In this way, the control function did not drift from the ideal due to plant wear or changes in the nature of the transported materials. The genetic algorithm was used to minimize energy consumption. In principle, the controller could be tuned to optimize any fitness function, such as plant wear or product degradation, providing that these aspects can be measured on-line. Commercial equipment is enabling routine measurements to be obtained with significant benefit to those wishing to gain detailed analysis of flow structure.[14]

Fluidized Beds

An obvious extension of the measurements described above is the use of ECT to follow the complex dynamic interactions between gas and solids in fluidized-bed reactors, dryers, and

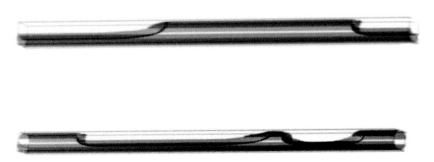

FIGURE 7.2   Real-time visualization of granular flows in pneumatic conveying using electrical capacitance tomography. [From Ostrowski, K. L., Luke, S. P., Bennett, M. A., and Williams, R. A., *Powder Technol.*, 102, 1, 1999. With permission.]

risers.[1,5] Figure 7.4 shows experimental measurements of the solid volume fraction at the base of a 150-mm-diameter bed adjacent to the distributor plate.[15] Images can be acquired for different flow regimes, determined by the superficial gas velocity (u). For both bubbling ($u < 0.2$ m/s) and slugging ($0.2 < u < 0.73$ m/s), fluidization bubbles appear near the wall of the pipe. The bubble diameter increases with increasing air velocity, but the flow patterns are not markedly different. In the transition from slugging to the turbulent regime ($u > 0.73$ m/s), the pipe cross section is largely occupied by gas, and bubbles are not observed attached to the walls. Detailed time transients can be obtained which allow the bubble size and transit velocity to be estimated. ECT enables the dynamic characteristics of the bed to be assessed and then analyzed, using chaos theory approaches or other statistical methods. The correlation fluctuations in time and space can be deduced either within a plane and between multiple planes spaced up the fluidized bed. These have found widespread application, as reviewed above.[6]

Powder Mixing

For laboratory mixing and scale-up studies, it is possible to label a particle with a positron-emitting tracer and to follow the trajectory of an individual particle in three dimensions using a positron-emission tomography camera.[16] The method of positron-emission particle tracking (PEPT) relies on detecting the pairs of back-to-back gamma rays produced by the annihilation of positrons, emitted from a radioactive tracer, with electrons. Typically, the volume that can be proved is up

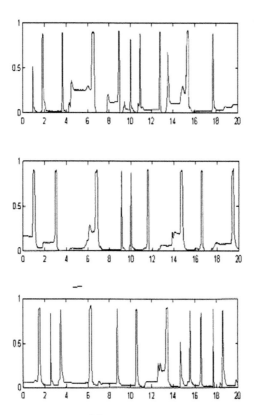

FIGURE 7.3   Examples on measurements of fluctuations of the relative volume fraction of solids across at a single plane during dense-phase conveying (slugging) with time, for three different dense-phase flows. [From Ostrowski, K. L., Luke, S. P., Bennett, M. A., and Williams, R. A., *Chem. Eng. J.*, 3524, 1, 1999. With permission.]

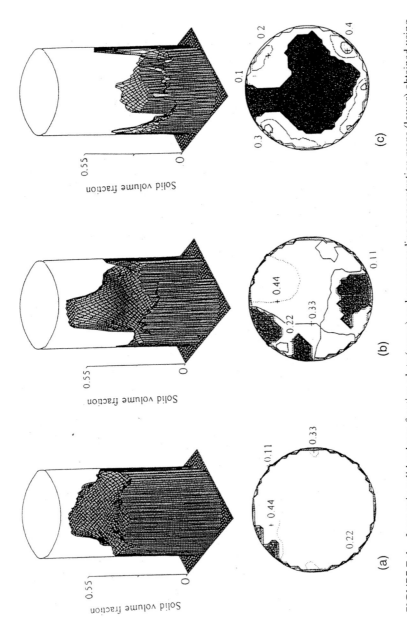

FIGURE 7.4 Isometric solid volume fraction plots (upper) and corresponding concentration maps (lower) obtained using real-time ECT in the vicinity of the distributor of a silica-gas fluidized bed for three flow regimes: (a) bubbling fluidization, (b) slugging regime, and (c) in the transitional regime from the slugging to turbulent regime.

to $500 \times 500 \times 250$ mm³ with a voxel resolution of approximately $0.5 \times 1.5 \times 1.5$ mm³.[17] Figure 7.5 shows a sample result of a two-dimensional slice through a laboratory bead mill), from which detailed information of bead velocities and slip effects can be deduced.[18] Similar methods were also used to follow the size-dependent segregation of particles in a dense suspension[4]; particle motion in granulators, extruders, dryers, conveyors, and so forth and for validation of Distinct Element Modeling (DEM) predictions.[19]

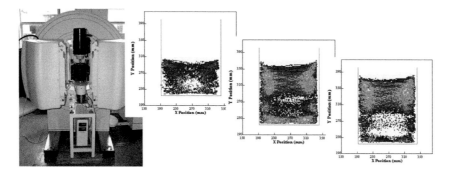

FIGURE 7.5 PET studies of a laboratory stirred bead mill (left) showing bead distribution and occupancy changes on increasing stirring speed (200, 600, 800 rpm).

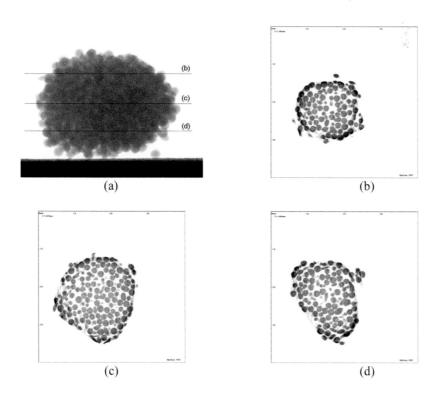

FIGURE 7.6 X-ray microtomograms of granules, showing (a) transmission image and (b–d) sections showing particle positions. [After Golchert, D. J., Moreno, R., Ghadiri, M., Litster, J., and Williams, R. A., *Adv. Powder Technol.*, 15, 447, 2004. With permission.]

Powder and Granule Structure

The use of bench-top X-ray microtomography systems is now commonplace, delivering spatial resolution down to a few microns for three-dimensional scanning of small samples (a few mm³) over a few hours.[7] These methods can provide detailed information on granule microstructure and packing of assemblies such as granules,[20] as shown in Figure 7.6, and structural modifications such as drying.[21] Further development in higher-resolution systems can be anticipated.

REFERENCES

1. Williams, R. A. and Beck, M. S., Eds., *Process Tomography–Principles, Techniques, and Applications,* Butterworth-Heinemann, Oxford, 1995, p. 581.
2. West, R. M, Jia, X., and Williams, R. A., *Chem. Eng. J.,* 77, 1–2, 31, 2000.
3. Williams, R. A. and Xie, C. G., *Part. Part. Syst. Charact.,* 10, 252, 1993.
4. Scott, D. M. and Williams, R. A., Eds., *Frontiers in Industrial Process Tomography,* AIChE/Engineering Foundation, New York, 1995.
5. Dyakowski, T., Jeanmeure, L. F. C., and Jaworski, A., *Powder Technol.,* 112, 174, 2000.
6. Mosorov, V., Sankowski, D., Mazurkiewicz, L., and Dyakowski, T., *Meas. Sci. Technol.,* 13, 1810, 2002.
7. Williams, R. A. and Jia, X., *Adv. Powder Technol.,* 14(1), 1, 2003.
8. McKee, S. L., Bell, T., Dyakowski, T., Williams, R. A., and Allen, T., *Powder Technol.,* 82, 105, 1995.
9. Ostrowski, K. L., Luke, S. P., Bennett, M. A., and Williams, R. A., *Powder Technol.,* 102, 1, 1999.
10. Ostrowski, K. L., Luke, S. P., Bennett, M. A., and Williams, R. A., *Chem. Eng. J.,* 3524, 1, 1999.
11. Jaworski, A. and Dyakowski, T., in *Proceedings of the Second World Congress on Industrial Process Tomography,* Virtual Centre for Industrial Process Tomography, Leeds/Manchester, 2001, p. 353.
12. Jaworski, A. and Dyakowski, T., *Powder Technol.,* 125, 279, 2002.
13. Neuffer, D., Alvarez, A., Owens, D. H., Ostrowski, K. L., Luke, S. P., and Williams, R. A., in *Proceedings of the First World Congress on Industrial Process Tomography* (Buxton), Virtual Centre for Industrial Process Tomography, Leeds/Manchester, 1999, p. 71.
14. Hunt, A., Pendleton, J. D., and White, R. B, in *Proceedings of the Third World Congress on Industrial Process Tomography,* Virtual Centre for Industrial Process Tomography, Manchester/Leeds, 2003, p. 281.
15. Wang, S. J., Dyakowski, T., Xie, C. G., Williams, R. A., and Beck, M. S., *Chem. Eng. J.,* 56, 95, 1995.
16. Parker, D. J., Hawkesworth, M. R., Broadbent, C. J., Fowles, P., Fryer, T. D., and McNeil, P.A., *Nucl. Instrum. Methods,* A348, 583, 1994.
17. Parker, D. J. et al., *Nucl. Inst. Meth. Phy. Res. A,* 477, 540, 2002.
18. Conway-Baker, J., Barley, R. W., Williams, R. A., Jia, X., Kostuch, J. A., McLoughlin, B., and Parker, D. J., *Minerals. Eng.,* 15, 53, 2002.
19. Seville, J. P. K., in *Proceedings Particulate Systems Analysis,* Harrogate, UK, 2003.
20. Golchert, D. J., Moreno, R., Ghadiri, M., Litster, J., and Williams, R. A., *Adv. Powder Technol.,* 15, 447, 2004.
21. Leonard, A., Blacher, P., Marchot, P., Pirard, J.-P., and Crine, M., *Proceedings of the Third World Congress on Industrial Process Tomography,* Virtual Centre for Industrial Process Tomography, Manchester/Leeds, 2003, p. 730.

Part III

Working Atmospheres and Hazards

3.1 Health Effects Due to Particle Matter

Isamu Tanaka and Hiroshi Yamato

University of Occupational and Environmental Health, Kitakyushu, Fukuoka, Japan

3.1.1 INTRODUCTION

There are many kinds of dusts and other airborne particles in the total environment in which humans work and live. They vary from time to time in number and chemical and physical properties. In the work environment, the health hazards resulting from the exposure to them is of particular importance because they are a common feature and their biological effects are often insidious. In order to prevent or minimize impairment of the health of workers, it is essential to know the concentrations and diameters of harmful substances, their chemical and physical characteristics, and their biological actions. We must be aware not only of the biological effects but also of the mechanisms underlying and controlling their occurrence.

This chapter discusses principles relating to the lung response to exposure to particulate matter in the work environment.

3.1.2 RESPIRATORY SYSTEM

The primary purpose of the respiratory organ is to act as a gas exchange mechanism. Figure 1.1 shows a schematic anatomy of the lung. Air inspired through the nose or the mouth enters the larynx and then the trachea, which is divided into two main bronchi. The bronchial tubes repeatedly divide and diminish in diameter with bifurcation, the smallest tubes being called bronchioles. The diameter of the main bronchi and terminal bronchioles are 1.2 cm and 0.5 mm, respectively. These dimensions and numbers have major implications with respect to deposition and clearance of particles entering the lung while entrained in respired air. Terminal bronchioles lead into the respiratory bronchioles and then into the respiratory space which is composed of a number of alveoli. A very thin layer of venous blood is circulated through the pulmonary capillaries, which are arranged in a network over the surface of approximately 300 million air-containing alveoli. The total effective gas exchange surface of the alveoli has been estimated at 100 m². The alveolocapillary wall separating the blood from the gas phase is on average 0.55 μm in thickness. During the passage of venous blood over the surface of this barrier, the rapid movement of oxygen from the air into the blood and of carbon dioxide out of the blood into the air phase is carried out by gas diffusion.

The trachea and the bronchial tubes are covered with ciliated cells on the epithelium. A carpet of mucus is continuously moved upward by ciliated action, where the mucus is then swallowed or expectorated. This system, known as the mucociliary escalator, is the mechanism whereby particles deposited on the airway surface are removed.

Beyond the terminal bronchioles the surface lining of the airways is no longer ciliated. Each alveolus is lined mainly by type I pneumocytes. Type II pneumocytes apparently secrete surfactant, a material that affects the surface tension of the ultrathin liquid layer on the epithelium. Alveolar macrophages, large free cells 10–50 μm in diameter, are located within the alveolar

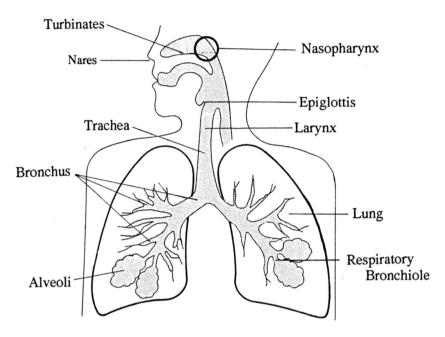

FIGURE 1.1   Schematic structure of the respiratory tract.

region. These are large mobile cells that are the principal phagocytic cell. They also are involved in immunologic reactions.

3.1.3   PENETRATION AND DEPOSITION OF PARTICLES IN THE RESPIRATORY TRACT

Particles penetrate into the human body through the respiratory system, causing a local effect on the lungs and other parts of the system and then passing into the blood stream and causing systemic effects.

Very soluble substances are easily absorbed from all parts of the respiratory tract. Therefore the site of deposition is of less importance for systemic effects.

For insoluble particles, the site of deposition in the respiratory system is of fundamental importance. It depends on the aerodynamic properties of the particle, its shape, dimensions of the airways, and the pattern of breathing.

Particles vary markedly in shape and density, and both these factors play a role in the behavior of particles in the respiratory tract. For our purposes we consider all particles as being spheres of unit density, with the understanding that there could be some variation between particles with respect to speed of settling, depending on their shape and density. This equivalent spherical diameter of unit density is called the aerodynamic diameter.

The dynamics of inhaled particles in the respiratory tract consider particles in terms of the aerodynamic diameter, to reach a reasonable agreement between theoretical predictions and experimental observations. When considering the deposition of particles, the respiratory system is divided into three regions:

1. Nasal-pharyngeal: This begins at the nose and goes to the level of the larynx or epiglottis.
2. Tracheobronchial: This next region consists of the trachea and the bronchial tubes down to the terminal bronchioles.
3. Pulmonary: The third region is regarded as the gas exchange space of the lungs.

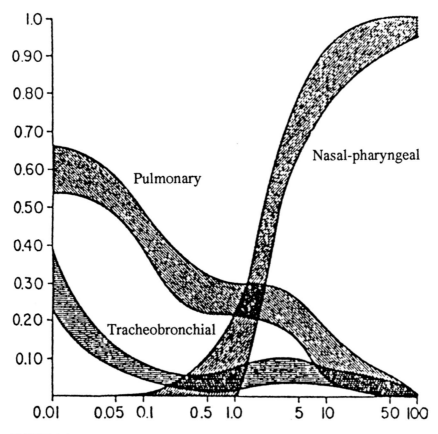

FIGURE 1.2   Regional deposition fraction in the respiratory tract. Each shaded area indicates the variability of deposition for a given mass media aerodynamic diameter (μm) in each region.

Figure 1.2 indicates the range of variability of deposition for a given aerodynamic diameter in each lung compartment.

The particles are deposited in the lung by the following four mechanisms:

1. Sedimentation: The deposition is proportional to the particle speed of settling (proportional to aerodynamic diameter squared) and to the time for settling.
2. Internal impaction: The deposition is also proportional to the particle speed of settling and air velocity, especially where the flow line changes direction.
3. Brownian displacement: The deposition of small particles of less than 0.1 μm in size is governed by diffusional force.
4. Interception: The deposition is provided by the geometric dimensions of a particle following an air stream line.

3.1.4   FATE OF DEPOSITED PARTICLES

The fate of the particles deposited in the respiratory tract is dependent on their diameter and solubility and the site of their deposition, following the defense mechanism in the respiratory system for inhaled particles.

1. Nasal filtration: Particles larger than 10–15 pm are mainly trapped in the nose by filtration by nasal hairs and in the throat and upper airways by impaction. Ciliated epithelium-covered

mucus in the nose contributes to the removal of particles, which are then swallowed. These swallowed particles are significant when dealing with materials that cause systematic intoxication.

2. Mucociliary clearance: The trachea and bronchi, down to the terminal bronchioles, are lined with a ciliated epithelium covered by a mucous layer. The cilia are in continuous and synchronized motion, which causes the mucous layer to have a continuous upward movement at a speed usually in the range of 10–30 mm/min. Particles deposited on the ciliated epithelium are moved toward the epiglottis by the mucociliary escalator, and then swallowed into the gastrointestinal tract or expectorated from the organism within a relatively short time.

3. Coughing and sneezing: These reactions transport the particles toward the upper respiratory tract. Coughing and sneezing create strong currents of air, which help to clear the airways.

4. Phagocytosis: The epithelium of the pulmonary spaces is not ciliated. However, insoluble particles deposited in this region are engulfed by alveolar macrophage cells. The engulfed particles move up to the ciliated epithelium and then are removed upward by the mucociliary escalator and out of the respiratory system. They may also remain in the pulmonary space or enter the lymphatic system. Particles containing free crystalline silica are cytotoxic; that is, they destroy the alveolar macrophage cells.

5. Dissolution: The particles deposited in nonciliated airways are cleared by dissolution.

3.1.5   HEALTH EFFECTS OF INHALED PARTICLES

Some particles cause various biological effects.

1. Pneumoconioses: The pneumoconioses constitute lung diseases that result from the inhalation of certain types of particles deposited in the pulmonary spaces. The alveolar macrophages are part of the defense system in this region. They ingest particles but may be killed or damaged in the process, which results in fibrogenic responses in lung tissue. Some particles introduce such responses, which lead to loss of elasticity in the lung and impair gas diffusion. There is a long interval known as the latent period between first exposure and development of the disease. Silicosis is the most important of the pneumoconioses. Its risk is linked to the percentage of free crystalline silica of inhaled particles. The typical pathological changes are determined by the cytotoxic and fibrogenic capacity of the silica particles. Other pneumoconiosis may be produced by deposition in the nonciliated airways of excessive amounts of, for example, asbestos, coal mine dust, beryllium, kaolin, barium, tin, iron oxide, talc, and graphite.

2. Systemic poisoning: The inhaled particles penetrate into the circulation and internal organs of the body after dissolution. Manganese, lead, cadmium, and the other heavy metals are examples of toxic systemic agents occurring in particulate form.

3. Cancer: Arsenic, chromate, policyclic aromatic hydrocarbons, nickel, radioactive particles, and asbestos have a carcenogenic potential leading to lung cancer. Soluble carcinogens may be a risk to both lungs and other organs.

4. Irritation: Exposure to irritant particles may lead to tracheitis and bronchitis, pneumonitis, and pulmonary edema. There are many irritant particles in the work environment, such as fumes of Cd, Be, V, and Zn, pesticides, acid mists, and fluorides.

5. Allergic reactions: The deposition and retention of inhaled sensitizing substances may produce allergic reactions or other sensitivity responses. Synthetic organic compounds (amine compound, plastics), many vegetable dusts (bagasse, corn, cotton, flax, flour, straw, tea, tobacco, wood), and metal dusts (nickel, chromium) may produce inhalant allergy, asthma, hay fever, or urticaria.

6. Metal fume fever: This is a condition resulting from exposure to freshly generated metal fumes (zinc oxides, magnesium oxide). The main symptoms include chills, fever, muscle pains, nausea, and weakness. It is an acute condition, of brief duration, and with no after effects.
7. Infection: Particles containing fungi, viral, or bacterial pathogens may play a role in the transmission of infectious diseases.

3.1.6 THRESHOLD LIMIT VALUE

Regarding chemical substances present in inhaled air as suspensions of particles, the potential hazard depends on particle size as well as mass concentration.

The threshold limit value (TLV) refers to the airborne concentration of particles and represents conditions under which it is believed that nearly all workers may be repeatedly exposed day after day without adverse health effects.

The TLV is based on available information from industrial experience and from experimental human and animal studies. Health impairments include those that shorten life expectancy, compromise physiological function, impair the capability for resisting other toxic substances or disease processes, or adversely affect reproductive function or developmental processes.

The size-selective TLV of particles has been recommended for crystalline silica for many years in recognition of the well-established association between silicosis and respirable mass concentration. The size-selective TLV of particles is usually divided into two types:

1. Inhalable particles: These materials are hazardous when deposited anywhere in the respiratory tract. The aerodynamic diameter of the inhalable particles is less than 200 μm. The concentration of the inhalable particles is measured as the total dust by an air filter sampler with the inlet air speed of 0.5–0.8 m/s.
2. Respirable particles: The materials are hazardous when deposited in the gas exchange region. The aerodynamic diameter of respirable particles is less than 10 μm. The mass concentration of the respirable particles is measured by the air sampler with an elutriater or cyclone.

The Japan Society for Occupational Health has recommended TLVs for airborne particles in the work environment by using the occupational exposure limits as shown in Table 1.1.

TABLE 1.1   Occupational Exposure Limits for Dusts

I   Dust with more than 10% crystalline silica

$$\text{Respirable dust:} \quad M = \frac{2.9}{0.22Q+1}\,\text{mg/m}^3, \quad \text{Total dust:} \quad M = \frac{12}{0.23Q+1}\,\text{mg/m}^3$$

$$\text{M: OEL,} \quad \text{Q:} \quad \text{content of crystalline silica \%}$$

II   Dust with less than 10% of free crystalline silica

Class 1:   OEL: Respirable dust 0.5 mg/m^3, total dust 2 mg/m^3
             Activated charcoal, alumina, aluminum, bentonite, diatomite, graphite, kaolinite, pagodite, pyrites, pyrite, cinder, talc

Class 2:   OEL: Respirable dust 1 mg/m^3, total dust 4 mg/m^3
             Bakelite, carbon black, coal, cork dust, cotton dust, iron oxide, grain dust, joss stick material dust, marble, portland cement, titanium dioxide, wood dust, zinc oxide

Class 3:   OEL: Respirable dust 2 mg/m^3, total dust 8 mg/m^3
             Limestone, inorganic and organic dusts other than classes 1 and 2

The toxicity of all particles generated in the work environment cannot evaluated. For new particles developed and produced, the nature of the database associating exposure and disease is inconclusive. If a disease is observed in laboratory animals exposed to new particles, this would need to be viewed as a positive indicator of the potential health hazard for humans. Especially, inhalation animal studies are credible because of the similarity of human exposures.

The health effects of airborne particles in the work environment should be investigated in order to prevent impairment.

REFERENCES

1. Task Group on Lung Dynamics, *Health Phys.*, 12, 173–207, 1966.
2. Japan Society for Occupational Health, *J. Occup. Health*, 44, 267–282, 2002.
3. World Health Organization, *WHO Offset Publication, No. 80,* WHO, Geneva, 1984, p. 75.

3.2 Respiratory Protective Devices for Particulate Matter

Isamu Tanaka and Hajime Hori
University of Occupational and Environmental Health, Kitakyushu, Fukuoka, Japan

3.2.1 INTRODUCTION

Respiratory protective devices (respirators) are used for protecting the respiratory system from inhalation of particulate matter. However, the use of respirators is the last choice for occupational hygiene management. The first step for management is to improve working environments, for example, to enclose dust-generation sources, to change the method or process of work, and to use local and general ventilation systems. Respirators should be used in an environment where the workplace is still hazardous even though improvements have been made, or where improvements are difficult or impossible.

3.2.2 TYPES OF RESPIRATORS

In principle, respirators are classified into two types: air-purifying respirators and atmosphere-supplying respirators. The air-purifying respirator removes contaminants from the ambient air, and the atmosphere-supplying respirator provides air from a source other than the surrounding atmosphere. Both types can be further subclassified by the type of inlet covering and the mode of operation. Figure 2.1 shows the subclassifications of respirators for particulate matter.

3.2.3 AIR-PURIFYING RESPIRATORS

Classification of Air-Purifying Respirators

Air-purifying respirators are further classified as nonpowered and powered respirators.

Nonpowered Respirators

Air in the atmosphere enters the filter in a manner that corresponds to workers' breathing. Only tight-fitting coverings are available for nonpowered respirators, because the pressure in the covering may be negative and contaminants could then come through the covering during inhalation. Nonpowered respirators are generally small and are easily maintained. They restrict the wearer's movements the least and may present little physiological strain to the wearer.

They should not be used in atmospheres containing less than 18% oxygen nor in atmospheres immediately dangerous to life or health.

Powered Air-Purifying Respirators

The powered air-purifying respirator (PAPR) uses a blower to pass contaminated air through an element that removes the contaminants and supplies the purified air to a respiratory inlet covering. The covering may be a facepiece, helmet, or hood.

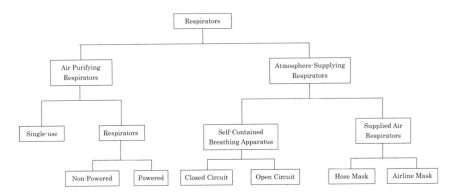

FIGURE 2.1   Classification of respirators.

This type of respirator comes in several different configurations. One configuration consists of the air-purifying element(s) attached to a small blower which is worn on the belt and is connected to the respiratory inlet covering by a flexible tube. Another type consists of an air-purifying element attached to a stationary blower powered by a battery or an external power source and connected by a long flexible tube to the respiratory inlet covering. The third type consists of a helmet or facepiece to which the air-purifying element and blower are attached. Only the battery is carried on the belt.

Filtration Mechanisms and Lifetime

All particulate filtering respirators use a filter to remove the contaminant. Many kinds of filters are available, such as nonwoven fiber material, natural wool or synthetic fiber felt, compressed natural wool or synthetic fiber felt, resin-impregnated felt, loose-packed fibrous glass, and a high-efficiency filter made of a flat sheet of material.

Particulate filters are of two types, that is, absolute and nonabsolute filters. Absolute filters use screening to remove particles from the air; that is, they exclude the particles that are larger than the pores. However, this type of filter has large breathing resistance due to small pores. Therefore, most respirator filters are nonabsolute filters. In nonabsolute filters, pore sizes are larger than the particles to be removed, but the particles are captured and removed by combinations of interception, sedimentation, inertial impaction, diffusion, and electrostatic force. The contribution of each factor to the filtration mechanism depends on the flow rate through the filter and the size of particle.

The airflow resistance of a particulate-removing respirator filter element increases as the quantity of particles retained in the filter increases. This resistance increases the breathing resistance offered by a nonpowered respirator and may reduce the rate of airflow in a powered respirator. Filter element plugging by retained particles may also limit the continuous use time of a respirator. Between uses, reusable respirators should be packaged to reduce exposure to conditions that cause filter degradation, such as high humidity.

In powered air-purifying respirators, since there is a constant flow through the air-purifying element instead of a flow only during inhalation, the useful service lifetimes of the elements are shorter than those of comparable ones attached to a negative pressure respirator. In order to overcome this problem, some powered air-purifying respirators have a spring-loaded exhalation valve assembly.

Filter Efficiency

For air-purifying respirators, the filtration efficiency (FE) of a filter is defined as follows.

$$\text{FE} = \frac{C_o - C_i}{C_o} = 1 - \frac{C_i}{C_o} \tag{2.1}$$

TABLE 2.1   Description of Filter Classes Certified under 42 CFR 84

Class of Filter	Efficiency (%)	Test Agent	Maximum Loading (mg)	Type of Contaminant	Service Time
N-series		NaCl	200	Solid and water-based particulates (i.e., nonoil aerosols)	Nonspecific
N100	99.7				
N99	99				
N95	95				
R-series		DOP oil	200	Any	One work shift
R100	99.7				
R99	99				
R95	95				
P-series		DOP oil	Stabilized efficiency	Any	Nonspecific
P100	99.7				
P99	99				
P95	95				

Note: The three categories of resistance to filter efficiency degradation are N (not resistant to oil), R (resistant to oil), and P (oil proof).

Where C_i is defined as the concentration of a contaminant inside the respirator facepiece cavity, and C_o is defined as the contaminant concentration outside the respirator facepiece. The filtration efficiency is applied only to component parts of the respirator system.

Table 2.1 shows the filter efficiency for nonpowered, air-purifying, particulate-filter respirators certified under 42 CFR Part 84 in the United States.[1]

3.2.4   ATMOSPHERE-SUPPLYING RESPIRATORS

Supplied-Air Respirators

Airline Respirators

Airline respirators use compressed air from a stationary source delivered through a hose under pressure. The respiratory inlet covering may be a facepiece, helmet, hood, or complete suit. These are available in demand, pressure-demand, and continuous-flow configurations. In a demand or negative pressure–type regulator, air flows into the facepiece only on demand by the wearer. On the other hand, a pressure-demand or positive pressure regulator has a modified regulator and special exhalation valve to maintain positive pressure in the facepiece at all times, so that any leakage should be outward. Therefore, a pressure-demand airline respirator provides very good protection.

The airline respirator may be used for long continuous periods. The disadvantage of an airline respirator is the loss of the source of respirable air supplied to the respiratory inlet covering. Also, the trailing air supply hose of the airline respirator severely restricts the wearer's mobility.

Hose Masks

Hose masks supply air from an uncontaminated source to the respiratory inlet covering through a strong, large-diameter hose. Two types are available. One has a hand- or motor-operated air blower that pushes low-pressure air through the hose to the respiratory inlet covering. The other type of hose mask has no blower and requires the wearer to inhale through the hose. The hose mask without blower has many advantages, namely, its long use periods, simple construction, low bulk, easy maintenance, and minimal operating cost. An advantage of the hose mask with blower is its minimal resistance to breathing.

The contaminated air may leak in if the covering fits poorly, because the air pressure inside the respiratory inlet covering of the hose mask with no blower is negative during inhalation. Therefore, hose masks, with and without blower, are certified only for use in atmospheres not immediately dangerous to life or health. The trailing air supply hose of the hose mask also severely limits a worker's mobility.

Self-Contained Breathing Apparatus

The distinguishing feature of a self-contained breathing apparatus (SCBA) is that the wearer need not be connected to a stationary breathing gas source, such as an air compressor. Therefore, it allows comparatively free movement over an unlimited area. Instead, enough air or oxygen for up to 4 hours, depending on the design, is carried by the wearer. This limited service life makes SCBAs unsuitable for routine use for long continuous periods. Two types are available.

Open Circuit

An open-circuit SCBA exhausts the exhaled air into the atmosphere. A cylinder of high-pressure compressed air supplies air to a regulator that reduces the pressure for delivery to the facepiece. This regulator also serves as a flow regulator by passing air to the facepiece on demand. Most open-circuit SCBAs have a service life of 30 to 60 minutes.

Two types of open-circuit SCBA are available: demand and pressure demand. These are very similar in basic operation to airline respirators, except that the air is supplied from a portable air source instead of a stationary one.

Closed Circuit

The exhaled air is rebreathed after carbon dioxide has been removed and the oxygen content restored by a compressed or liquid oxygen source or an oxygen-generating solid. These devices have a longer service life than the open-circuit SCBA, that is, they are designed for 1 to 4 hour use in oxygen-deficient atmospheres. Because negative pressure is created in the facepiece of a nonpositive pressure apparatus during inhalation, there is increased leakage potential. Therefore, negative pressure closed-circuit SCBAs should be used in atmospheres not immediately dangerous to life or health only. For use in oxygen-deficient atmospheres over long periods, closed-circuit SCBAs are also satisfactory.

3.2.5   PROTECTION FACTOR

The protection factor (PF) of a respirator is defined as follows:

$$PF = \frac{C_o}{C_{in}} \qquad (2.2)$$

where, C_{in} is defined as the concentration of contaminant the inside of the covering. C_o is defined as the contaminant concentration outside the respirator facepiece. PF assessments are made almost exclusively on man/respirator systems. It is important to recognize that on a man/respirator system, C_{in} depends on a complicated function of many individual sources of penetration (e.g., air-purifying element penetration, exhalation valve penetration, face seal penetration, and other inboard

TABLE 2.2   Protection Factor Values against Particulate Exposures Assigned by NIOSH

Type of Respirators	Assigned PF
Air purifying respirators	5–50
Powered air purifying respirators	25–50
Atmosphere-supplying respirators	
Supplied air/continuous flow	25–50
Supplied air/SCBA/demand	10–50
Supplied air/SCBA/pressure demand	1000–10000

penetrations) and those environmental conditions that would affect penetration. To deal with the multiple methods for determining and applying protection factors, a number of definitions have been proposed.[2] Table 2.2 shows protection factors assigned by NIOSH.[3]

3.2.6   NOTES FOR USING RESPIRATORS

A worker should be aware of the following:

1. Air-purifying respirators cannot be used if the oxygen concentration in the atmosphere is less than 18%.
2. Respirators should be selected on the basis of hazards to which the worker is exposed. Coverings should be fitted for the worker's face.
3. Respirators should be regularly inspected.
4. Worn and deteriorated parts should be replaced.
5. Respirators should not be allowed to leak contaminants through gaps between the worker's face and the coverings.
6. Respirators should be kept clean and disinfected.
7. Respirators should be stored in a dry, clean place.
8. Used filters should be disposed of by a method that does not scattering the dust.
9. The worker should be instructed and trained in the proper use of respirators and their limitations.
10. An assigned person who is responsible for mask management should instruct workers how to use respirators.

REFERENCES

1. National Institute for Occupational and Safety and Health, *NIOSH Guide to Industrial Respiratory Protection,* Bollinger, N. J. and Schutz, R. H., Eds., U.S. Department of Health and Human Services, Cincinnati, OH, 1987.
2. National Institute for Occupational and Safety and Health, *NIOSH Guide to the Selection and Use of Particulate Respirators Certified Under 42 CFR 84,* NIOSH Publication No. 96-101, U.S. Department of Health and Human Services, Cincinnati, OH, 1996.
3. Myers, W. R., Lenhart, S. W., Campbell, D., and Provost, G., *Am. Ind. Hyg. Assoc. J.,* 44, B25, 1983.

3.3 Spontaneous Ignition and Dust Explosion

*Tatsuo Tanaka**

Hokkaido University West, Sapporo, Japan

3.3.1 SPONTANEOUS IGNITION OF POWDER DEPOSITS

In spite of the absence of fire sources, an ignition or subsequent dust explosion can occur in a coal or grain dust deposit. This is called spontaneous ignition and takes place as the result of an accumulation of heat of reaction, raising the local temperature, which, in turn, promotes the rate of reaction to give off more heat. When the deposit temperature exceeds a certain limit, the powder deposit is ignited after a certain time. The former is the ignition temperature, which is defined as the highest ambient temperature at which no ignition can take place, and the latter is called the induction time. These two characteristics should be estimated for safety purposes for a specific powder deposit.

Basic Differential Equation

Consider a cylindrical powder deposit of an infinite length placed at time 0 under the ambient temperature T_a. According to thermal theory, a combustible dust deposit has two heat fluxes: one is the heat generation caused by the chemical reaction, and the other is the heat loss dissipated to the surroundings. If the reaction rate is expressed by the zero-order equation of Arrhenius, the heat balance is taken as

$$\rho C\left(\frac{\partial T}{\partial t}\right) = \left(\frac{k}{r}\right)\left[\frac{\partial}{\partial r}\left(r\frac{\partial T}{\partial t}\right)\right] + Qf\exp\left(-\frac{E}{RT}\right) \tag{3.1}$$

where r is the radial distance (m) in the cylindrical coordinates, t is the storage time, ρC is the volumetric specific heat (J/m^3K), k is the thermal conductivity of powder (W/mK), $T(r, t)$ is the temperature of the deposit (K), Q is the heat of reaction (J/mol), f is the frequency factor ($mol/s/m^3$), E is the activation energy (J/mol), and R is the universal gas constant (J/mol K). This equation is solved numerically under the following initial and boundary conditions by using the specific data of cork as

I.C. $T = T_r (= 293 \text{ K})$, at $t = 0$

B.C. $-k\partial T/\partial r = h(T - T_a)$ at $r = R_1$

where T_r is the room temperature, R_1 is the radius of the cylinder, and h is the heat transfer coefficient from the side wall of the deposit.

For example, the calculated temperatures[1] are compared with Leuschke's[2] experimental data of the same conditions under $T_a = 414$ K, as in Figure 3.1a and 3.1b. We see that good agreement is obtained between the theory and the experiment where R_1 and the length are 0.16 m, and the subscripts m, h, and b are at the center, halfway, and near the surface of the deposit, respectively. The temperature profiles immediately before the ignition vary with the ambient temperature, as indicated in Figure 3.2 and Figure 3.3, the calculation and the experiment being compared at 414 K and 444 K, respectively. When

* Retired

349

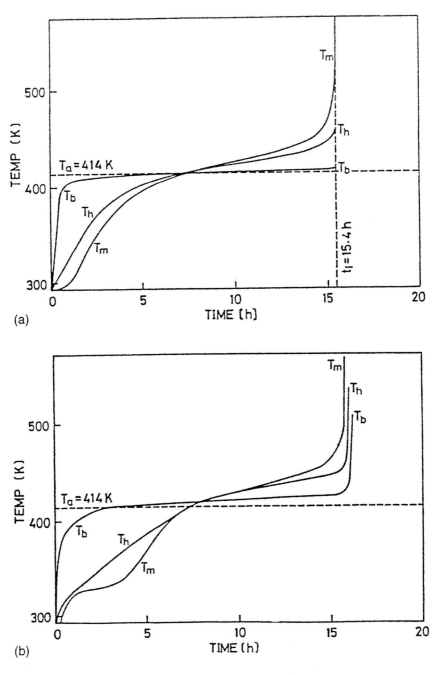

(a)

(b)

FIGURE 3.1   (a) Heat propagation simulated in cork dust deposit in the lapse of time at $T_a = 414$ K. [From Liang, H., Tanaka, T., and Nakajima, Y., *J. Soc. Powder Technol. Jpn.*, 23, 326, 1986. With permission.] (b) The corresponding experimental data by Leuschke. [From Liang, H., Tanaka, T., and Nakajima, Y., *J. Soc. Powder Technol. Jpn.*, 23, 326, 1986. With permission.]

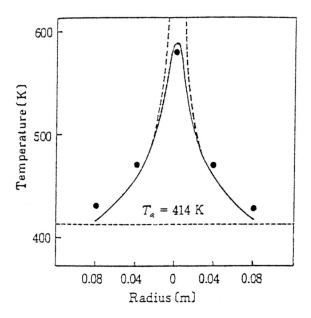

FIGURE 3.2 Temperature profile simulated immediately before ignition ($T_a = 414$ K). [From Liang, H., Tanaka, T., and Nakajima, Y., *J. Soc. Powder Technol. Jpn.*, 23, 326, 1986. With permission.]

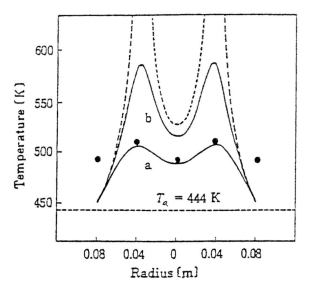

FIGURE 3.3 Temperature profile simulated immediately before the ignition ($T_a = 444$K). [From Liang, H., Tanaka, T., and Nakajima, Y., *J. Soc. Powder Technol. Jpn.*, 23, 326, 1986. With permission.]

the ambient temperature is set to 408 K, the calculated deposit temperature maintains constant. The measured ignition temperature has been determined to be 412 K.

Leuschke[3] varied the volume of the cork dust deposit to examine the ignition temperature, which is compared with the calculation by Equation 3.1, as shown in Figure 3.4. Good agreement suggests that simulation is possible by use of Equation 3.1 to examine various factors affecting the ignition

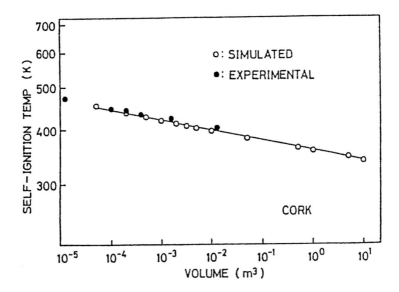

FIGURE 3.4 Simulated dependence of self-ignition temperature on the dust volume compared with experimental data. [From Liang, H. and Tanaka, T., *Kagaku Kogaku Ronbunshu*, 13, 63–70, 1987. With permission.]

temperature as well as the induction time. For this purpose, Equation 3.1 is rewritten in a dimensionless form, together with the initial and boundary conditions:

$$\frac{\partial^2 \theta}{\partial z^2} + \frac{1}{z}\frac{\partial \theta}{\partial z} - \frac{\partial \theta}{\partial \tau} = -SG \exp\left(-\frac{G}{1+\theta/G}\right)$$

(3.2)

$$\text{I.C. } \theta = -GD \text{ at } \tau = 0$$

$$\text{B.C. } -\partial \theta / \partial z = B\theta \text{ at } z = 1$$

where $\theta = E(T - T_a)/RT_a^2$, $z = r/R_1$, $\tau = kt\rho CR_1^2$, $B = hR_1/k$ (= Biot number),

$S = QfR_1^2/kT_a$, $G = E/RT_a$, and $D = (T_a - T_r)/T_a$.

Evaluating Ignition Temperature and Induction Time

From the above definitions, the dimensionless induction time τ_i is related to τ_i by $\tau_i = kt_i/\rho CR_1^2$ by corresponding to a reasonably high value of the dimensionless temperature θ, and the fire spreads immediately in all directions once the bed is ignited. τ_i can thus be regarded independent of θ and z. Therefore, θ_i becomes a function of S, G, D, and B. The numerical calculations give interrelationships between θ_i and for G various values of the parameters S, D, and B and represent that the critical value of G_c, to give $\tau_i \to \infty$ becomes a function of only S in the cases where B is larger than 25. This means that the heat generated equals the heat lost to the surroundings (i.e., the steady state), where $\partial \theta/\partial \tau$ is equal to 0 in Equation 3.2. This is the critical condition that has been dealt with by past workers who tried to solve the basic equation in an analytical way, namely, the right-hand side in Equation 3.2 was approximated at steady state by Frank-Kamenetskii,[4] because for $\theta/G \ll 1$

$$-SG \exp\left(-\frac{G}{1+\theta/G}\right) \approx -SG \exp\left[-G(1-\theta)\right]$$

$$= -SG \exp(-G) \exp(\theta) = -\delta \exp(\theta)$$

(3.3)

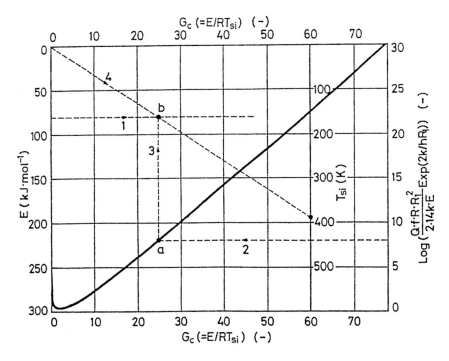

FIGURE 3.5   Nomogram for calculating the ignition temperature. [From Liang, H. and Tanaka, T., *Kagaku Kogaku Ronbunshu,* 13, 63–70, 1987. With permission.]

where $\delta = SG \exp(-G)$. Therefore, in the steady state we have

$$\frac{d^2\theta}{dz^2} + \left(\frac{1}{z}\right)\frac{d\theta}{dz} \approx -\delta\exp(\theta) \tag{3.4}$$

As far as the above equation for the cylindrical coordinates is concerned, it has an analytical solution θ of for all δ up to δ_c, the latter of which was given by Frank-Kamenetskii to be $2\exp(-2/B)$, B being the Biot number. According to our calculations, where no approximation be made, there is obtained $\delta_c = 2.14 \exp(-2/B)$. If δ_c is exceeded, there is no steady-state solution of Equation 3.2, suggesting the runaway reaction. After all, we have an equation including all original variables concerned for obtaining the ignition temperature, T_{si}.

$$\left(\frac{QfR_1^2 E}{kRT_{si}^2}\right)\exp\left(-\frac{E}{RT_{si}}\right)\exp\left(\frac{2k}{hR_1}\right) = 2.14 \tag{3.5}$$

Likewise, the Frank-Kamenetskii number for spherical coordinates and for the plane slab are respectively 3.32 and 3.51, as mentioned later. A nomograph is presented in Figure 3.5,[5] which makes it possible to obtain the ignition temperature of a given cylindrical deposit based on Equation 3.5. Starting from the right ordinate of the calculated value, draw a horizontal line 2 which intersects the reference curve at point a. Draw a horizontal line 1 from the left ordinate of the activation energy and obtain an intersecting point b with the vertical line passing the point a, and connect b with the original point, which can be extended straight forward to the temperature axis as denoted by line 4, where we can get the corresponding ignition temperature T_{si}.

On the other hand, the dimensionless induction time is somewhat complicated but is correlated in the following empirical expression:

$$\tau_i = K\left(\frac{G}{S}\right)^m \qquad (3.6)$$

where K and m depend on the dimensionless groups, D and W [$=R(T_a-T_{st})/E$]. For instance, when G/S is greater than $10^{0.2/(b-a)}$, $\tau_i = 10^{0.4}(G/S)^a$; when G/S is smaller than $10^{0.9/(b-a)}$, then $\tau_i = 10^{1.1}(G/S)^a$; when G/S is in between, then $\tau_i = 10^{0.2}(G/S)^b$, wherein $a = 2.32 \times 1.55^{-\log D} \times 3.93^{\log W}$, $b = 2.02^{(1-\log D)} \times 4.77^{\log W}$ and $c = 3.67 \times 1.59^{-\log D} \times 4.20^{\log W}$. The calculated induction time by use of Equation 3.6 together with the known values of cork dust deposit are shown in Figure 3.6a against the volume of the deposit. Unfortunately, there are no data for cork, but similar data are available for tobacco. The estimation seems to be reasonable from Figure 3.6b.

Examination of the Steady-State Solution

If δ is smaller than δ_c, the system might be safe; that is, no spontaneous ignition would take place. More rigorous insight can be given for a plane slab of the length L. The steady-state equation corresponding to Equation 3.4 becomes

$$\frac{d^2\theta}{dz^2} + \delta\exp(u) = 0 \qquad (3.7)$$

where z is the distance from the bottom of the slab divided by L. The analytical solution is obtained, under $z = 0$ and 1, $\theta = 0$, and $z = 1/2$, $\theta = \theta_m$ (the maximum temperature), as

$$\frac{-2\ln\left(\sqrt{\exp\theta_m} - \sqrt{\exp\theta_m - 1}\right)}{\sqrt{\exp\theta_m}} = \sqrt{\frac{\delta}{2}} \qquad (3.8)$$

The left-hand side of the above equation becomes maximum when $\theta_m = 1.1868$, and, accordingly, δ on the right-hand side is equal to 3.51. This is the critical value δ_c, beyond which there is no solution of Equation 3.7; that is, the slab becomes a runaway system. However, if δ is less than δ_c (e.g., $\delta = 2$), then Equation 3.7 has two solutions, $\theta_m = 0.32895$ and 2.8955, the latter of which, however, is an unstable solution, and the system might become a runaway state according to the circumstances. Consequently, even when δ is less than δ_c, the system is not necessarily safe and stable. In fact, the temperature profile within the slab is much higher for the latter case, and if the profile be exceeded for some reason, the system cannot hold a thermal stability. In either way, aeration quenching appears an effective countermeasure against the disaster.

Effect of Oxygen Diffusion on Thermal Stability in a Powder Bed

Before discussing air quenching, the thermal stability influenced by the diffusion of oxygen in a plane slab should be considered. So far, the basic equation has been solved under the assumption that the oxygen concentration is constant everywhere due to the sufficient makeup. However, if the diffusion of oxygen is not enough, then the aeration of fresh air will enhance the rate of reaction, leading eventually to a hazardous situation. The heat balance equation for a one-dimensional plane slab can be written as[6]

$$\rho C\left(\frac{\partial T}{\partial t}\right) = k\frac{\partial^2 T}{\partial x^2} + Q\left(\frac{c}{c_0}\right)^n f\exp\left(-\frac{E}{RT}\right) \qquad (3.9)$$

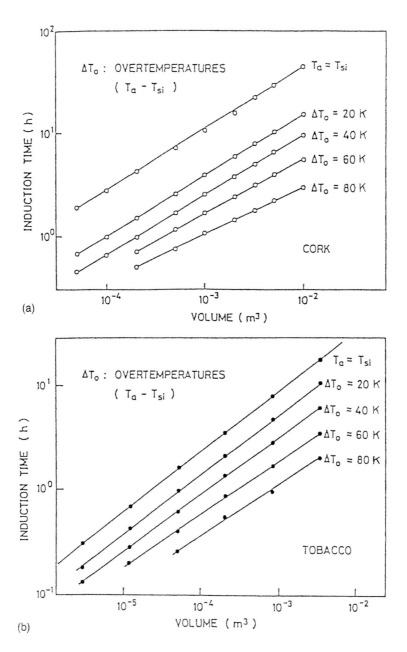

FIGURE 3.6 (a) The induction time simulated for different overtemperatures varied with the deposit volume. [From Liang, H., Tanaka, T., and Nakajima, Y., *J. Soc. Powder Technol. Jpn.*, 23, 326, 1986. With permission.] (b) The experimental induction time to ignition by Leuschke for tobacco. [From Liang, H., Tanaka, T., and Nakajima, Y., *J. Soc. Powder Technol. Jpn.*, 23, 326, 1986. With permission.]

where x is the distance from the bottom of the slab, c is the concentration of oxygen as a function of x and t, c_0 is the oxygen concentration outside the powder bed, and n is the order of reaction. In regard to the oxygen diffusion through the bed, we can write

$$\varepsilon\left(\frac{\partial c}{\partial t}\right) = D^*\left(\frac{\partial^2 c}{\partial x^2}\right) - \left(\frac{c}{c_0}\right)^n f \exp\left(-\frac{E}{RT}\right) \tag{3.10}$$

where ε is the porosity of the bed and D^* is the effective diffusivity of oxygen. Equation 3.9 and Equation 3.10 are rewritten in the nondimensional forms by utilizing the dimensionless terms quoted in Equation 3.2:

$$\left(\frac{\rho C L^2}{k}\right)\frac{\partial \theta}{\partial t} = \frac{\partial^2 \theta}{\partial z^2} + \delta\phi^n \exp\left(\frac{G\theta}{G+\theta}\right) \tag{3.11}$$

$$\left(\frac{\varepsilon L^2}{D^*}\right)\frac{\partial \phi}{\partial t} = \frac{\partial^2 \phi}{\partial z^2} - \left(\frac{\Lambda}{G}\right)\delta\phi^n \exp\left(\frac{G\theta}{G+\theta}\right) \tag{3.12}$$

where the length L is used in place of R_1 for z and δ, and $\phi = c/c_0$, an important dimensionless group $\Lambda = (kT_a/D^*c_0 Q)$, and δ is the Frank-Kamenetskii number defined by $(QL^2 G/kT_a)f \exp(-G)$. At steady state, combining Equation 3.11 and Equation 3.12, we have

$$\frac{d^2\phi}{dz^2} = -\left(\frac{\Lambda}{G}\right)\left(\frac{d^2\theta}{dz^2}\right) \tag{3.13}$$

The following boundary condition is assumed here that the bottom of the bed is closed to air flow and the ambient temperature is T_a; $\theta = 0$, $d\phi/dz = 0$; at $z = 0$; and $\theta = 0$, $\phi = 1$ at $z = 1$, from which we get

$$\frac{d^2\theta}{dz^2} = -\delta\left[1 - \left(\frac{\Lambda}{G}\right)\{\theta + (1-z)\theta'_0\}\right]^n \exp\left(\frac{G\theta}{G+\theta}\right) \tag{3.14}$$

where $\theta'_0 = d\theta/dz$ at $z = 0$. The boundary condition is $\theta = 0$ at $z = 0$ and 1.

Figure 3.7 shows the calculated θ at $z = 1$ as a function of θ'_0 for some sets of parameters, from which we see that there is a single steady-state solution for $\delta = 4$, but when $\delta = 4.5$, three solutions exist. At $\delta = 5.0387$, the two solutions come closer to give a multiple solution, beyond which there is obtained only one steady-state solution. Therefore, three different solutions are possible in a certain range of δ. Of the solutions, the ones labeled B turn out to be unstable solutions; as stable solutions, we have the lower-temperature mode (A) and the higher-temperature mode (C) in general, as shown in Figure 3.8. The arrows indicate the direction of the distribution shift as δ increases. On the other hand, the oxygen distributions are depicted in Figure 3.9, in which for the lower-temperature mode, the oxygen concentration is higher, whereas for the higher-temperature mode, the oxygen is extremely low in the major part of the bed. Here, the smoldering combustion can be assumed. When δ is smaller than $\delta_c = 5.0387$, the lower-temperature mode certainly exists, but this mode coincides with the unstable solution at $\delta = 5.039$, being likely to move toward the higher-temperature mode, as is seen from Figure 3.8 (A_3, $B_3 \rightarrow C_3$). At any rate, the first type of critical state can be called the Frank-Kamenetskii type, and the second type called the oxygen-deficient type.

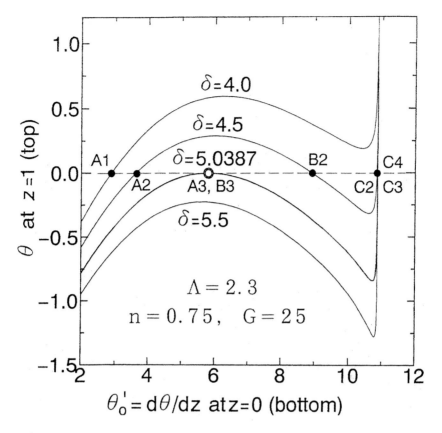

FIGURE 3.7   The calculated θ at $z = 1$ as a function of θ_0'. [From Nakajima, Y., Furusawa, S., Liang H., and Tanaka, T., *J. Soc. Powder Technol. Jpn.*, 30, 247–254, 1993. With permission.]

Stability Criteria and Type Discrimination

Assumption is made here that the thermal stability is secured in a powder bed if (1) the lower-temperature mode exists and/or (2) the maximum temperature rise in the higher-temperature mode is smaller than $0.2T_a$. The latter number means that about 60°C can be allowed for the temperature rise. As the result of the numerical integration of Equation 3.14 for a variety of the parameters concerned, Λ, G, and n, Figure 3.10 shows the critical δ_c values thus determined, in which the bottom broken line refers to the Frank-Kamenetskii theory ($\delta_c = 3.51$). Note that a remarkable difference between δ_c and the bottom value is due to the pronounced effect of the limited rate of oxygen diffusion on the thermal stability. The critical state of the Frank-Kamenetskii type lies on the left portion of the curve of a constant G divided by an open circle. Once δ_c is exceeded, the system is runaway toward the higher-temperature mode and eventually results in smoldering or even explosion if the bed remains uncontrolled. On the other hand, the right portion of the curve indicates the oxygen-deficient type, in which the temperature rises up gradually and continuously from $0.2T_a$ if δ is greater than δ_c. From the theory, however, the maximum temperature rise never exceeds T_a/Λ.

For these reasons, aeration quenching is effective for preventing spontaneous ignition if the bed is of the Frank-Kamenetskii type, but rather dangerous for the oxygen-deficient type. The discrimination can be done by use of values of G, n, and Λ, as shown in Figure 3.10. It is fortunate that the scale of the bed and the frequency factor of reaction are not necessary for the discrimination.

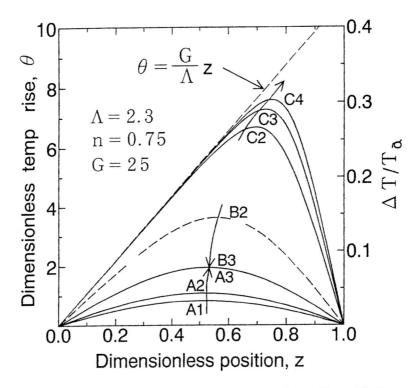

FIGURE 3.8 Temperature distributions for solutions given in Figure 3.7. [From Nakajima, Y., Furusawa, S., Liang H., and Tanaka, T., *J. Soc. Powder Technol. Jpn.*, 30, 247–254, 1993. With permission.]

Effect of Aeration Quenching on Criticality for Spontaneous Ignition

As discussed above, aeration is expected to be effective for the Frank-Kamenetskii type. A quantitative analysis is briefly given in this section for the aeration contributing to the thermal stability of the powder bed and a new risk arising from the aeration will be mentioned.[7]

The heat balance equation in a one-dimensional plane slab with aeration is written with x being the distance from the bottom of the slab:

$$\rho C\left(\frac{\partial T}{\partial t}\right) = k\left(\frac{\partial^2 T}{\partial x^2}\right) - G_a C_a\left(\frac{\partial T}{\partial x}\right) + Qf\,\exp\left(-\frac{E}{RT}\right) \tag{3.15}$$

where G_a is the mass velocity of air through the slab and C_a is the specific heat of air. This equation can be reduced to a dimensionless expression as

$$\frac{\partial \theta}{\partial \tau} = \frac{\partial^2 \theta}{\partial z^2} - \beta\left(\frac{\partial \theta}{\partial z}\right) + \delta \exp\left(\frac{\theta}{1+\theta/G}\right) \tag{3.16}$$

wherein $\tau = tk/\rho CL^2$, $z = x/L$, $\theta = (T-T_a)G/T_a$, and $\beta = G_a C_a L/k$ (the characteristic term for aeration). As the results of dynamic characteristics analysis, the following condition holds for the thermal stability:

$$\delta < \pi^2 + \left(\frac{\beta^2}{4}\right) = \pi^2 + \frac{(G_a C_a L/k)^2}{4} \tag{3.17}$$

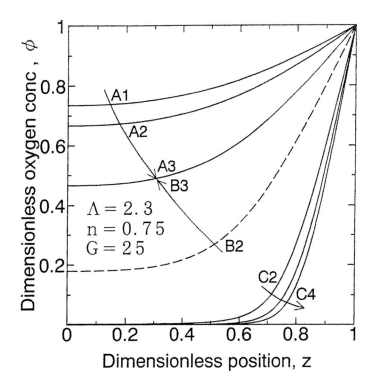

FIGURE 3.9 Oxygen concentration distributions for solutions given in Figure 3.7. [From Nakajima, Y., Furusawa, S., Liang H., and Tanaka, T., *J. Soc. Powder Technol. Jpn.*, 30, 247–254, 1993. With permission.]

The above condition should be consistent with that proposed by Frank-Kamenetskii when $\beta = 0$ ($G_a = 0$). The critical value calculated is equal to 9.87 ($=\pi^2$) which is three times greater than that calculated with $\delta_c = 3.51$. Taking δ_m at the maximum temperature T_m, in place of δ at T_a, gives

$$\delta_m = \delta \left(\frac{G}{G+\theta_m}\right)^2 \exp\left(\frac{G\theta_m}{G+\theta_m}\right) < \pi^2 + \left(\frac{\beta^2}{4}\right) \tag{3.18}$$

where θ_m is the maximum dimensionless temperature rise [i.e., $G(T_m-T_a)/T_a$]. When θ_c is the critical temperature rise, θ_m should be replaced by θ_c in Equation 3.18 to give

$$\delta_c = \left[\pi^2 + \left(\frac{\beta^2}{4}\right)\right]\left[\frac{G+\theta_c}{G}\right]^2 \exp\left(-\frac{G\theta_c}{G+\theta_c}\right) \tag{3.19}$$

Calculating δ_c by using $\theta_c = 1.187$ of Equation 3.8 when $\beta = 0$, we have $\delta_c = 3.49$ (if $G = 25$ is assumed), which agrees with 3.51, the criticality constant.

To obtain θ_m and θ_c for various values of G, β and δ, Equation 3.16 is numerically integrated at steady state, $\partial\theta/\partial\tau = 0$ with the boundary conditions that $\theta = 0$ at $z = 0$, and $d\theta/dz = 0$ at $z = 1$. Figure 3.11 shows the calculated θ_0 at $z = 0$. The values of θ_1 corresponding to $\theta_0 = 0$ are the proper values to fulfill the steady-state solution of Equation 3.16. Within a limited range of δ, there are two values, the lower one of which is stable and the other is unstable. There seems to be a critical value of $\delta_c = 215$, beyond which no equilibrium solution can exist for this condition ($G = 20$, $\beta = 200$).

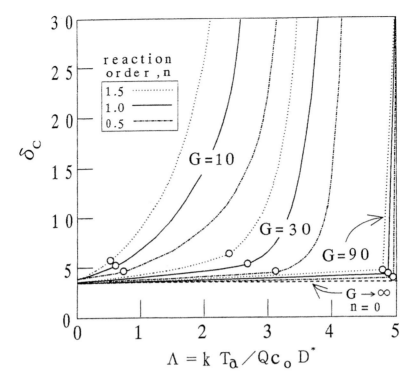

FIGURE 3.10 Critical values of varied with G, n, and Λ. [From Nakajima, Y., Furusawa, S., Liang, H., and Tanaka, T., *J. Soc. Powder Technol. Jpn.*, 30, 247–254, 1993. With permission.]

Obviously the lower value of θ_1 in Figure 3.11 is equal to θ_m, the maximum temperature rise obtained for a single set of the parameters, G, β, and δ. The calculated results are summarized by the following empirical equation as a function of G, β, and δ:

$$\theta_m = -\ln\left(1 - \frac{\delta \exp(-1.9/G)}{\beta}\right) \exp\left(\frac{3.8}{G}\right) \tag{3.20}$$

When $\theta_m = \theta_c$, $\delta = \delta_c$ in the above equation, so that δ_c and θ_c are solved using Equation 3.18 and Equation 3.20 simultaneously, as shown in Figure 3.12 and Figure 3.13. It can be noted that δ_c is of the same order of β, which amounts to a few hundred, assuming the aeration superficial velocity of 1 cm/s with a slab a few meters thick. This outstanding difference in δ_c from 3.51 without aeration suggests an essential improvement in the thermal stability of the powder layer.

Risk Accompanied by Aeration

Because of the unstable solution, there is a possibility for spontaneous ignition to occur when some disturbances cause the temperature to exceed the stable solution. For protection, the following restorable temperature distribution should be maintained even if δ is smaller than δ_c:

$$\hat{\theta} = \frac{G(T - T_a)}{T_a} = -\ln(1 - \alpha z); \quad \alpha = 1 - \exp(-\theta_c) \tag{3.21}$$

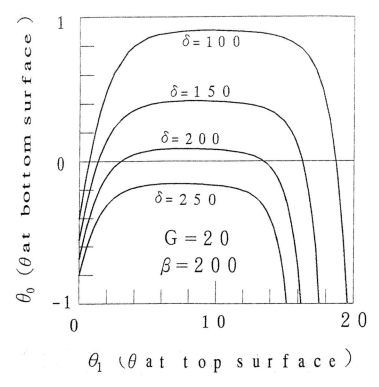

FIGURE 3.11 Calculated bottom temperature θ_0 plotted against θ_1. [From Nakajima, Y., Furusawa, S., Liang, H., and Tanaka, T., *J. Soc. Powder Technol. Jpn.*, 30, 483–489, 1993. With permission.]

where θ is the restorable temperature. θ_c can be found in Figure 3.13. It is recommended for safety purposes to monitor and regulate the temperature distribution along the axis of the slab in response to any unexpected disturbance, so that the temperature may be kept well below the restorable limit.

Finally a remark will be given about the physicochemical properties needed for the calculations above. The order of reaction (oxidation), n, is reported to be from 0.4 to 0.8 for coal and 0.7 for sawdust. G ranges from 10 to 100: 10 for coal, 40 for sawdust and cork, 60 for cellulose, and 50–100 for nitrocompounds. D^* can be calculated from $\varepsilon^{2/3} D_o$ where D_o is the diffusivity of oxygen by 1.8 $\times 10^{-5}(T/273)^{1.75}$ (m²/s). Accordingly, $\Lambda (= kT_a/D^*c_0 Q)$ ranges from 1 to 1.3 for sawdust, 2–2.5 for cork, 2.5–3 for coal, and the highest is 5. The heat evolution Qf(J/m³ s) is plotted against the activation energy E (kJ/mol) in Figure 3.14 for various materials.[8]

3.3.2 DUST EXPLOSION MECHANISM AND PREVENTION

A lot of combustible solid materials are handled in industries, in which extreme care should be taken for fine powder against dust explosion, which is a phenomenon of the ignition of a particle followed by flame propagation between solid particles. The ignitability and the explosibility are represented by measurements for ignition temperature, minimum ignition energy, flame propagation velocity, pressure development in a closed vessel, and so on. Although a number of experiments have been conducted with various equipment, theoretical analysis remains difficult and insufficient, which is due to the extreme complexity of factors involved in the mechanism of dust explosion. Here, a simple dust cloud model will be described, aiming at the theoretical prediction of various phenomena of dust explosion, which is compared with experimental data, leading to a sufficient understanding of the mechanisms.

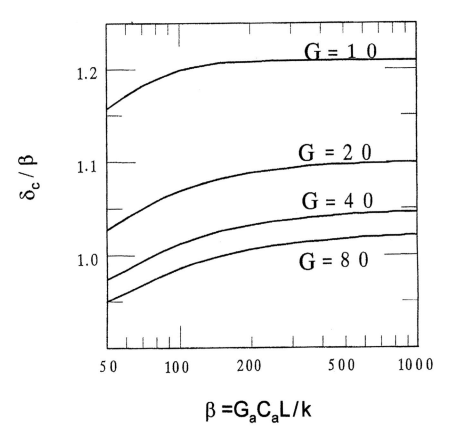

FIGURE 3.12 Critical δ as a function of G and β for aeration quenching. [From Nakajima, Y., Furusawa, S., Liang, H., and Tanaka, T., *J. Soc. Powder Technol. Jpn.*, 30, 483–489, 1993. With permission.]

Dust Cloud Model

The uniform dispersion postulated is illustrated in Figure 3.15, where the explosion begins with ignition of the central particle of diameter D_p, and the flame propagates radially with spherical symmetry. If the flat flame propagation in ducts is to be considered, a one-dimensional configuration is appropriate. Particles are distributed on each spherical surface with one particle at the center. Distances between two neighboring surfaces and between the center and the second surface are all L; L is the representative distance, burned or unburned, in the radial direction. Therefore,

$$L = \left(\frac{\rho_s}{c_d}\right)^{1/3} D_p \tag{3.22}$$

where c_d is the dust concentration and ρ_s is the density of the solid. The number of particles on the nth spherical surface, $N(n)$, can be calculated from

$$N(n) = 24n^2 - 48n + 26 \tag{3.23}$$

Flame propagation from burning particles to the next unburned ones is assumed to occur when the unburned particles are heated by convection and radiation from the flame and the temperature of

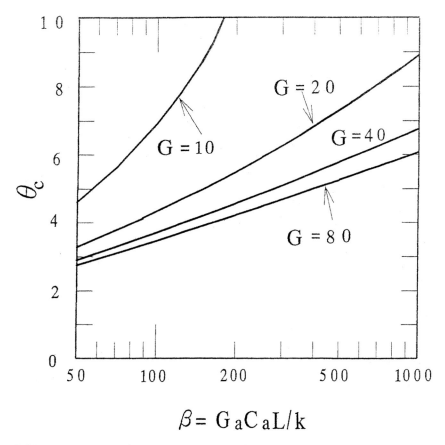

$$\beta = G_a C_a L / k$$

FIGURE 3.13 Critical θ as a function of G and for aeration quenching. [From Nakajima, Y., Furusawa, S., Liang, H., and Tanaka, T., *J. Soc. Powder Technol. Jpn.*, 30, 483–489, 1993. With permission.]

particles reaches the ignition point. Suppose the flame has just propagated to the $(n-1)$th spherical surface. The flame front radius b is assumed to be

$$b = (n-1)L + R_b \qquad (3.24)$$

where R_b is the flame radius inherent to a single burning particle. For liquid droplets, R_b is given by Miesse[10] as

$$R_b = \left(\frac{D_p}{4}\right)\left[1 + \sqrt{1 + \left(\frac{2k^*}{D_p}\right)}\right] \qquad (3.25)$$

where k^* is an equilibrium constant ranging from 1 to 10 cm for liquids. Similar values of k^* are assumed for solids. For calculating the temperature of unburned particles in the nth surface at time t, a dynamic heat balance is taken as

$$\left(\frac{\pi}{6}\right)D_p^3 \rho_s C_{ps}\left(\frac{dT_s}{dt}\right) = \pi D_p^2\left[h(T_g - T_s) + \frac{1}{2}F\varepsilon_p\varepsilon_f\sigma T_f^4\right] \qquad (3.26)$$

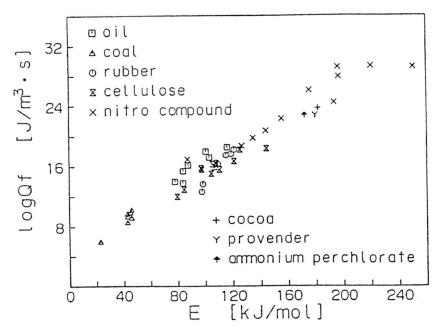

FIGURE 3.14 Relationship between heat generation rate and the activation energy for some materials. [From Liang, H. and Tanaka, T., *Funsai (Micromeritics)*, 34, 42, 1990.]

where C_{ps} is the specific heat of particles, h is the heat transfer coefficient, T_s and T_g are the temperatures of particles and gas at a distance nL from the center at time t, respectively, F is the view factor, ε_p and ε_f are the emissivity of the solid and flame, respectively, and σ is the Stefan–Boltzmann constant. T_f is the flame front temperature. Integrating the above equation gives

$$T_s = \exp(-At)\left[A \int_0^1 \exp(At) T_g(nL,t) dt + \left(\frac{B}{A}\right)[\exp(At)-1] + T_{s0} \right] \qquad (3.27)$$

where $A = 6h/(C_{ps}\rho_s D_p)$ and $B = 3F\varepsilon_p\varepsilon_f\sigma T_f^4/C_{ps}\rho_s D_p$. T_{s0} is T_s at time 0.

Solving the thermal diffusion equation with respect to the spherical coordinates, we have the temperature of gas $T_g(r,t)$ using the boundary and initial conditions: $T_g = T_f$ at $r = b$ and $T_g = T_{i0} =$ constant at $t = 0$.

$$T_g(r,t) = (T_f - T_{i0})\left(\frac{b}{r}\right) erfc\left(\frac{r-b}{2\sqrt{\kappa t}}\right) + T_{i0} \qquad (3.28)$$

where κ is the heat diffusivity. Substituting T_g into Equation 3.27, the temperature of particle can be calculated.

Let Δt be the time interval for the flame to propagate from the $(n-1)$th surface to the nth surface, calculated from t and $T_s =$ ignition temp. using Equation 3.27. Figure 3.16 shows the calculated Δt against n, whereas Δt plotted against c_d, the solid concentration, is depicted in Figure 3.17, indicating that Δt is of the order of 10^{-3} s, from which one can estimate the burning velocity to be less than 1 m/s as well as the flame propagation velocity for deflagration to be less than 100 m/s, all of which roughly agree with our past experience. The latter is calculated based on the gas expansion due to an increase in moles and heat release as a result of particle combustion.[9]

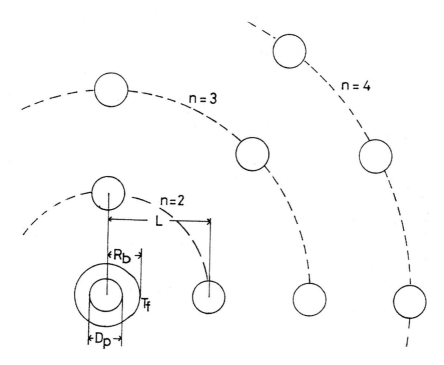

FIGURE 3.15 Dust cloud model. [From Nomura, S. and Tanaka. T., *Kagaku Kogaku Rombunshu,* 5, 47–51, 1979. With permission.]

Assuming that the rate of combustion of a single particle is controlled by oxygen diffusion, it is reported by Yagi and Kunii[11] that $D_p(t)$, the particle diameter at time t after ignition at $t = 0$, is given by

$$\frac{t}{\tau} = 1 - \frac{D_p(t)}{D_{p0}} \tag{3.29}$$

Accordingly, the mass of a particle burned by time t, $m(t)$, is given by

$$m(t) = m_0 \left\{ 1 - \left[1 - \left(\frac{t}{\tau} \right) \right]^3 \right\} \tag{3.30}$$

where τ is the time needed for complete combustion of a particle and is proportional to the square of the initial diameter as

$$\tau = K_{d0} D_{p0}^2 \tag{3.31}$$

where K_{d0} is called the burning constant, which is about 2000 s/cm² for solids on the average, according to Essenhigh and Fells.[12]

When the flame propagates to the nth spherical surface at time t after ignition at the center at $t = 0$, the total mass of particles burned, $M(t)$, is obtained as the sum of the mass of particles burned in each spherical surface in Figure 3.15, namely,

$$M(t) = m_0 \sum_{i=1}^{n} N(i) \left[1 - \left(1 - \frac{t - (i-1)\Delta t}{\tau} \right)^3 \right] \tag{3.32}$$

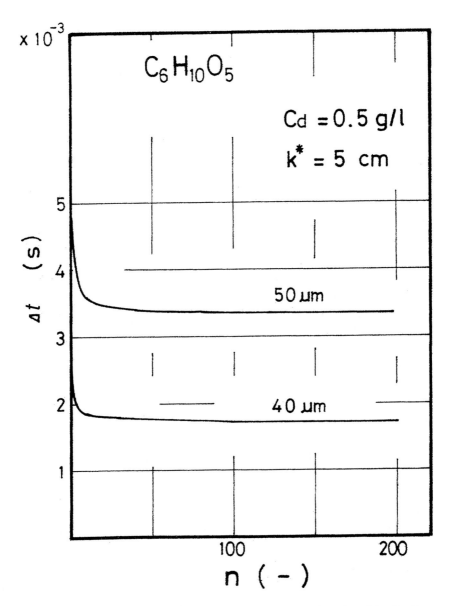

FIGURE 3.16 Calculated time interval variation with the number of particles for flame propagation. [From Nomura, S. and Tanaka, T., *Kagaku Kogaku Rombunshu*, 5, 47–51, 1979. With permission.]

Substituting Equation 3.23 into Equation 3.32 and approximating gives

$$\bar{M}(t) = \frac{M(t)}{M_0} = \left(\frac{8m_0}{M_0}\right)\left(\frac{t}{\Delta t}\right)^3 \qquad (3.33)$$

where M_0 is the maximum mass of particles burned in a closed vessel, which is proportional to the vessel volume V_0 as

$$M_0 = aV_0 \qquad (3.34)$$

where a is a proportionality constant depending on the dust and oxygen concentrations. These terms should be used later.

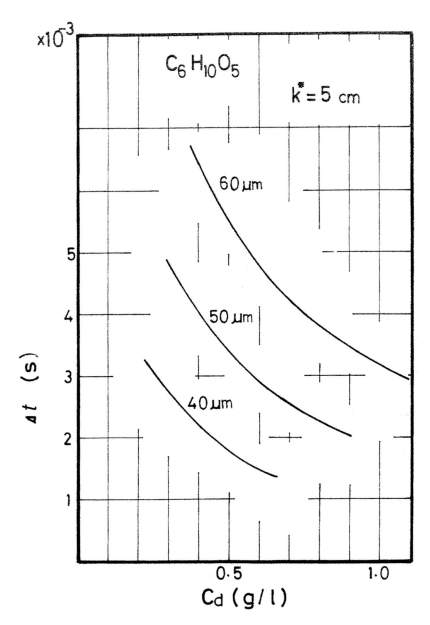

FIGURE 3.17 Calculated time interval between the neighboring particles for flame propagation varied with dust concentration. [From Nomura, S. and Tanaka, T., *Kagaku Kogaku Rombunshu*, 5, 47–51, 1979. With permission.]

Upper and Lower Explosible Limit Concentrations

In order for a flame to propagate throughout a dust cloud, the dust concentration must lie within an explosible region bounded by the upper and lower limits. These limits are defined respectively as the minimum and maximum dust concentrations to sustain flame propagation. For safety, combustible powders should be handled in a region outside the two limits.

The lower limit can be considered to be the concentration calculable from Equation 3.22 by substituting the maximal distance that is obtained from the condition that the temperature of particle 2 reaches its ignition point when particle 1 is just burned out. Equation 3.27 through Equation 3.31 are simultaneously solved by the use of $b = R_b$ and $r = L$ to give the lower-limit explosible concentration as shown in Figure 3.18. Calculations are carried out for several k^* values ranging from 1 to 10, and the experimental data appear reasonable against the particle size, so long as the appropriate value of k^* can be specified.

Near the upper limit, however, oxygen is short, and the flame propagation is strongly affected by the oxygen diffusion. For simplicity of calculation, a situation is assumed where the total oxygen present in a dust cloud is equally allocated to all of the particles in the system. Accordingly, the flame propagation from a burning particle to the neighboring unburned particles should be made before the allocated oxygen to the first particle is entirely dissipated. The critical condition is mathematically given by the above derivation of the lower limit, taking the role of oxygen into consideration. The time t_1 needed for exhausting the allocated oxygen is given by[13]

$$\frac{\tau_1}{\tau} = 1 - \left[1 - \left(\frac{c_g M_p}{c_d N_1}\right)\right]^{1/3} \tag{3.35}$$

where c_g is the oxygen concentration in the air, M_p is the molecular weight of the particle, and N_1 is the number of moles of oxygen by combustion of 1 mole of the particle. Figure 3.19 compared the calculation (solid line) with experiment (dotted line) for some materials. General agreement is obtained, in which we see that the concentration of oxygen lower than 10% would sustain no flame propagation, suggesting the basis for inerting to prevent explosion.

Ignition Temperature and Minimum Ignition Energy

As measures of explosion sensitivity, both ignition temperature and ignition energy are determined experimentally by use of a specific furnace and an apparatus preparing the ignition spark electrode. Cassel and Liebman[15] discussed theoretically the ignition temperature using the rates of heat generation on a particle surface and of heat loss, as denoted by H and U, respectively, and the ignition temperature defined as the minimum air temperature for initiating combustion is obtained by solving simultaneously the two relationships $H = U$ and $dH/dT_s = dU/dT_s$, where T_s is the solid temperature, and they neglected the effect of radiation to give rise to disagreement with experimental data. Analysis by Mitsui and Tanaka,[13] taking the radiant heat from particles to the wall of apparatus in addition to the convective heat into account, is carried out by the same procedure to give Figure 3.20, in which the ignition temperatures are plotted against the particle diameter for some metals and plastics. The different trends of the curves arise from the different emissivities of the particles. With regard to the minimum ignition energy, Kalkert and Schecker[16] proposed the following theoretical equation, ρ_a being the air density:

$$W = (4\pi\kappa)^{3/2} \rho_a C_a \left[\left(\frac{\ln 2}{12}\right)\left(\frac{\rho_s C_{ps}}{k}\right)\right]^{3/2} T_f D_p^3 \tag{3.36}$$

where W is the minimum ignition energy (J) and T_f is the flame temperature. Proportionality between the measured W and the cube of particle size D_p is found for polyethylene to prove the above equation.

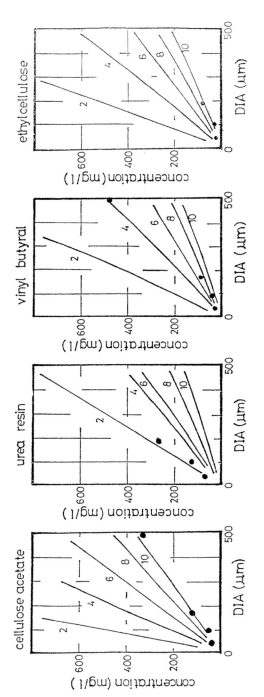

FIGURE 3.18 Comparison between calculated and experimental data of minimum explosible limit for various materials. [From Nomura, S. and Tanaka. T., *Kagaku Kogaku Rombunshu*, 5, 47–51, 1979. With permission.]

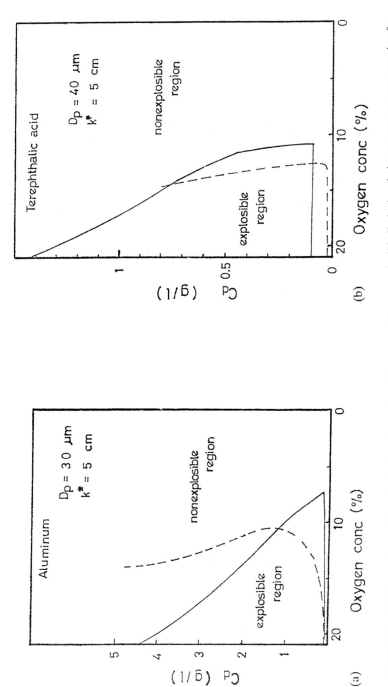

FIGURE 3.19 (a) Comparison of the calculated and experimental upper and lower explosible limits in relation to oxygen concentration for aluminum. [From Nomura, S., Torimoto, M., and Tanaka, T., *J. Soc. Chem. Eng. Jpn.*, 45, 327, 1981. With permission.] (b) Comparison of the calculated and experimental upper and lower explosible limits in relation to oxygen concentration for terephthaic acid. [From Nomura, S., Torimoto, M., and Tanaka, T., *J. Soc. Chem. Eng. Jpn.*, 45, 327, 1981. With permission.]

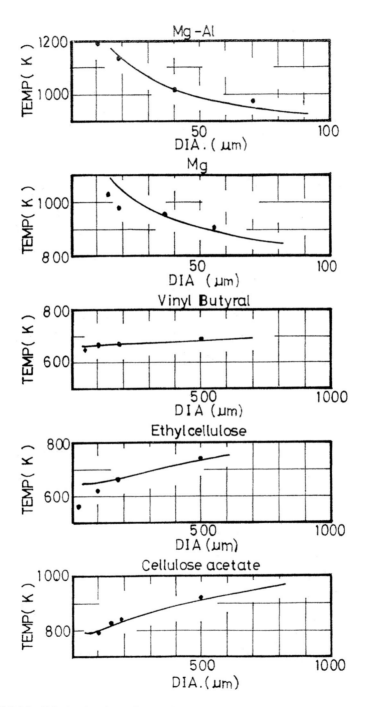

FIGURE 3.20 Calculated and experimental ignition temperatures plotted against particle diameter for some materials. [From Mitsui, R. and Tanaka, T., *IEC Process Des. Dev.*, 12, 384, 1973.]

Pressure Development on Closed and Vented Vessels

One of the most important explosion characteristics is the rate of pressure rise in a closed vessel. Using spherical vessels, a scale-up relationship between the maximum rate of pressure rise $(dP/dt)_{max}$ and the vessel volume V_0, which is called the "cubic law," was obtained experimentally:

$$V_0^{1/3}\left(\frac{dP}{dt}\right)_{max} = K_{st} \tag{3.37}$$

where K_{st} is a constant. However, no theoretical background is given to the above empirical equation about whether K_{st} is a universal constant or dependent on some experimental variables, without which wide applicability cannot be expected.

Adiabatic change is assumed in a closed vessel of volume V_0. After ignition at the vessel center, the reduced pressure $\bar{P}(= P/P_0)$ starts to rise up due to the gas mole increase and the heat released by combustion. The rate of pressure rise at time t is

$$\left(\frac{d\bar{P}}{dt}\right) = \gamma \bar{P}^{1-(1/\gamma)}\left(\bar{P}_m^{1/\gamma} - 1\right)\left(\frac{d\bar{M}(t)}{dt}\right) \tag{3.38}$$

where $\bar{P}_m = P_{max}/P_0$ with $P_{max} = RT_{g0}n_{c0}/V_0$, wherein T_{g0} is the maximum gas temperature, n_{c0} is the resultant total number of gas moles, and γ is the ratio of the specific heats. This equation can be integrated to give

$$\bar{P} = \left[\left(\bar{P}_m^{1/g} - 1\right)\bar{M}(t) + 1\right]^g \tag{3.39}$$

The relationship between P and t for spherical vessels calculated from Equation 3.39 is shown in Figure 3.21 by using Equation 3.33. The maximum rate of pressure rise is obtained at the point (P_{max}, t_0) where t_0 is the time when the total mass of particles burned $M(t_0)$ becomes unity. From Equation 3.33

$$t_0 = \left(\frac{M_0}{8m_0}\right)^{1/3}\Delta t \tag{3.40}$$

and

$$\frac{d\bar{M}(t)}{dt} = \left(\frac{24m_0}{M_0\ t}\right)\left(\frac{t}{\Delta t}\right)^2 \tag{3.41}$$

Combining Equation 3.34 and Equation 3.38 through Equation 3.41, we have the following relation:

$$V_0^{1/3}\left(\frac{dP}{dt}\right)_{max} = \left(\frac{\gamma D_{p0}}{\Delta t}\right)\left(\frac{36\pi\rho_s}{a}\right)^{1/3}P_m\left(1 - \bar{P}_m^{-1/\gamma}\right) \tag{3.42}$$

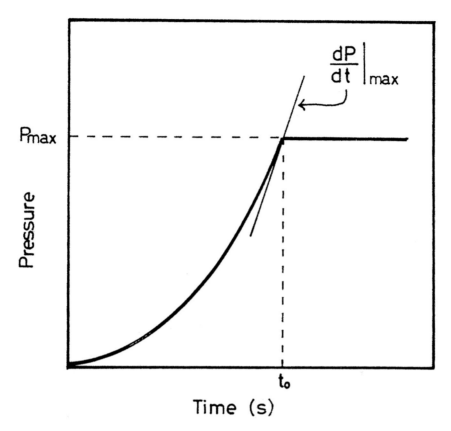

FIGURE 3.21 Calculated pressure development in the lapse of time. [From Nomura, S. and Tanaka. T., *Kagaku Kogaku Rombunshu*, 5, 47–51, 1979. With permission.]

This equation proves the validity of the empirical cubic law given by Equation 3.37, where K_{st} is obviously represented by the right-hand side of Equation 3.42 and is constant for a fixed dust concentration and particle size. Figure 3.22 compares the theory with the experimental data of Ishihama and Enomoto[17] for starch dust. The predicted maximum rate of pressure rise is somewhat greater than the experiment, as the theory is based on the assumption of adiabatic change.

The most common and economical approach to explosion protection is to incorporate relief vents. The area of the vent for a particular vessel must be large enough to discharge the combustion products quickly to prevent pressures exceeding the strength of the plant vessel. Suppose the dust explosion takes place in a vented vessel with volume V_0 and vent area S. The macroscopic energy balance is applied to the system and the following approximate equation is given for the rate of pressure rise:

$$\frac{d\bar{P}}{dt} = \gamma \bar{P}^{1-(1/\gamma)} \left(\bar{P}_m^{1/\gamma} - 1 \right) \left(\frac{d\bar{M}(t)}{dt} \right) - \left(\frac{\gamma}{V_0} \right) \bar{P}^{1-(1-\gamma)} \times \left[\bar{P}_m^{1/\gamma} \left(\frac{w_b}{\rho_m} \right) + \left(\frac{w_n}{\rho_0} \right) \right] \tag{3.43}$$

The first term on the right-hand side is the rate of pressure rise due to the combustion with the same form as that of Equation 3.38, and the second term is the reduction of the developed pressure by discharge of burned and unburned gases through the vent. w denotes the mass flow rate, and the sub-

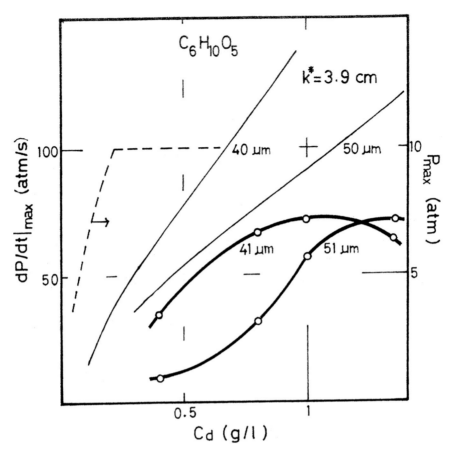

FIGURE 3.22 Calculated maximum rate of pressure rise compared with experimental data by Ishihama and Enomot. [From Nomura, S. and Tanaka. T., *Kagaku Kogaku Rombunshu*, 5, 47–51, 1979. With permission.]

scripts b and u denote "burned and unburned." When the pressure becomes maximum, as denoted by P_v, $d\bar{P}/dt = 0$. The first term could be replaced by the maximum rate of pressure rise in a closed vessel and the second term is approximated using the coefficient of discharge and the vented area S as follows:

$$0 = \left(\frac{d\bar{P}}{dt}\right)_{max} - \left(\frac{S}{V_0}\right)\left(\frac{P_0}{r_0}\right)^{1/2} f(\gamma) g(\bar{P}_v) \tag{3.44}$$

where $f(\gamma)$ and $g(\bar{P}_v)$ are given by Nomura and Tanaka[18] for $\bar{P} < 1/0.53$

$$f(\gamma) = g\left(\frac{2\gamma}{\gamma-1}\right)^{1/2}, \; g(\bar{P}_v) = \bar{P}_v^{1-(1/\gamma)}\left(\bar{P}_v^{1-(1/\gamma)} - 1\right)^{1/2}$$

and for $\bar{P} > 1/0.53$

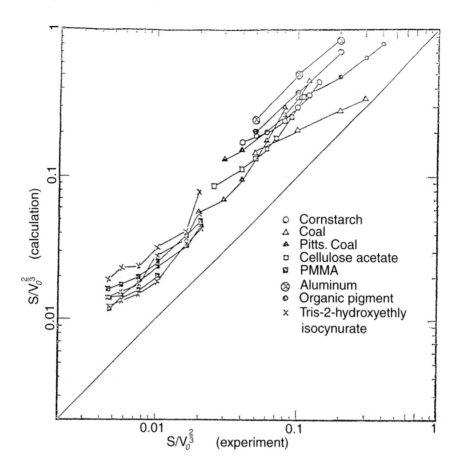

FIGURE 3.23 Comparison of calculated $s/v_0^{2/3}$ with experimental $s/v_0^{2/3}$. [From Nomura, S. and Tanaka, T., *J. Chem. Eng. Jpn.*, 13, 309–313, 1980. With permission.]

$$f(\gamma) = \gamma \left[\gamma \left(\frac{2}{\gamma+1} \right)^{(\gamma+1)/(\gamma-1)} \right]^{1/2} , \; g\left(\overline{P}_v\right) = \overline{P}_v^{(3\gamma-1)/2\gamma} \tag{3.45}$$

Substituting the cubic law in Equation 3.37 leads to a new concept of a vent ratio:

$$\frac{S}{V_0^{2/3}} = \left[\left(\frac{P_0}{K_{st}} \right) \left(\frac{P_0}{\rho_0} \right)^{1/2} f(\gamma) g\left(\overline{P}_v\right) \right]^{-1} \tag{3.46}$$

The new dimensionless vent ratio, $S/V_0^{2/3}$, is estimated from K_{st} for a given material $\overline{P}_v$ and the pre-determined maximum pressure related directly to the strength of the plant. It seems dubious that the conventional vent ratio, S/V_0 (i.e., the dimensional parameter) provides a sound basis for designing a vent area for various scales of equipment. As to the vented explosions, many experiments were carried out with various scales of test apparatus and various materials.[19–21] A set of reported $\overline{P}_v$ and K_{st}

values are substituted into Equation 3.46 to calculate the theoretical $S/V_0^{2/3}$ which is plotted against the experimental one as depicted in Figure 3.23. The theoretical data, being a little above the corresponding straight line, are favorable for the purpose of the safety design. For all of the calculations throughout this section, the SI unit system should be used, although most equations are dimensionally sound. For instance, P_0 and P_{max} are in Pa, V_0 in m³, S in m², K_{st} in Pa/m s, and so on.

Measurements of Dust Explosion Characteristics

Explosibility tests have been developed in many countries, particularly in Europe and the United States. Although a number of similarities are present in the tests, a current lack of standardization has resulted in the variations in the test procedures in each country.[22] In Japan, the Association of Powder Process Industry and Engineering standardized a test method for dust explosibility in 1991,[23] developing two types, which are mainly used for determining the explosibility of a test powder and the lower explosible limit concentration. The first type is called the Vertical Tube Apparatus, which is characterized by the dust cloud formed by means of blown-up compressed air. The second type, called the Tapping Sieve Type Apparatus, features the sieve tapping mechanism for obtaining more homogeneous dust dispersion. As to the determination of the severity of dust

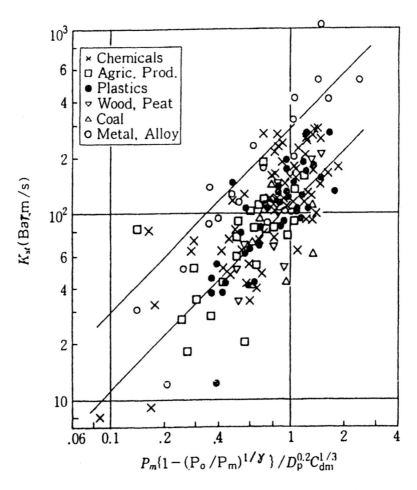

FIGURE 3.24 Estimating K_{st} from other characteristics in comparison with experimental data for a number of materials. [From Tanaka, J., *Soc. Powder Technol. Jpn.*, 32, 237–239, 1995. With permission.]

explosion, investigation is being made for the standardization with a 1-m³ cylindrical apparatus having a length-to-diameter ratio close to unity. The cubic law described in the preceding section turns out to hold for the spherical volume greater than 20-l. Therefore, a 20-l sphere test apparatus has also been investigated for use as an alternative for determining the maximum pressure as well as the maximum rate of pressure rise. The Research Institute of Industrial Safety, Japan,[24] published a guide to test a method for explosion pressure and the rate of pressure rise for combustible dusts, in which are recorded a number of experimental data for wood, coal, agricultural products, plastics, chemicals, metals, and so on. The documented parameters are particle size, lower explosible limit, maximum pressure, K_{st} value, and others. These data can be used for the confirmation of the theoretical derivation, Equation 3.42, as indicated in Figure 3.24, in which D_{p0}, P_m and ρ_s are utilized along with c_{dm} (minimum explosible limit) in place of a (= effective combustible dust concentration defined in Equation 3.34). Δt is a function of D_{p0} as shown in Figure 3.16. Except for metals and alloys, the experimental K_{st} appears to lie along a straight line corresponding to the calculated K_{st},[25] because the densities of most materials range from 0.5 to 1.5. On the other hand, the ρ_s of metal and alloy ranges from 2 to 8 (g/cm³), so that the data are placed a little above, in proportion to $(\rho_s)^{1/3}$ according to Equation 3.42. The unit of K_{st} in Figure 3.24 is expressed conventionally by bar m/s, as indicated in the original table.

Explosion Suppression and Protection

Reduction of oxygen concentration is significant for dust explosion prevention as discussed in relation to the explosible limits in Equation 3.36. This is termed "inerting," and the oxygen is frequently replaced by nitrogen, carbon dioxide, and flue gas. Inerting is also made by adding inert dusts such as $CaCO_3$, $CaSO_4$, $NaHCO_3$, and so forth, which are called diluents, reducing the explosibility of the combustible dust. Relief venting is also described theoretically in Equation 3.46. The vent should be located in the vicinity of an expected ignition position and various types of diaphragms for vent cover made of flexible plastics, metal foil, and so on are available. An automatic explosion suppression system should also be considered, which consists of a detector responding to the pressure developed to transfer the signal to a controller unit, which actuates the extinguisher supply to the fire to suppress explosion.

REFERENCES

1. Liang, H., Tanaka, T., and Nakajima, Y., *J. Soc. Powder Technol. Jpn.*, 23, 326, 1986.
2. Leuschke, G., *Int. Symp. Loss Prevent. Safety*, 2, 647, 1980.
3. Leuschke, G., *Inst. Chem. Eng. Symp. Ser.*, 68, 1981.
4. Frank-Kamenetskii, D. A., in *Diffusion and Heat Transfer in Chemical Kinetics*, 2nd Ed., Transl. Appleton, Plenum Press, New York, 1969, p. 388.
5. Liang, H. and Tanaka, T., *Kagaku Kogaku Ronbunshu*, 13, 63–70, 1987.
6. Nakajima, Y., Furusawa, S., Liang, H., and Tanaka, T., *J. Soc. Powder Technol. Jpn.*, 30, 247–254, 1993.
7. Nakajima, Y., Furusawa, S., Liang H., and Tanaka, T., *J. Soc. Powder Technol. Jpn.*, 30, 483–489, 1993.
8. Liang, H. and Tanaka, T., *Funsai (Micromeritics)*, 34, 42, 1990.
9. Nomura, S. and Tanaka. T., *Kagaku Kogaku Ronbunshu*, 5, 47–51, 1979.
10. Miesse, C. C., in *Sixth Symposium (International) on Combustion*, Combustion Institute, Pittsburgh, 1957, p. 732.
11. Yagi, S. and Kunii, D., in *Fifth Symposium (International) on Combustion*, Combustion Institute, Pittsburgh, 1955, p. 231.
12. Essenhigh, R. H. and Fells, H., *Discuss. Faraday Soc.*, 30, 208, 1960.
13. Nomura, S., Torimoto, M., and Tanaka, T., *J. Soc. Chem. Eng. Jpn.*, 45, 327, 1981.
14. Mitsui, R. and Tanaka, T., *IEC Process Des. Dev.*, 12, 384, 1973.
15. Cassel, H. M. and Liebman, I., *Combust. Flame*, 3, 467, 1959.

16. Kalkert, N. and Schecker, H. G., *Chem. Ing. Tech.*, 51, 1248, 1979.

17. Ishihama, W. and Enomoto, H., in *Fifteenth Symposium (International) on Combustion*, Combustion Institute, Pittsburgh, 1974, 479.

18. Nomura, S. and Tanaka, T., *J. Chem. Eng. Jpn.*, 13, 309–313, 1980.

19. Hartmann, I. and Nagy, J., *Ind. Eng. Chem.*, 49, 1743, 1957.

20. Donat, C., *CEP Tech. Manual Loss Prevent.*, 11, 87, 1977.

21. Schwab, R. S. and Othmer, D. F., *Chem. Process Eng.* (April), 165, 1964.

22. Firld, P., *Dust Explosion*, Elsevier, Amsterdam, 1982, p. 46.

23. Association of Powder Process Industry and Engineering, *Test Method for Dust Explosibility*, ASP 002-1991, 1991, p. 1.

24 Research Institute of Industrial Safety, Japan, *Guide to Test Method for Explosion Pressure and Rate of Pressure Rise for Combustible Dusts*, RIIS-TR-94-1, 1994, p. 24.

25. Tanaka, J., *Soc. Powder Technol. Jpn.*, 32, 237–239, 1995.

Index